Radiation Damage in DNA: Structure/Function Relationships at Early Times

Edited by

Alfred F. Fuciarelli

Biology and Chemistry Department
Battelle, Pacific Northwest Laboratory
PO Box 999, MSIN P7-56
Richland, WA 99352

and

John D. Zimbrick

National Research Council
National Academy of Sciences
2101 Constitution Avenue, N.W.
Suite 342
Washington, D.C. 20418

BATTELLE PRESS
Columbus • Richland

DISCLAIMER

PACIFIC NORTHWEST LABORATORY
operated by
BATTELLE MEMORIAL INSTITUTE
for the
UNITED STATES DEPARTMENT OF ENERGY
under Contract DE-AC06-76RLO 1830

Library of Congress Cataloging-in-Publication Data

Radiation damage in DNA : structure/function relationships at early times / edited by
Alfred F. Fuciarelli and John D. Zimbrick.
p. cm.
"The fourth in a series of international workshops was held at the Salishan Lodge in Gleneden Beach, Oregon, October 1-5, 1994" —
Pref.
ISBN 0-935470-90-5
1. DNA—Effect of radiation on—Congresses. 2. Radiogenetics—Congresses.
I. Fuciarelli, Alfred F., 1960- . II. Zimbrick, John D., 1938- .
QP624.R33 1995
574.87'3282--dc20 95-7178
CIP

Printed in the United States of America

Battelle Press
505 King Avenue
Columbus, Ohio 43201-2693
614-424-6393
1-800-451-3543
FAX: 614-424-3819

PREFACE

The fourth in a series of international workshops was held at the Salishan Lodge in Gleneden Beach, Oregon, October 1-6, 1994. Since its inception at Mülheim in 1988, this series has been highly successful in bringing together an interdisciplinary group of physicists, chemists and biologists to focus on the central theme of radiation-induced damage to DNA. Each of the three previous workshops has built upon the groundwork laid previously, and has provided a forum wherein the latest cutting-edge science is used to shed new and more revealing light on the DNA damage arena. This workshop continued in that tradition with its expansion to include presentations and discussions on the structural biology of damaged DNA including DNA damaged by monochromatic light; the relevance of these perturbed structures to the many facets of DNA repair; the technical problems associated with preparation and/or availability of suitable damaged DNA templates for study; and the new technologies available from state-of-the-art mass spectrometry, nuclear magnetic resonance spectroscopy, and pulsed electron paramagnetic resonance spectroscopy to study structural properties of damaged DNA oligomers and time-resolved free radical processes. Besides the experimental studies just described, the workshop also included a session on computer modeling of damaged DNA structures, and a discussion on how computer modeling and experimental studies can complement each other. Lastly, the workshop included sessions on structure/function relationships at higher levels of DNA organization and how these relationships might vary depending on this level of organization.

The organizers and participants are most grateful to the sponsors for making this workshop possible. These are: Radiation Research Society; Department of Energy; National Cancer Institute; Life Sciences Center, Battelle, Pacific Northwest Laboratories; Beckman Instruments, Inc.; and Hewlett-Packard, Inc. We also thank Ms. Carolyn Woodall and the staff of the Salishan Lodge for their excellent service and facilities. Finally, we thank Mary S. Lipton, Ellen C. Sisk and Anita Zimbrick for providing invaluable assistance on logistics and secretarial support for the workshop.

Contents

Chapter 1

Models for Energy Deposition in DNA

ENERGY LOSS BY ELECTRONS IN DNA

Simon M. Pimblott and Jay A. LaVerne
Notre Dame Radiation Laboratory
University of Notre Dame
Notre Dame, Indiana 46556

Abstract

The dipole oscillator strength distribution (DOSD) for dry, solid DNA was constructed from experimentally derived optical constants and X-ray attenuation cross-sections. Various energy loss parameters, including the stopping power, the inelastic mean free path and the inelastic collision cross-section, were calculated from the DOSD and stochastic Monte Carlo simulations were performed to evaluated the energy loss distribution for energetic electrons. Large differences were found between the predictions for DNA and those previously reported for gaseous media, raising serious questions about the usefulness of calculations of radiation-induced damage in DNA based on cross-sections for gaseous media.

Introduction

An understanding of the transfer of energy from an energetic electron to DNA is of fundamental importance in the elucidation of the mechanisms of direct radiation-induced damage to DNA [1]. Several elaborate Monte Carlo calculations have been made using cross-sections for gaseous water as approximations for those for DNA [2-4]. In principle, these studies should provide detailed information about radiation-induced damage to DNA [5]. Experimental studies on the range, and on the energy loss, of electrons in gaseous media [6] and experiments performed using thin organic films and water ices [7,8] show that information for gases cannot be extrapolated to condensed media in a simple way. Consequently, the data obtained in the Monte Carlo simulations are of limited value because of the questionable cross-sections used. No investigation of the energy loss of energetic electrons in DNA which incorporates realistic cross-sections for DNA has been presented.

This study describes the construction of the dipole oscillator strength distribution (DOSD) for DNA and its use in the investigation of the energy loss by energetic electrons. In the following section, the DOSD is obtained from the optical constants, n and k, for DNA [9] and from atomic X-ray attenuation cross-sections [10]. Expressions relating the energy loss properties of electrons to the DOSD of the medium are presented in Section 3. A Monte Carlo simulation method for modeling the distribution of energy loss events produced by an energetic electron is also briefly discussed. Section 4 compares the predictions of calculations for DNA with results obtained for liquid and gaseous water [11]. The final section is a brief summary.

Dipole oscillator strength distribution

To construct the DOSD for DNA, two different sources of information are employed. Inagaki *et al.* [9] have derived the optical constants, n and k, of calf thymus DNA from critical-angle measurements and from transmission measurements using thin films for photons with energies between 2 and 82 eV. The refractive index, n, and the extinction coefficient, k, of a medium are related to the dielectric constants, ε_1 and ε_2, by the relationships [9]

$$\varepsilon_1 = n^2 - k^2 \tag{1}$$

and

$$\varepsilon_2 = 2nk. \tag{2}$$

The energy loss function, $Im(-1/\varepsilon(\omega))$ is defined by the dielectric response function,

$$\varepsilon(\omega) = \varepsilon_1 + i\varepsilon_2, \tag{3}$$

and is related to the dipole oscillator strength distribution by [12]

$$f(\omega)/\omega = 2mIm(-1/\varepsilon(\omega))/(h^2e^2n) \tag{4}$$

where $f(\omega)$ is in eV^{-1}, m and e are the electron mass and charge, respectively, h is Plank's constant and n is the number density of molecular units (nucleotides in this instance). This formalism and the data of Inagaki *et al.* [9] allow the calculation of the DOSD for energies less than 82 eV.

At higher energies than 82 eV, the DOSD is fabricated using the atomic photon attenuation cross-sections [13] for the atoms comprising the nucleotides. The DOSD of a medium is directly proportional to its photon attenuation cross-section, σ_1 [13]. When σ_1 has the units of Mb (= 10^{-22} m^2) the constant of proportionality is 9.111×10^{-3}. At high energies, molecular cross-sections can be approximated from the component atomic cross-sections using the "mixture rule" [14], for instance for water, $\sigma_{H_2O} = 2\sigma_H + \sigma_O$. In consequence, the DOSD for DNA at high energies can be built up from the individual atomic photon attenuation cross-sections of the atoms comprising its nucleotides.

Table 1. Molecular parameters for typical DNA nucleotides.

Nucleotide	Molecular formula	Molecular weight	Valence electrons	Core electrons
T-A	$C_{20}H_{27}N_7O_{13}P_2$	651	238	92
C-G	$C_{19}H_{26}N_8O_{13}P_2$	652	238	92

A typical DNA nucleotide consists of either a thymine-adenine (T-A) or a cytosine-guanine (C-G) base pair plus two phosphate groups and two sugars. The molecular parameters for these nucleotides are listed in Table 1 and the resulting sums for the DOSD of DNA are

T-A pair : $$f_{DNA}(\omega) = 20f_C(\omega) + 27f_H(\omega) + 7f_N(\omega) + 13f_O(\omega) + 2f_P(\omega) \quad (5)$$

and

C-G pair : $$f_{DNA}(\omega) = 19f_C(\omega) + 26f_H(\omega) + 8f_N(\omega) + 13f_O(\omega) + 2f_P(\omega). \quad (6)$$

The choice of a T-A or a C-G base-pair is found to have no noticeable effect on the DOSD from 100 eV to 1 MeV.

The DOSD constructed using the available experimental optical constants and atomic attenuation cross-sections has a gap between 82 eV and 100 eV. A linear interpolation was used to bridge between the two sets of predictions. The constructed DOSD for DNA is shown in Figure 1. The dipole oscillator sum for the effective number of electrons that can receive an energy transfer is

$$Z_{eff} = \int_{\omega_o}^{\omega_{max}} f(\omega)d\omega \tag{7}$$

and that for the mean excitation energy is

$$ln(I_{eff}) = \int_{\omega_o}^{\omega_{max}} f(\omega)ln(\omega)d\omega / Z_{eff}. \tag{8}$$

The parameters ω_0 and ω_{max} are the minimum and maximum permissible energy transfer, respectively. As ω_{max} increases Z_{eff} and I_{eff} are found to approach the values of 330 and 77.9 eV, respectively.

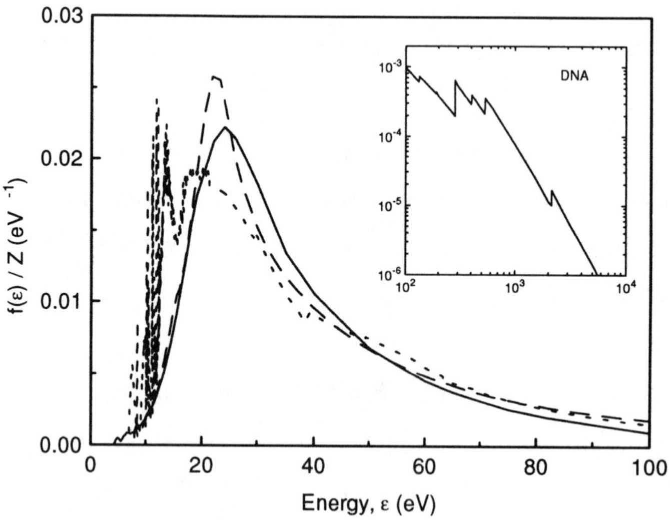

Figure 1. Normalized dipole oscillator strength distributions for DNA, liquid water, gaseous water. Key : (solid line) DNA, (dashed line) liquid water, (dotted line) gaseous water.

Energy Loss by Energetic Electrons

The energy loss properties of energetic electrons can been expressed in terms of the DOSD [11, 15-17]. This formulation relies on the quadratic extrapolation of the dielectric response function into the energy-momentum plane [18], and has the advantage that all the information necessary to describe the property of interest, including the effect of phase, is contained within the DOSD [16].

The expressions for the stopping power, S, the inelastic mean free path, Λ, and the cumulative inelastic collision cross-section, σ(ε,E), for an energetic electron are [17]

$$S = 2\chi n\int_{\varepsilon_o}^{\varepsilon_{max}} f(\varepsilon)G(\varepsilon / E)d\varepsilon, \tag{9}$$

$$\Lambda^{-1} = \chi n\int_{\varepsilon_o}^{\varepsilon_{max}} \{f(\varepsilon)L(\varepsilon / E) / \varepsilon\}d\varepsilon \tag{10}$$

and

$$\sigma(\varepsilon,E) = \left(\Lambda^{-1}(E) + (\chi / E)\int_{\varepsilon_o}^{\varepsilon} f(\gamma)M(\gamma / E,\varepsilon / E)d\gamma\right) / n, \tag{11}$$

with $\chi = \pi e^4 / E$. The functions G, L and M depend upon ε and the kinetic energy of the electron, E, and are given explicitly in reference [17]. (To emphasize the distinction between energetic electrons and photons, ε rather than ω is used to denote the energy variable.)

The track of energy deposition events produced by the passage of an energetic electron through a medium is a stochastic phenomenon [19]. Elaborate Monte Carlo simulation techniques have been developed for modeling the detailed structure of electron tracks [20-24], however, these calculations require a large number of cross-sections, which are frequently unavailable [23]. In contrast, simulation for the number density of the energy loss events is straightforward and uses only the normalized cumulative inelastic cross-section [11, 17, 25],

$$Y = \sigma(\varepsilon,E) / \sigma(\varepsilon_{max},E) \tag{12}$$

which is available from the DOSD via equation (13). The technique, has been described in detail elsewhere [11] and involves only the energy degradation of the primary electron, and all of its secondaries of energy greater than 5 keV, down to 5 keV. Each simulation typically considers 10^5 different primary electron tracks.

Results and Discussion

The dipole oscillator strength distribution for DNA is compared with those for liquid and gaseous water [11] in Figure 1. The peak of the distribution for DNA is at 24 eV which is higher in energy than the peaks for liquid water (21-22 eV) and gaseous water (19 eV). The DOSD for DNA is somewhat similar to that for liquid water, but has a broader peak. There are considerable differences between the DOSDs for the condensed media and that for gaseous water. As the energy loss properties of electrons are related to the DOSD of the medium, the DOSDs suggest significant differences will be found in the energy loss properties of DNA and gaseous water.

The density-normalized stopping powers and inelastic mean free paths for electrons in DNA, liquid water and gaseous water are compared in Figure 2. At low energies, there are very clear discrepancies between the predictions for solid DNA and liquid water, and those for gaseous water; the difference in the stopping power between DNA and gaseous water is greater than 10% for electron energies less than 1 keV.

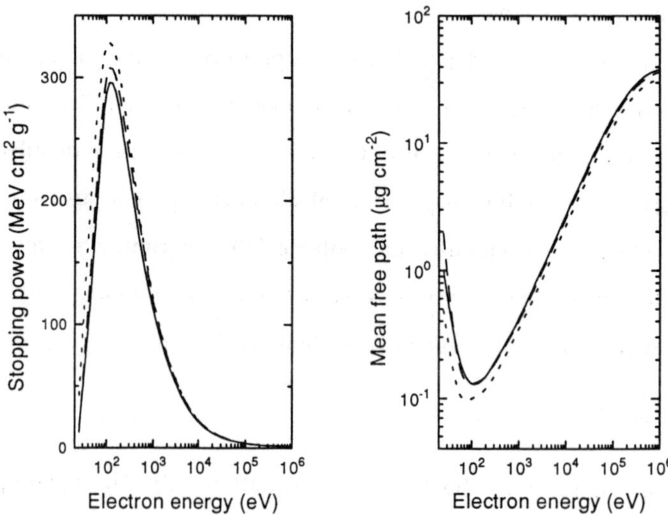

Figure 2. Effect of electron energy on the stopping power and inelastic mean free path of DNA, liquid water and gaseous water. Key as in Figure 1.

The inelastic mean free path, L, describes the mean distance an electron travels between inelastic collisions. A more important parameter is the range of an electron, which is the distance that it travels as its energy degrades from $E+\Delta E$ to E. The csda range is calculated by integrating the inverse of the stopping power from $E+\Delta E$ to E. The effect of incident energy on the csda range of an electron is shown in Figure 3. The ranges in DNA and in liquid water are very similar for all energies, but very different from those in gaseous water at low energies.

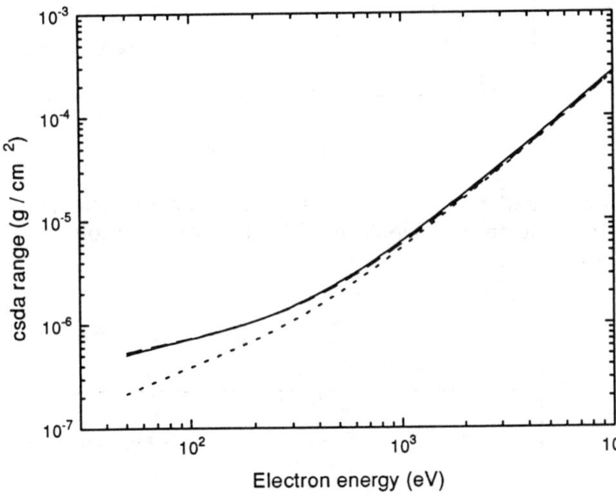

Figure 3. Effect of incident energy on the csda range of electrons in DNA, liquid water and gaseous water. Key as in Figure 1.

The normalized Y cross-sections, equation (12), for 1 MeV electrons in DNA, liquid water and gaseous water are compared in Figure 4. There are some small differences between the Y cross-sections for DNA and liquid water, however, these are insignificant compared to those between these media and gaseous water. The normalized cross-sections for the gaseous water are considerably larger than those of DNA and of liquid water for energy losses smaller than 50 eV.

The effect of medium on the number density of energy loss events for a 1 MeV electron track is examined in Figure 5. The most probable event in DNA, and in liquid water, has an energy of 22 eV which is much larger than in gaseous water

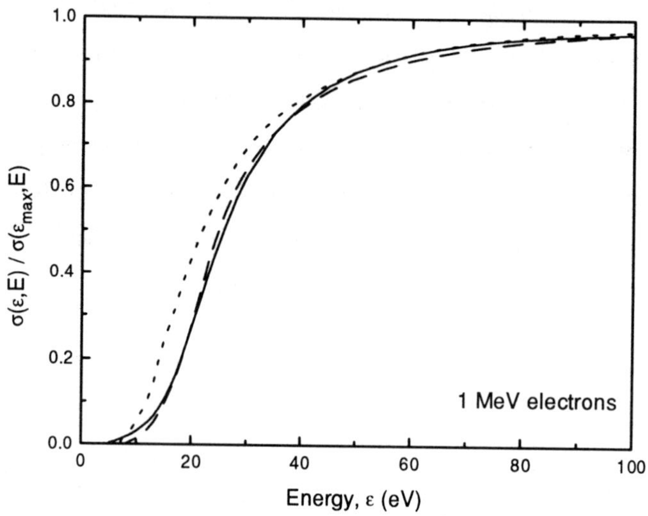

Figure 4. Normalized cumulative cross-section for an inelastic collision for a 1 MeV electron in DNA, liquid water and gaseous water. Key as in Figure 1.

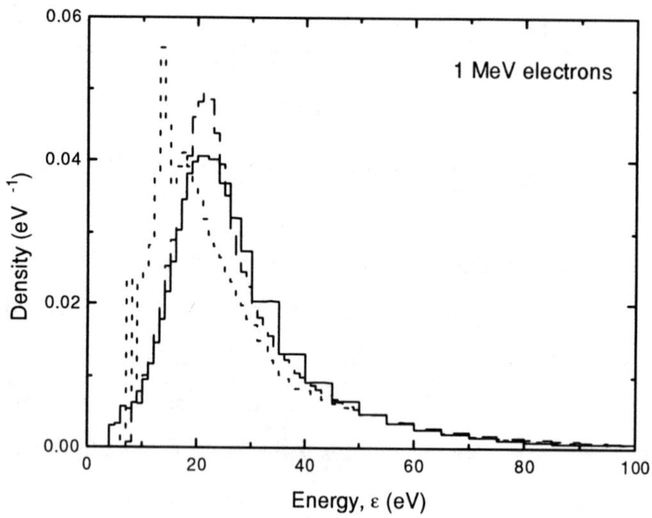

Figure 5. Density of energy loss events for the track of a 1 MeV electron in DNA, liquid water and gaseous water. Key as in Figure 1.

(15 eV). The mean track averaged energy losses are 39 eV, 38 eV and 34 eV in DNA, liquid water and gaseous water, respectively.

Summary

An experimentally based dipole oscillator strength distribution was constructed for DNA and was used to calculate a number of the energy loss properties of energetic electrons in DNA. The predictions of these calculations were compared with similar calculations for liquid water and gaseous water. Very significant differences are found between the results for DNA and gaseous water, however, there is some similarity between DNA and liquid water. These differences suggest that the qualitative validity of any calculation using gaseous water as a model for DNA is questionable and imply that any quantitative predictions are probably in serious error.

Acknowledgment

The research described was supported by the Office of Basic Energy Sciences of the U.S. Department of Energy. This is Contribution NDRL-3723 of the Notre Dame Radiation Laboratory.

References

1. E.M. Fielden and P. O'Neill, ed., *The Early Effects of Radiation on DNA,* NATO ASI Series, Series H, Vol. 25, Springer-Verlag: Berlin 1990.
2. D.E. Charlton, D.T. Goodhead, W.E. Wilson, and H.G. Paretzke, The deposition of energy in small cylindrical targets by high LET radiations, *Radiat. Prot. Dosim.,* **13,** 123-125 (1985).
3. H. Nikjoo, D.T. Goodhead, D.E. Charlton, and H.G. Paretzke, Energy dseposition in small cylindrical targets by ultra-soft X-rays, *Physics in Medicine and Biology,* **34,** 691-705 (1989).
4. H. Nikjoo, D.T. Goodhead, and D.E. Charlton, Energy deposited in small cylindrical targets by monoenergetic electrons, *Int. J. Radiat. Biol.,* **60,** 739-756 (1991).
5. D.T. Goodhead, H.P. Leenhouts, H.G. Paretzke, M. Terrissol, H. Nikjoo, and R. Blaauboer, Track structure approaches to the interpretation of radiation effects in DNA, *Radiat. Prot. Dosim.,* **52,** 217-223 (1994).
6. J.A. LaVerne and A. Mozumder, Energy loss and thermalization of low energy electrons, *Radiat. Phys. Chem.,* **23,** 637-660 (1984).
7. A.M. Rauth and J.A. Simpson, The energy loss of electrons in solids, *Radiat. Res.,* **22,** 643-661 (1964).

8. L. Sanche, Primary interactions of low energy electrons in condensed matter, in *Excess Electrons in Dielectric Media,* C. Ferradini and J.-P. Jay-Gerrin, Editor, CRC Press, Boca Raton, 1991.
9. T. lnagaki, R.N. Hamm, E.T. Arakawa, and L.R. Painter, Optical and dielectric properties of DNA in the extreme ultraviolet, *J. Chem. Phys.,* **61,** 4246-4250 (1974).
10. W.J. Veigele, Photon cross-sections from 0.1 key to 1 MeV for elements Z =1 to Z = 94, *Atomic Data Tables,* **5,** 51-111 (1973).
11. S.M. Pimblott, J.A. LaVerne, A. Mozumder, and N.J.B. Green, Structure of electron tracks in water. 1. Distribution of energy deposition events, *J. Phys. Chem.,* **94,** 488-495 (1990).
12. J.C. Ashley, Stopping-power of liquid water for low energy electrons, *Radiat. Res.,* **89,** 25-31 (1982).
13. R.D. Hudson, Critical review of ultraviolet photoabsorption cross sections for molecules of astrophysical and aeronomic interest, *Rev. Geophys. Space Phys.,* **9,** 305-406 (1971).
14. G.D. Zeiss, W.J. Meath, J.C.F. MacDonald, and D.J. Dawson, Dipole spectrum of water vapor and its relation to the energy loss of fast charged particles, *Radiat. Res.,* **63,** 64-82 (1975).
15. A. Mozumder and J.A. LaVerne, Range and range straggling of low energy electrons: General considerations and application to N_2, 0_2 and H_20, *J. Phys. Chem.,* **89,** 930-936 (1985).
16. J.A. LaVerne and A. Mozumder, Effect of phase on the stopping and range distribution of low-energy electrons in water, *J. Phys. Chem.,* **90,** 3242-3247 (1986).
17. S.M. Pimblott and J.A. LaVerne, Energy loss by electrons in gaseous saturated hydrocarbons, *J. Phys. Chem.,* **95,** 3907-3914 (1991).
18. J.C. Ashley, Interaction of low energy electrons with condensed matter : stopping powers and inelastic mean free paths from optical data, *J.Electron Spectrosc. Relat. Phenom.,* **46,** 199-214 (1988).
19. H.G. Paretzke, Radiation track structure theory, in *Kinetics of Nonhomogeneous Processes. A Practical Introduction for Chemists, Biologists, Physicists, and Materials Scientists,* G.R. Freeman, Editor, Wiley-Interscience, New York, 1987.
20. M. Zaider, D.J. Brenner, and W.E. Wilson, The applications of track calculations to radiobiology. 1. Monte Carlo simulation of proton tracks, *Radiat. Res.,* **95,** 231-247 (1983).
21. J.E. Turner, R.N. Hamm, H.A. Wright, R.H. Ritchie, J.L. Magee, A. Chatterjee, and W.E. Bolch, Studies to link the basic radiation physics and chemistry of liquid water, *Radiat. Phys. Chem.,* **32,** 503-510 (1988).
22. H.G. Paretzke, J.E. Turner, R.N. Hamm, R.H. Ritchie, and H.A. Wright, Spatial distributions of inelastic events produced by electrons in gaseous and liquid water, *Radiat. Res.,* **127,** 121129 (1991).
23. M.A. Hill and F.A. Smith, Calculation of initial and primary yields in the radiolysis of water, *Radiat. Phys. Chem.,* **43,** 265-280 (1994).
24. M. Terrissol and A. Beaudre, Simulation of space and time evolution of radiolytic species induced by electrons in water, *Radiat. Prot. Dosim.,* **31,** 175-177 (1990).
25. A. Mozumder and J.L. Magee, Theory of radiation chemistry. VII. Structure and reaction in low-LET tracks, *J. Chem. Phys.,* **45,** 3332-3341 (1966).

RADICAL PRODUCTION IN BIOLOGICAL SYSTEMS

J. R. Johnson and G. Akabani
Health Protection Department
Life Sciences Center
Pacific Northwest Laboratory
Box 999, Richland
WA, 99352, USA

Introduction

This paper describes our effort to develop a metric for radiation exposure that is more fundamental than adsorbed dose and upon which a metric for exposure to chemicals could be based. This metric is based on the production of radicals by the two agents.

Radicals produced by radiation in biological systems are commonly assumed to be the same as those produced in water despite the presence of a variety of complex molecules. This may explain why the extensive efforts to describe the relationship between energy deposition (track structure) and molecular damage to DNA, based on the spectrum of radicals produced, have not been successful in explaining simple biological effects such as cell killing. Current models (Goodhead, 1992) assume that DNA and its basic elements are immersed in water-like media and only model the production and diffusion of water-based radicals and their interaction with DNA structures; these models lack the cross sections associated with each macro-component of DNA and only treat water-based radicals. It has been found that such models are not realistic because DNA is not immersed in pure water (von Sonntag, 1987).

A computer code capable of simulating electron tracks, low-energy electrons, energy deposition in small molecules, and radical production and diffusion in water like media has been developed. This code is still in at a primitive stage and development is continuing. It is being used to study radical production by radiation, and radical diffusion and interactions in simple molecular systems following their production. We are extending the code to radical production by chemicals to complement our PBPK modeling efforts (Jarvis et al, 1992; Thrall et al, 1992). It therefore has been developed primarily for use with radionuclides that are in biological materials, and not for radiation fields.

Computer Code

The modular approach of the code is shown in Figure 1, and the modules are discussed below.

The Monte Carlo code is divided in two basic modules in order to have a consistent approach to assess energy deposition patterns and chemical outcome on biological systems. The first module is based on physical models for electron transport (track structure); we use the basic principles of single scattering to assess energy deposition in simple molecules. A database was generated containing the cross sections for many well known molecular structures in DNA such as small proteins and amino acids.

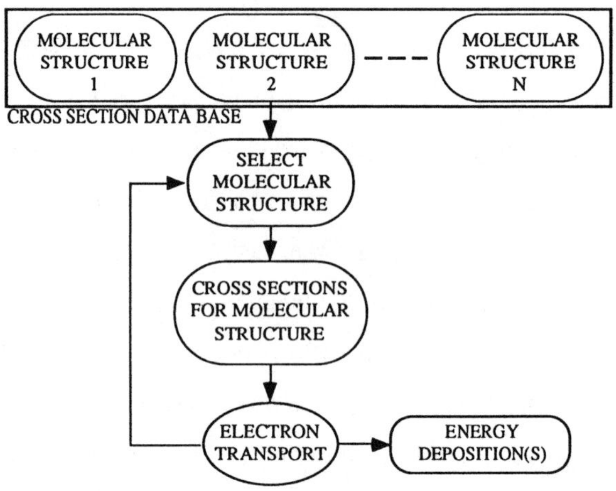

Figure 1. Physics module for the transport of electrons in molecular\ structures based on density concentration

Molecular Structure Data Base

The code requires the user to select a molecular structure from the data base. Molecular structures (simulated in computer code) have been obtained from several sources such as the International Nucleotide Sequence Database (Japan), Protein Data Bank (Brookhaven National Laboratory) or the National Center for Biotechnology Information (NIH). Even though there is a large number of molecules to consider, they can be simplified by reducing them by homology to a basic structure on which calculations can be done. This approach is consistent with our needs. The dynamics of the problem and macromolecules in DNA is too complex for our studies. These simplified molecules are then characterized in terms of probability of dissociation, recombination, and photon and fluorescence production.

Physics Module

This module (Figure 2) simulates the physical processes that can occur following ionization as electrons interact with atoms and molecular structure of the selected molecule. The most commonly observed damage to cells following irradiation is base damage and DNA strand breaks (Cleaver, 1992). This damage results from

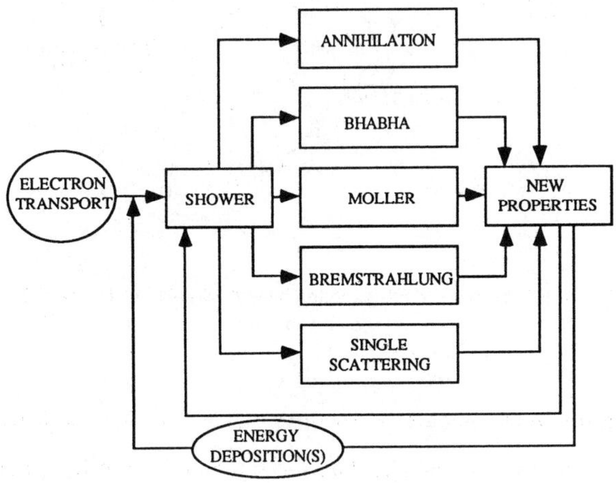

Figure 2. Flow chart of electron transport.

processes involving initial energy deposition (track structure), radical formation and radical reactions. By determining the energy deposition patterns and the simulation of the energy transfer to molecules, radical production and diffusion, and the dissociation of large molecules can be modeled. The results of this modeling is the generation of new molecular species and the correlation of molecular changes with DNA damage.

Chemistry Module

This module is shown schematically in Figure 3. DNA damage following exposure of cells to chemicals is similar, except for the initial energy deposition and radical production. Every specific molecule is characterized in terms of its probability of molecular dissociation, recombination, vibrational energy dissipation, fluorescence, and photon production. Every phenomenon has an associated probability which is randomly selected based on initial radical interaction and available energy.

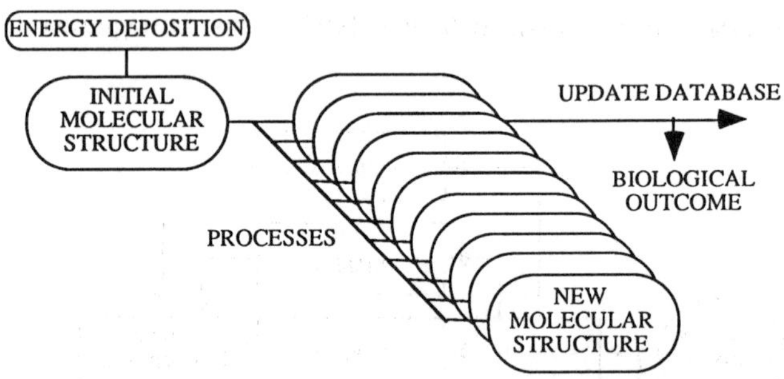

Figure 3. Chemical module for the determination of new molecules and their outcome.

When a molecular species is formed, it is checked against the existing database. If a new molecular species has been formed, it is included in the initial database to then allow it to become part of the overall molecular pool. This process is important with high dose rate exposures when large amounts of new molecular structures are generated in small time periods. In this case the chemical process is completely changed.

Discussion

The ICRP and the NCRP have both endorsed the "total detriment" (TD) concept for radiation protection in their most recent recommendations (ICRP, 1991; NCRP, 1993). This concept involves converting risks to a common unit so that they can be added and compared to other risks in society. The common unit used by these organizations is the "length of life lost" from fatal cancers and severe genetic effects. Non-fatal cancers are included by weighting them with a "quality of life" factor. This process results in their recommendations on dose limits being given in "Effective" dose units. This quantity is derived from "Absorbed Dose," but it is actually a risk unit.

Their charters do not include protection from exposures to toxic chemicals, but a natural and necessary extension of the TD concept in protection is to include them. At present, this is difficult because there is no equivalent concept to absorbed dose for chemical exposures on which "chemical risk" factors can be based. This extension will be more important if the preliminary finding of Claycamp and Luo (1994) that plutonium will cause chemically induced oxidative DNA damage is confirmed by others.

Both chemicals and radiation interact with DNA by the production of radicals, and this process could be a methodology of adding the risk from chemical exposures into the TD concept of protection.

Conclusion

The fundamental objective of this code development is to provide testable predictions on radical production by radiation and chemicals and their interactions with DNA. These tests will lay the foundations upon which a metric of risk from the separate or combined exposure to radiation and chemicals could be based. The code development is nearly complete, and work with others to develop and test predictions has begun.

Acknowledgement

Pacific Northwest Laboratory is operated for the U.S. DOE under contract DE-AC06-76RLO 1830.

References

Cleaver, J., Jen, J., Lutze, L., Wineger, R., And Mitchell, D. DNA and Chromosomal Structure Determine the Molecular Forms of Radiation Damage and Rates of Repair, Proceedings of the 27th Annual Meeting of the NCRP (Proceedings No. 13)

Claycamp, H.G. and Luo, D. (1994) Plutonium-Catalyzed Oxidative DNA Damage in the Absence of Significant Alpha-Particle Decay, Radiation Research, 137, pp 114-117

Goodhead, D.T. (1992) Radiation Tracks in Biological Materials: Initial Damage in Cells, DNA and Associated Structures. pp 25-37 in 'Genes, Cancer and Radiation Protection' Proceedings of the 27th Annual Meeting of the NCRP(Proceedings No. 13)

ICRP 1991, 1990 Recommendations of the International Commission on Radiological Protection, ICRP Publication 60, Annals of the ICRP, Volume 21(1-30) Pergamon Press, Elmsford, New York

Jarvis, M.F., Thrall, K.D., Hui, T.E., Johnson, J.R., Leonowich, J.A., Jarvis, T.T. and Mohatt, J.V. (1992) The Integrated Use of Instrumentation and PBPK Models to Establish, Presented at the 31st Hanford Symposium on Health and the Environment

NCRP 1993, Limitation of Exposure to Ionizing Radiation, NCRP Report No. 116 National Council on Radiation Protection and Measurements, Bethesda, Maryland.

Thrall, K.D., Hui, T.E., Johnson, J.R. and Jarvis, M.F. (1992) Use of Models to establish occupational exposure to Chemicals and/or Radioactivity, Presented at the 4th Annual Meeting of the International Society of Environmental Epidemiology (ISEE) and the International Society of Exposure Analysis (ISEA), Cuernavaca, Morelos, Mexico

von Sonntag, C. (1987) The Chemical Basis of Radiation Biology, Taylor & Francis, London-New York-Philadelphia

REACTION OF HYDROXYL RADICAL WITH B- AND Z-DNA

V. Michalik+, L. Tartier, M. Spotheim-Maurizot and M. Charlier,
Centre de Biophysique Moléculaire, CNRS, rue Charles Sadron,
450 71 Orléans CEDEX 2, France

Introduction

The endogenous processes running in any aerobic organism produce an enormous amount of oxidative damage most of which is finally a result of the hydroxyl radical attack on DNA /1/. The ·OH radicals are also responsible for the majority of effects of ionizing radiation on DNA in aqueous solution /2/ and are probably responsible for a significant part of DNA damage induced in irradiated cells /3/. However, up to now the mechanisms of formation of DNA damages by the ·OH radical attack are only poorly understood. For example, it is not clear if ·OH radical attack depends on the DNA polymeric structure, if it is sequence specific and to what extent the ·OH induced base radicals contribute to the formation of frank strand breaks (FSB), mainly due to attack of sugars. It is excepted now that depending on sequence, topological constraints and environmental factors DNA can adopt different forms such as A, B, C, Z double helices, triple and quadruple helices /4/. Since there are large structural differences between the DNA in B- and Z-forms one may expect differences in the ·OH radical reaction with the two forms as well as differences in the processes subsequent to ·OH radical attack which lead to strand breakage. We have performed a modelling of ·OH radical attack on DNA in the B- and Z-form and compared the theoretical results with the experimental patterns of strand breakage in B- and Z-DNA. The calculated probabilities of ·OH radical attack on sugars and bases in DNA of a given sequence and form were compared to the experimental yields of FSB, i.e. revealed in a neutral pH, and alkali revealed breaks (ARB) in B- and Z-DNA irradiated in dilute aqueous solution. The possible mechanisms underlying the pathways from ·OH radical attack to strand breakage are discussed.

Method

To simulate the ·OH radical reaction with DNA we have used a model described in detail elsewhere /5/. It follows the similar models described in literature /6-9/ used to investigate the radiation induced strand breakage of B-DNA duplex. The model is based on the Monte Carlo method, the model of Smoluchowski, the model of the regular DNA atomic structure and the knowledge of reaction rate constants for ·OH radical reaction with DNA constituents. The model scheme is shown in Figure 1.

The atomic structure of regular B- and Z-DNA can be obtained from the X-ray diffraction data of Arnott /10/. The DNA is represented as an associated volume of all atom volumes given by the scaled van der Waals radii /11/. For the purposes of present work we have generated two DNA structures. The first one is 3'-$(GC)_{12}$CTAGGAGATCTC-5' in the B-form and the second structure has the same sequence but the first part is in the Z-form and the second part in the B-form. Since we are not able to simulate the region of the junction between the two forms we have generated $(GC)_{12}$ in the Z-form, CTAGGAGATCTC in the B-form and then this two structures were coupled and used together for the simulations of ·OH radical attack. The generated atomic DNA structure is enveloped by a water cylinder

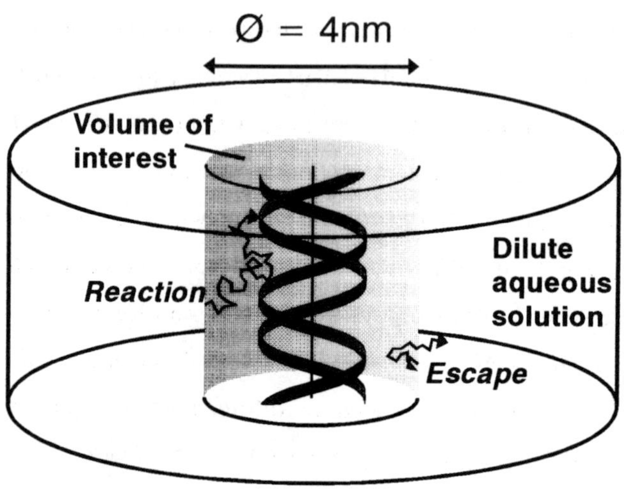

Figure 1. Scheme of the model. DNA placed in dilute aqueous solution is enveloped by a cylinder representing the volume of interest. ·OH radicals start their diffusive motion randomly from the surface of the volume of interest and can either react with DNA or escape from the volume.

4 nm in diameter representing the volume of interest in which the diffusion of ·OH radicals is simulated by the Monte Carlo technique. They start their diffusive motion at a random location from the cylinder coat in a random direction and then the simulation is carried out by a number of steps. At each step the ·OH radical is displaced in a random direction by 1 Å. The simulation of diffusion is terminated either by the ·OH radical reaction with the DNA or by its escape from the cylinder surrounding the DNA.

The simulation of hydroxyl radicals reaction with the DNA molecule is based on the model of Smoluchowski /12/. The ·OH radical which diffused to a distance less than or equal to a reaction radius from the reaction site of a given moiety will react with a unit probability. The reaction radii can be calculated from the rate constants /13/ for ·OH reaction with the DNA constituents. The calculated radii are: 1.2 Å for deoxyribose, 2.9 Å for adenine and cytosine, 3.0 Å for thymine and 4.3 Å for guanine. The considered reaction sites in DNA are C(6) and C(5) in pyrimidines, C(4), C(8) and C(5) in purines and all carbon positions in sugar moiety /2,14/. Accordingly, the ·OH radical which diffused to a distance less than or equal to 3.0 Å from the C(5) or C(6) carbon of thymine will react with this DNA constituent with a unit probability. Similarly, each ·OH radical being at a distance less than or equal to 1.2 Å from any carbon of sugar will react with the sugar.

Results

The results are presented in the form of the specific and overall probabilities of ·OH reaction with the sugar or base moiety. The specific probability refers to a particular nucleotide. The overall probability is an average value of the specific probabilities. As shown in Table I the respective overall probability of ·OH reaction with the sugar or base moiety is strongly dependent on the DNA form. The bases are more susceptible to ·OH radical attack in the Z form of DNA than in the B form. It correspond to the structural differences between the two DNA forms. In the B form of DNA the rise is 3.4 Å and the major and minor grooves are of nearly the equal depth. In the Z-form there is a higher rise per base pair (3.7 Å) and the major "groove" is convex with exposed cytosine C_5 and guanine N_7 and C_8 atoms.

Consequently, the bases are more accessible to ·OH radicals in the Z-form than in the B-form. Another factor influencing the proportion of ·OH radicals reacting with the bases and sugars is the base composition of DNA. For the GC rich B-form reactions with the bases represent 76% of all ·OH reactions with DNA. For B-DNA with the GC/AT ratio equal to one, reactions with the deoxyribose moiety represent 32% of all ·OH reactions with DNA. This is due to the relatively great reaction radius of the guanine moiety which thus reacts with ·OH radical more efficiently than do other bases.

DNA SPECIFICATION	·OH ATTACK [%]		DISTRIBUTION OF OH ATTACK ON DNA A:T:C:G		
	bases	sugars	nucleotides	bases	sugars
B-DNA double helix GC/AT = 1	68	32	1:1:1:2	1:1:1:3	1:1:1:1
B-DNA double helix GC/AT = ∞	75	25	0:0:3:7	0:0:1:4	0:0:1:1
Z-DNA double helix	92	8	0:0:1:13	0:0:1:20	0:0:1:2

Table 1. Calculated probabilities of ·OH radical reaction with sugar and base moieties in DNA of B- and Z-form.

Another factor modulating the attack of hydroxyl radical on DNA is the sequence of bases /5/. As shown in Figures 2 and 3 there is a base and sequence dependent variation in the number of ·OH reactions with a nucleotide. We have found that the number of ·OH attacks on a particular nucleotide varies between 0.6-1.7 times the average level of the ·OH reaction witha nucleotide in B-DNA duplex. For deoxyribose in B-DNA duplex the differences are considerably smaller, *i.e.* between about 0.75 and 1.25 times the average level of ·OH reaction with the sugars. For Z-DNA the pattern is simple and has the same shape for the ·OH reaction both with nucleotides and deoxyriboses. In the former case the differences are quite big, i.e. in the range from 0.1 to 2.0 while in the latter case they are smaller, *i.e.* between 0.6 and 1.3 times the average level of hydroxyl radical attack on nucleotides and sugars, respectively. In both cases guanine sites are more attacked than the cytosine sites and the difference is much greater for bases than for sugars.

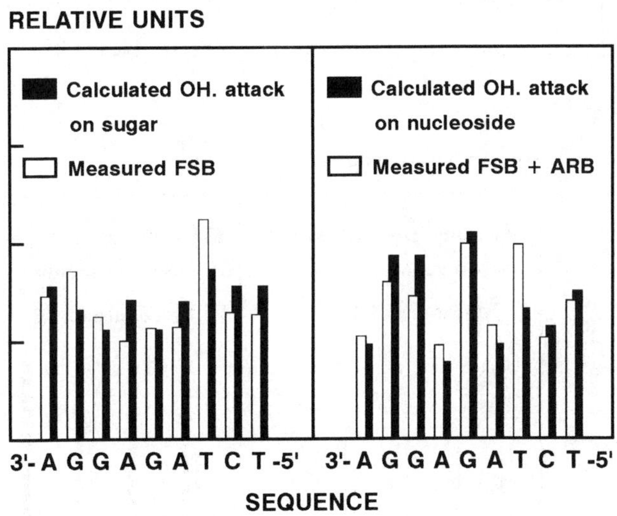

Figure 2. Patterns of ·OH radical attack on nucleotides and sugars in B-DNA compared with the patterns of strand breakage measured after gamma radiolysis /15/ in samples treated by piperidine revealing both frank strand breaks (FSB) and alkali revealed breaks (ARB) and neutral pH revealing only FSB. The height of each bar is proportional to the area of the band which correspond to a given nucleotide. It represents the observed frequency of breakage at the considered site.

In the same figures the predicted values are compared with experimental results /15/.In the experiments a plasmid with inserted Z-prone sequence was used. When plasmid is in the native supercoiled form the Z-prone sequence is in the Z-form. When the plasmid is relaxed it is in the B-form. Both the supercoiled and relaxed plasmids were irradiated, then a 92 base-pair-long fragment containing the Z-prone sequence was obtained after digestion by restriction enzymes. The fragment was labelled and deposited on a sequencing gel to identify either FSB or FSB and ARB. The ·OH radical reaction with nucleotides is compared with the measured cleavage of DNA treated by piperidine (alkaline conditions (ARB+FSB) and the damage on deoxyribose is compared with the yields of FSB. We can conclude that there is a good qualitative agreement between both sets of data for region in the B form as well as in the Z form. In spite of the fact that the quantitative agreement is much better for the part in B-form than for the part in Z-form the calculated values for guanosines are higher than the measured values in both forms.

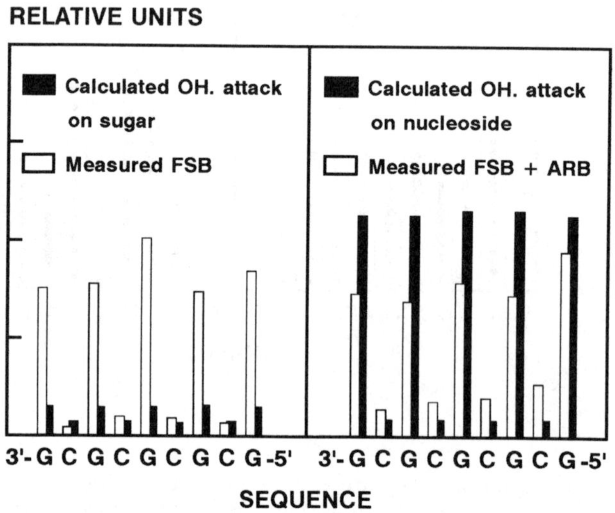

Figure 3. Patterns of ·OH radical attack on nucleosides and sugars in Z-DNA compared with the patterns of strand breakage measured after gamma radiolysis /15/. The part in B-form (shown in Figure 2) was used to normalize the data for the absolute comparison.

Discussion

The good agreement between modelling of attack on sugars and experimental data obtained in neutral conditions show that ·OH induced sugar radicals are mainly responsible for FSB and that ·OH induced base radicals contribute only slightly to the formation of these damages in B-DNA. The agreement between the modelling of the attack at both sugars plus bases and the experimental data obtained after alkaline treatment show that base radicals contribute to the formation of ARB in B-DNA. These conclusions are consistent with our previous study on the radiosensitivity of single- and double-stranded B-DNA /5/. We have found again here, but on a shorter sequence, the already reported variation of radiosensitivity of nucleotides. The major determinants of this variation are the character of the chemical reaction of hydroxyl radical with different DNA components together with their specific position in the structure (sequence).

The fact that the calculated values for guanosines are systematically greater than the measured ones can be explained as follows: a) the rate constant for guanine used in calculations is too big, b) the probability to be revealed after the alkali treatment as a strand break is for guanine base damages different than for other base damages, and c) a transfer of damage from guanine base to its sugar or adjacent bases modifies the measured patterns. For B-DNA each of these factors may play a role, but not a significant one, since the observed differences are not big. For Z-DNA the factors a) and b) could be responsible for the observed differences between the measured FSB+ARB and the calculated ·OH attack on nucleosides but cannot explain the differences found for the measured FSB and the calculated ·OH attack on sugars. Thus for Z-DNA the third factor seems to be important.

It is known that the guanosines are in Z-DNA in *syn* orientation /4/. This gives rise to interatomic contacts which can facilitate the transfer of damage from the base to the corresponding sugar. Moreover, the deoxyguanosine base is stacked with the cytosine in 3' direction but not at all with the cytosine in 5' and this can also facilitate a transfer of damage to the cytosine on 3' side. The involvement of such a transfer of ·OH damage from base to sugar has already been suggested in the formation of strand breaks /16/ and the mechanism might be an intra-DNA H-abstraction /17,18/. If we assume that about 30% of the guanine base radicals produced by ·OH reaction with Z-DNA are transferred to the associated deoxyribose and that about 10% are transferred to the cytosine in 3' direction, the calculated values fit much better with the measured ones (Figure 4).

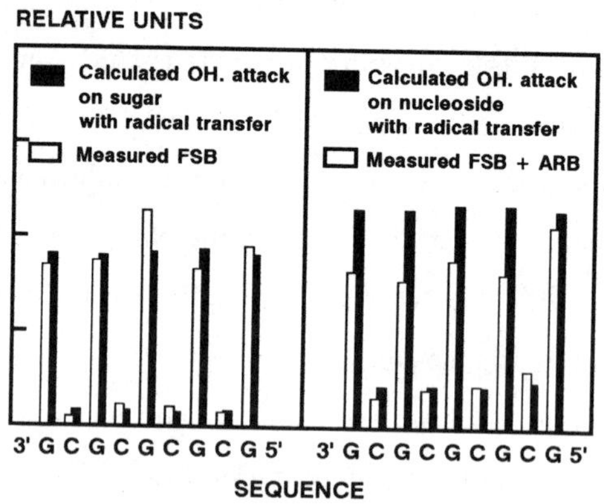

Figure 4. The calculated patterns of ·OH radical attack on nucleosides and sugars in Z-DNA when a radical transfer from guanine base to its sugar and the cytosine on 3' side is assumed, are compared with the measured patterns of strand breakage /15/.

It is rather surprising to see that from the chemical point of view a relatively simple model of ·OH radical interaction with DNA can very well predict sequence specific, non-trivial patterns of DNA cleavage for B-DNA and also qualitative differences in the radiolysis of B- and Z-forms of DNA in aqueous solution. The comparison of theoretical results with the experimental ones justifies the above described modelling approach to be used for studying a DNA damage induced by other reactive radicals in aqueous solution.

Acknowledgement

This work was supported by the Association pour la Recherche sur le Cancer (ARC) and the Ligue Nationale contre le Cancer. V.M. was supported by a fellowship from the Ministere de l'Enseignement Supérieur et de la Recherche and by grant 204/93/2451 from the Grant Agency of Czech Republic and by grant 135102 from the Academy of Sciences of Czech Republic.

References

1. B. Demple. and L. Harrison, Repair of oxidative damage to DNA: Enzymology and biology. *Ann.Rev.Biochem.* **63**, 915-948 (1994)
2. C.von Sonntag, *The chemical basis of radiation biology*, Taylor & Francis, London, 1987
3. J.F.Ward, DNA damage produced by ionizing radiation in mammalian cells: Identities, mechanisms of formation and reparability. *Progr.Nucl.Acid Res.Mol.Biol.* **35**, 95-125 (1988)
4. W.Saenger, *Principles of nucleic acid structure*, Springer Verlag, New York, 1983
5. V.Michalik, M.Spotheim-Maurizot and M.Charlier, Calculation of hydroxyl radical attack on different forms of DNA. (1994) submitted
6. A.Chatterjee and J.L.Magee, Theoretical investigation of the production of strand breaks in DNA by water radicals. *Radiat.Prot.Dosim.* **13**, 137-140 (1985)
7. H.A.Wright, J.L.Magee, R.N.Hamm, A.Chatterjee, J.E.Turner and C.E.Klots, Calculations of physical and chemical reactions produced in irradiated water containing DNA. *Radiat.Prot.Dosim.* **13**, 133-136 (1985)
8. M.Terrissol and E.Pomplun, Computer simulation of DNA-incorporated ^{125}I Auger cascades and of the associated radiation chemistry in aqueous solution. *Radiat.Prot.Dosim..* **52**, 177-181 (1994)
9. M.Bardash and M.Zaider, A stochastic treatment of radiation damage to DNA from indirect effects. *Radiat.Prot.Dosim.* **52**, 171-176 (1994)
10. R.Chandrasekaran and S.Arnott, The structures of DNA and RNA helices in oriented fibers. In *Numerical data and functional relationship in science and technology, New series* VII, 1b, (W.Saenger, Ed.), pp.31-170. Springer Verlag, Berlin (1989)
11. V.Michalik and M.Begusová, Target model of nucleosome particle for track structure calculations and DNA damage modelling. *Int.J.Radiat.Biol.* (1994), in press
12. M.V.Smoluchowski, Drei Vortrage uber diffusion, Brownsche molekularbewegung und koagulation von kolloidteilchen. *Physik.Zeits.* **17**, 557,(1916)
13. G.V.Buxton, C.L.Greenstock, W.P.Helman and Ross, A.B., Critical review of rate constants for reactions of hydrated electrons, hydrogen atoms and hydroxyl radicals in aqueous solution. *J.Phys.Chem.Ref.Data* **17**, 513-586 (1988)
14. S.Steenken, Purine bases, nucleosides, and nucleotides: Aqueous solution redox chemistry and transformation reactions of their radical cations and e^- and OH adducts. *Chem.Rev.* **89**, 503-520 (1989)
15. L.Tartier, V.Michalik, M.Spotheim-Maurizot, A.R.Rahmouni, R.Sabattier and M.Charlier, γ-radiolysis of Z-DNA. (1994) submitted
16. P.O'Neill and E.M.Fielden, Primary free radical processes in DNA. *Adv.Radiat.Biol.* **17**, 53-120 (1993)
17. K.Hildebrand and D.Shulte-Frohlinde, ESR studies on the mechanism of hydroxyl radical induced strand breakage of polyuridilic acid. *Int.J.Radiat.Biol.*, **55**, 725-738 (1989)
18. L.R.Karam, M.Dizdaroglu and M.G.Simic, Intramolecular H atom abstraction from the sugar moiety by thymine radicals in oligo- and polydeoxynucleotides. *Radiat.Res.*, **116**, 210-216 (1988)

CALCULATION OF RANGE AND DISTRIBUTIONS OF DAMAGE TO DNA BY HIGH- AND LOW-LET RADIATIONS

Hooshang Nikjoo and David E. Charlton
MRC Radiobiology Unit, Chilton, Didcot, OX11 0RD, UK.
Concordia University, Physics Dept., Montreal Canada

Abstract

Data are presented for the distribution and complexity of damage in DNA by radiations of high- and low-LET. The results of two methods of calculation are described: (a) Tracks of alpha particles and electrons were examined for clusters of ionizations of differing complexity in spherical volumes of DNA size targets. (b) Patterns of energy depositions in sugar/phosphate moiety and the bases were converted to events which may or may not lead to DNA damage. The patterns were examined for the distribution of number of damage sites and the lengths of the damage segments in DNA. The results indicate that the majority of damage in short segments of DNA are simple and the frequency of more complex damage types is very low. The data suggest that biophysical modelling should be carried-out in higher order structures of DNA such as chromatin fibre.

Introduction

The aim of this study is to analyse the distribution of initial damaging events in DNA exposed to low- and high-LET radiations. For this purpose we have examined several properties of DNA damage which may characterise the differences between high- and low- LET radiations and have quantified the frequency and magnitudes with the severity of the damage. The damage event in the context of this paper is defined as either a physical event, such as the number of ionizations or as an amount of energy deposited in the target volume, or as a chemical event, produced by diffusing OH radicals, both of which can lead to damage in DNA such as strand breakage or base damage.

Ionizing radiations produce a very large numbers of physical and chemical events in biological environments (Hutchinson 1985). Very few of these initial events are of relevance to the final biological effects. There is now a general consensus that there exists a correlation between initial biophysical/chemical damage in the DNA and the final observed biological effects (Goodhead and Nikjoo 1989, Goodhead and Charlton 1986). To explain this correlation a number of hypotheses have been put forward to describe the likely consequences of the biological effects in terms of clustered damage in DNA (Goodhead 1988, 1994) or locally multiple damage sites or LMDS (Ward 1988, 1991). Both the 'clustered damage' and 'LMDS' describe in principle the same phenomena and proposed that the more severe damage to DNA has less likelihood of being repaired than the more prevalent simple damage. Attempts to quantify the concept of 'clustered damage' or 'LMDS' in DNA in terms of a quantity of energy deposited in the target volume or by numbers of ionizations in spheres of DNA and larger target sizes have been made by several authors. Monte Carlo track structure studies of energy deposition in targets similar to biological macromolecules have been carried out by Charlton *et. al.*(1989), Holley and Chatterjee (1990), Turner *et. al.*(1982), Zaider (1993), Paretzke (1987) and Nikjoo *et. al.*(1994a). Brenner and Ward (1992) and Michalik (1993) have carried out cluster analysis of ionizations for ion and electron tracks. In general, these authors find correlation between double strand breaks and clusters of ionizations containing at least two to five ionizations in sites of diameters 1 to 4 nm. Studies on the distribution of damage length in the hit section are few and indicate that the damage region is a short segment of the order of about 5 to 15 base pairs in length (Martin and Haseltine 1981, Glickman 1987).

Model of DNA

Previously, we used a volume model of DNA in our calculations (Charlton *et. al.*1989, Nikjoo *et. al.*1994). The volume model describes the regions of sugar phosphate and bases of a linear segment of double helix canonical B-DNA. Recently we developed an atomistic model of DNA including the first hydration shell (Umrania *et. al.*1994) for our work on biophysical modelling including early chemistry of radiation damage. Figure (1) shows the two models of DNA side by side. The atomistic model is a decamer B-DNA including the primary hydration shell. The hydration shell which is a knowledge-based model was generated to provide the most likely sites for the distribution of water molecules around the four

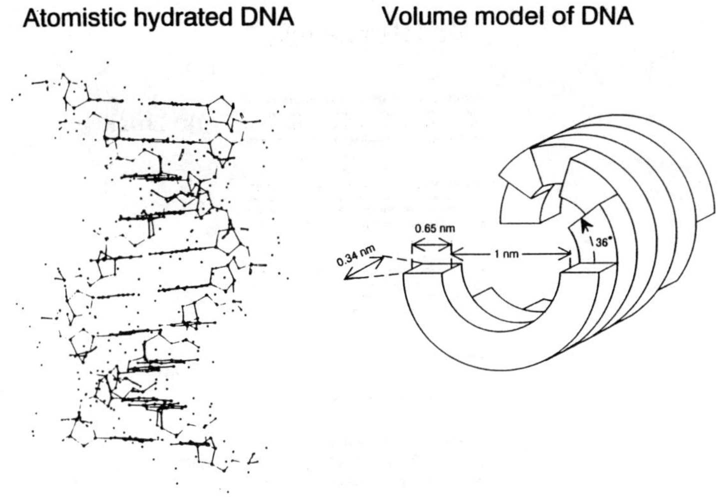

Figure 1. The volume model and the atomistic hydrated model of double-stranded DNA used in the calculations. The volume model shows the volume occupied by one sugar/phosphate chain. The atomistic hydrated decamer B-DNA shows the distribution of water molecules around the four bases, the sugar residues and the phosphate groups.

bases, the sugar residues and the phosphate groups in a canonical B-DNA conformation. The simple volume model was compared with the atomistic canonical hydrated DNA in terms of strand break formation and it was found that for the present calculations the two models produce very similar results. Therefore, it was decided to carry-out the present study with the volume model of DNA which is simpler to use.

<u>Model of DNA damage</u>

In line with our previous calculation, Figure (2) shows the various type of damage considered in this paper. In this paper 'Damage' is produced by direct energy deposition greater than 17.5 eV in sugar-phosphate and bases (Charlton and Humm 1987, Pomplun *et. al.*1992) or by diffusing OH radicals (Nikjoo *et. al.*1994).

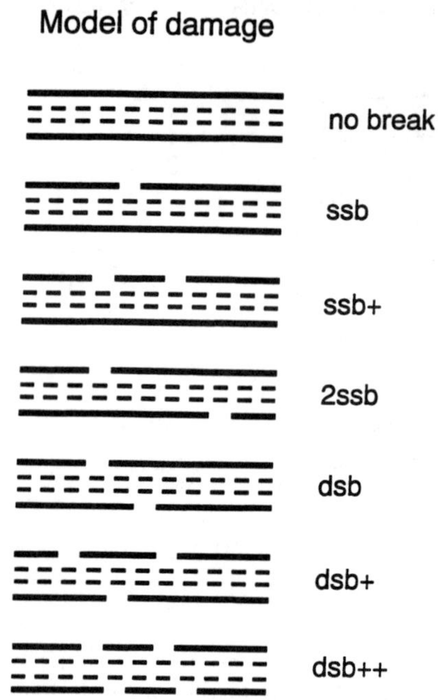

Figure 2. Model of DNA damage. Various possible damage types in sugar-phosphate of a double stranded DNA. A double strand break, dsb, is formed when two single strand breaks on opposite strands were separated by not more than 10 bp.

'Damage' in sugar-phosphate has been assumed to lead to a strand break and in bases to base damage. Strand break formations are considered to originate only from damage to sugar/phosphate moiety. A double strand break was assumed to be formed when two single strand breaks on opposite strands were within a maximum separation of 10 base pairs. If the distance between the two strand breaks was found to be greater than 10 base pairs then it was considered to be a '2ssb' damage. Although base damage has been included in models of damage no mechanism of strand break involving base damage has been invoked in this study.

Frequency of energy deposition in DNA

The first step in these calculations is simulation of electron and ion tracks molecular interaction by interaction. For alpha particles Monte Carlo track segments were obtained by the computer code PITS (Wilson *et. al.*1994). The code PITS (Positive Ion Track Simulation) has been written for simulating ion tracks in a variety of media, with or without secondary electron transport. The code was used to generate segments of tracks of alpha particles of energy 1.2 to 20 MeV in water. For the secondary electron cross section the input data for the code MOCA8B (Paretzke 1987) was used. For electrons used in this study the code CPA (Terrissol and Beaudre 1990) was used to generate the full slowing down of electrons in liquid water and the associated water radicals generated at time 10^{-12} s. The conceptual framework of finding the frequency of energy deposition in biological targets is to place the target molecule and the track of ion or electron in a virtual volume large enough to contain the entire track Figure (3). Then a search was made to see if any event was in the volume of the target.

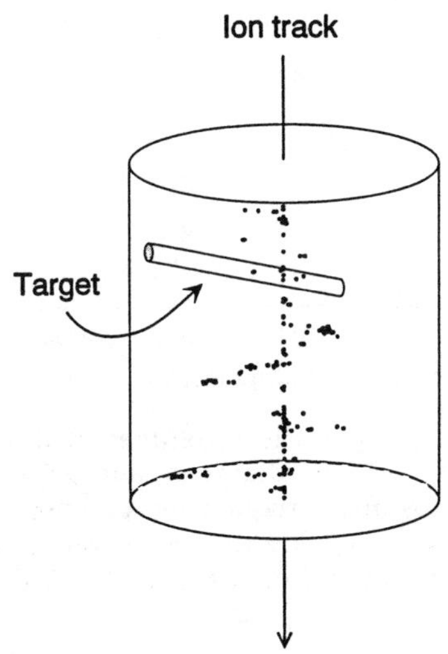

Figure 3. Schematic view of scoring of sites of energy deposition or ionizations in a segment of DNA by a track of 1 MeV proton. The cylinder defines a virtual volume which contains the entire track and its surfaces act as boundary for generation of random targets superimposed on the track.

The energy and location of the event was recorded for further analysis (Charlton *et. al.*1985, Charlton *et. al.*1992, Nikjoo *et. al.*1991). Figure (4) shows the frequency of energy depositions in a cylindrical target of diameter 2.3 nm and 18 nm in length similar to a linear segment of DNA 54 bp. The left ordinate show the absolute frequency of energy depositions in the cylindrical volume when randomly positioned and oriented in water and irradiated with 1 Gy of the reference radiation. The right ordinate show the number of such segments hit by the reference radiation. These data provide a general correlation between local energy depositions and observed biological effects. From these studies various hypothesis have been put forward on features of ionizing radiations (Goodhead and Nikjoo 1989, Goodhead 1994).

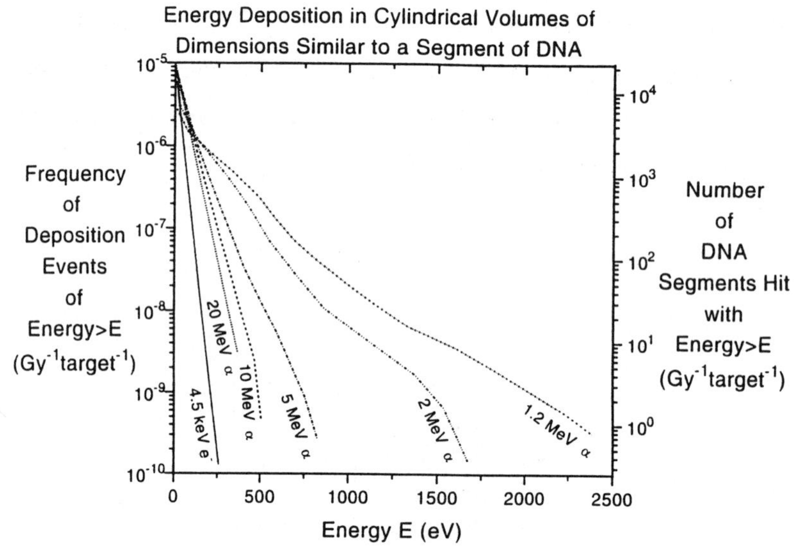

Figure 4. Energy deposition in cylindrical volumes of dimensions similar to a segment of DNA, 2.3 nm diameter and 18.4 nm long. The left ordinate shows absolute frequencies of energy depositions when DNA target exposed to 1 Gy of the reference radiation. The right ordinate shows the number of segments hit with energy greater than E per Gy of radiation.

Clustering of ionizations

Spatial distribution of ionizations in alpha-particle and electron tracks were obtained by the method of cluster analysis (Hartigan 1975) using the S-Plus K-means functions (Statistical Sciences, Inc.). Spheres of 2.3 nm diameter were centred around each cluster. Figure (5) show the frequency of clusters of ionizations in spheres of diameter 2.3 nm for 10 keV electrons and 20 MeV alpha-particle. These data show that for electrons the majority of ionizations are singletons and the frequency of more complex clusters of ionizations drops very rapidly. In contrast, the clusters for alpha-particles show higher frequency for clusters with larger numbers of ionizations. The frequency of more complex clusters increases with higher LET alpha particles and track ends.

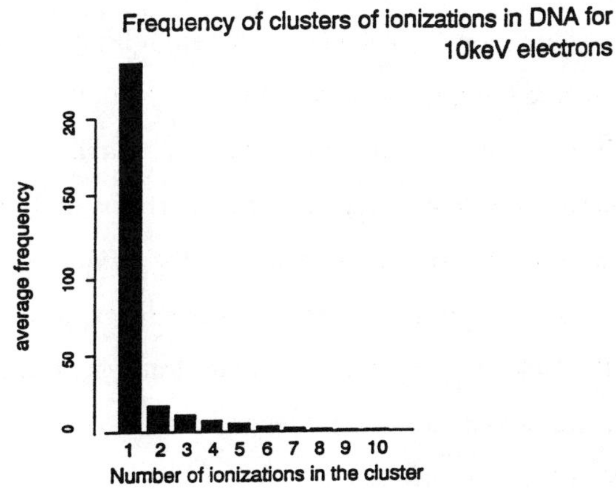

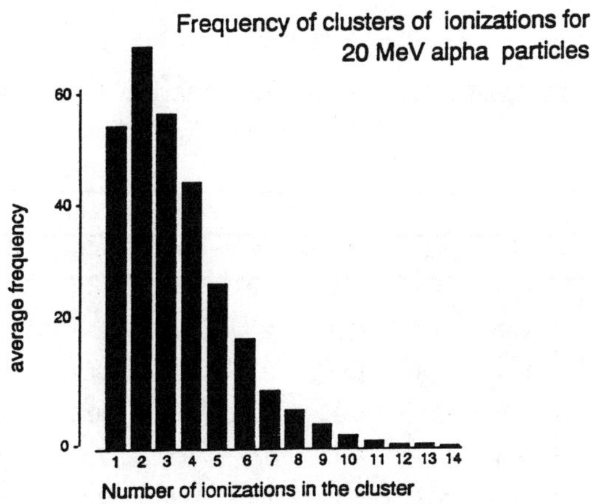

Figure 5. Frequency of clusters of ionizations in volumes 2.3 nm diameters for 10 keV electrons and 20 MeV alpha particles. For 10 keV electron 100 full slowing down tracks were sampled. For alpha-particles 10 microns of constant energy segment were sampled.

Distribution of energy depositions and damage in DNA

Detailed descriptions of energy depositions in sugar-phosphate and base regions of the volume model of DNA are presented in Figure (6). The top and bottom strings depict the double helix sugar-phosphate backbone and the centre string represents the bases of the DNA. The three vertical dots form a schematic representation of a nucleotide pair. Figure (6) shows a few examples of energy deposition in DNA for 10 MeV and 2 MeV alpha particles for damage resulting from direct energy deposition in DNA.

The distributions show variety and diversity in types of damage. Figure (6) show examples of hit regions by a 4.5 keV electron in liquid water. Here damage by OH radicals is included. Figure (6) shows that OH radicals are less localised in the DNA segment hit than hits from the direct damage. Table (1) shows a summary of frequencies of strand breakage data for alpha particles and electrons in terms of percentage of the total DNA segments hit by the reference radiation. The 'nb' column refer to those events which do not induce strand breakage in DNA. The final column of the table (1) (for complex double strand breaks, dsb++) clearly shows that as the LET increases an increasing fraction of the damage is of this type. For low energy electrons such events are rare.

Table (1)

Frequency distribution of strand breakage (%) in Sections of DNA
54 Nucleotides Long

	Model of damage						
Particle	nb	ssb	ssb+	2ssb	dsb	dsb+	dsb++
1.2 MeVa	32.6	18.8	11.4	1.0	0.8	16.5	18.9
2 MeVa	33.4	21.8	13.2	1.1	0.9	17.7	11.9
5 MeVa	38.4	28.8	12.5	0.9	0.6	14.9	3.9
10 MeVa	44.8	32.1	9.6	0.8	0.6	10.5	1.6
20 MeVa	51.6	32.0	6.9	0.6	0.5	7.4	1.0
4.5 keV e^-	64.1	27.8	3.5	1.4	0.6	2.4	0.2

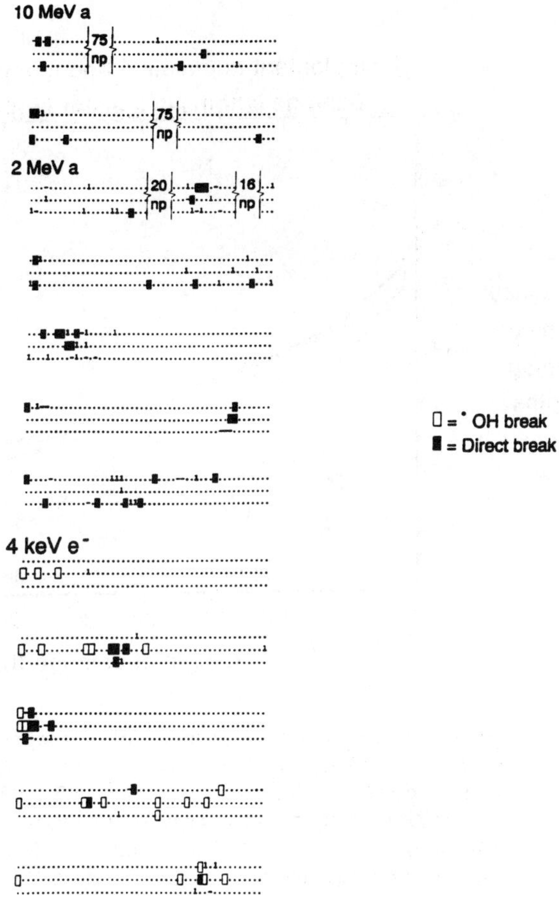

Figure 6. Various examples of sites hit by physical and chemical events in a track for alpha particles and electrons. The three horizontal strings represent the sugar-phosphate and the base regions of the DNA. The top and bottom pixels represent the two sugar-phosphate components of the base-pair and the middle one show the bases. The closed rectangle show the sites with hits leading to a damage by direct energy depositions. Sites with a '1' or '-' were hit with event(s) of various size which did not lead to a damage. The open rectangles show the sites of damage by diffusible hydroxyl radicals.

Frequency distribution of the number of damaged sites in the hit section of DNA and the length of the section were obtained from analysis of a large number of string data for each radiation energy. Here length is the number of base pairs between the extremities of the damage. Figure (7) show the complement-cumulative frequency

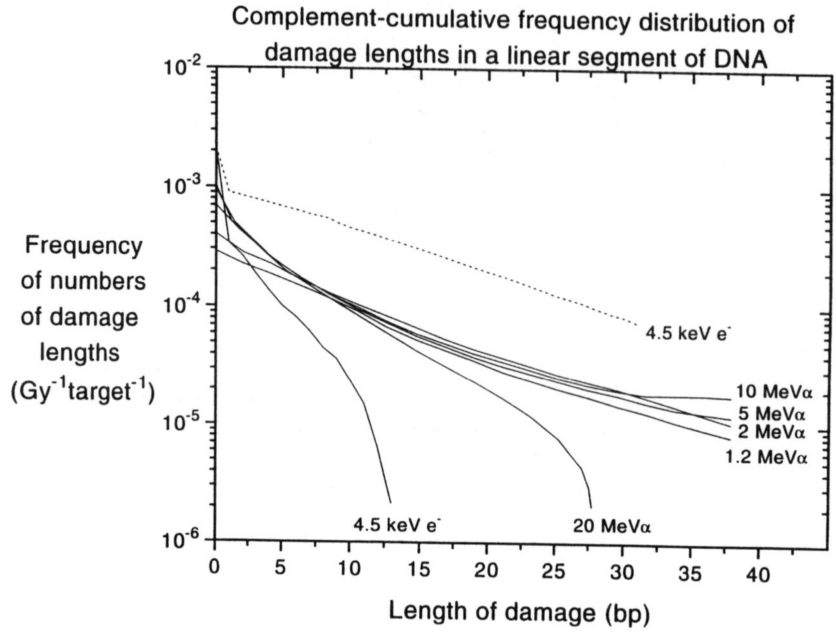

Figure 7. Absolute frequency of number of damaged lengths per Gy in a segment of DNA 18.4 nm, 54 bp, long. The solid lines show the distributions for alpha particles and 4.5 keV electrons for damaged hits arising from direct energy deposition events in DNA segment. The dotted line for 4.5 keV electron include damaged hits from both the direct energy depositions and the diffusible hydroxyl radicals.

distributions of damage lengths in a linear segment of DNA 54 base pairs long for alpha particles and electron. The alpha particles data show only the 'damage' for direct energy deposition while the data for 4.5 keV electrons show 'damage' from the diffusible radicals and the direct effect on the DNA. The effect of radical damage considerably increases the length of the damaged section. In Figures (7) and later in Figure (8) the dotted line refer to the composite damage contributed by direct and diffusing OH radicals in DNA. The distributions for the numbers of damaged sites within the hit region of the DNA are shown in Figure (8). While Figure (7) showed that the length of the damaged sections for alpha-particles were very similar, this figure show that the amount of damage within the length is strongly LET dependent.

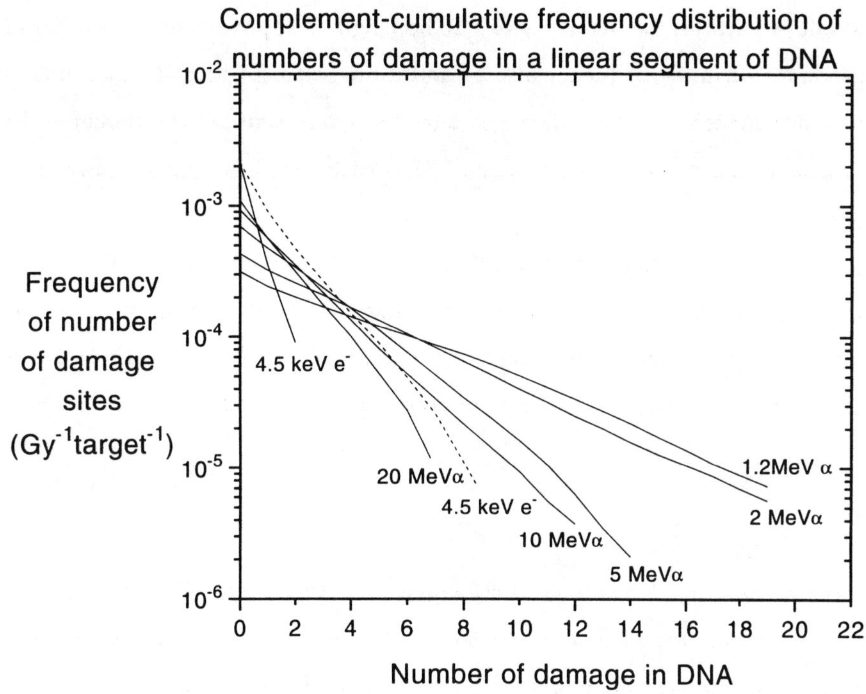

Figure 8. Comparison of frequency of number of damaged sites for various radiations. The solid lines show the distributions for alpha particles and 4.5 keV electrons for damaged hits arising from direct energy deposition events in DNA segment. The dotted line for 4.5 keV electron include damaged hits from both the direct energy depositions and the diffusible radicals.

Conclusions

We have presented data on the characteristics of damage in DNA by alpha particles and electrons. Data were presented on clustering of ionizations in DNA molecule for electrons and alpha particles, on distribution of damage length in linear segments of DNA and their number distributions. The data show that most events are simple both for low and high LET radiations and that the frequency of severe damage, even for high LET radiations is rare. For example, from Figure (5b) only about 20% of clusters in spheres with diameters 2.3 nm have more than 5 ionizations.

The frequency of the length of DNA damaged drops quickly with length. For 10 MeV alpha particles the frequency of a 10 bp length is approximately one order of magnitude less than the frequency of 1 bp and drops another order of magnitude at 20 bp. While the lengths of the damaged sites were very similar for different alpha particles, above 5 damaged sites the higher LET particles produced significantly more damage in the DNA.

Our data on frequency of lengths of damaged sites suggests small deletion lengths of DNA. However, there is evidence that genome structural organizations may influence the distribution of damage in DNA (Sapora *et. al.* 1991). Therefore, it is suggested to carry out the calculation in higher ordered structure of DNA such as chromatin.

Acknowledgements

The authors thank Philippa Middlehurst and Adrian Ford for helping with the preparation of this manuscript. The work was partially supported by the Commission of European Communities contract no. F13P-CT92 0027.

References

Brenner D J and Ward J F, 1992. *Constraints on energy deposition and target size of multiple-damaged sites associated with DNA double strand breaks*. Int. J. Radiat. Biol., **61**, 737-748

Charlton D E and Humm J L, 1988. *A method of calculating initial DNA strand breakage following the decay of incorporated* I^{125}. Int. J. Radiat. Biol. **53**, pp 353-365.

Charlton D E, Nikjoo H, Humm J L, 1989. *Calculation of initial yields of single- and double-strand breaks in cell nuclei from electrons, protons and alpha particles.* Init. J. Radiat. Biol., **56**, 1-19.

Charlton D E, Goodhead D T, Wilson W E, Paretzke H G. 1985. *The deposition of energy in small cylindrical targets by high LET radiation*. Radiat. Prot. Dos. **13**, 123-125.

Charlton D E, Nikjoo H, Goodhead D T, 1992. *Energy deposition in sub-microscopic volumes*. Radiation Research, A Twentieth-Century Perspective, volume II, pp 421-426, Academic Press San Diego.

Glickman B W, Drobetsky E A, de Boer J, Grosovksy A J, 1987. *Ionizing radiation induced point mutations in mammalian cells.* Radiat. Res. vol. 2, pp 562-567, Taylor & Francis London.

Goodhead D T, 1994. *Initial events in the cellular effects of ionizing radiations: clustered damage in DNA*. Init. J. Radiat. Biol., **65**, 7-17.

Goodhead D T 1988. *Spatial and Temporal distribution of energy.* Health Physics, **55**, 231-240.
Goodhead D T and Nikjoo H 1989. *Track structure analysis of ultrasoft X-rays compared to high- and low-LET radiations.* Int. J. Radiat. Biol., **55**, 513-529.
Goodhead D T and Charlton D E 1985. *Analysis of High-LET radiation effects in terms of local energy deposition.* Radiat. Prot. Dos., **13**, 253-258.
Hartigan J A, 1975. Clustering Algorithms, John Wiley & sons, New York.
Holley W R, Chatterjee A and Magee J L, 1990. *Production of DNA strand breaks by direct effects of heavy charged particles.* Radiat. Res. **121**, 161-168
Hutchinson F, 1985. *Chemical changes induced in DNA by ionizing radiation.* Progress in nucleic Acid Research and Molecular Biology, **32**, 115-154.
Martin R F and Haseltine W A, 1981. *Range of radio chemical damage to DNA with decay of iodine-125*, Science, **213**, 896.
Mikhalik V, 1993. *Energy deposition clusters in nanometer regions of charged-particle tracks*, Radiat. Res., **134**, 265-270.
Nikjoo H, Charlton D E, Goodhead D T, 1994a. *Monte Carlo track structure studies of energy deposition and calculation of initial DSB and RBE*, Adv. Space Res. Vol. 14, no 10, Pergamon Oxford, pp 161-180.
Nikjoo H, O'Neill P, Terrissol M, Goodhead D T, 1994b. Modelling of radiation-induced DNA damage: the early physical and chemical events. Int. J. Radiat. Biol., (in press).
Nikjoo H, Goodhead D T, Charlton D E, Paretzke H G, 1991. *Energy deposition in small cylindrical targets by monoenergetic electrons.* Int. J. Radiat. Biol., **60**, 739-756.
Paretzke H G, 1987, Radiation track structure theory, in: Kinetics of nonhomogenous processes, ed. G R Freeman, Wiley New York, pp 89-170
Pomplun E, Roch M, Terrissol M, 1992. *Simulation of strand break induction by DNA incorporated* I^{125}. in Biophysical Aspects of Auger Processes, eds. Howell R W, Narra V R, Sastry K S R, Rao D V., AAPM, New York. pp 137-152.
Sapora O, Belli M, Maione B, Pazzaglia S, Tabocchini M A, 1991. *The influence of genome structural organization on DNA damage and repairs in eukaryotic cells exposed to ionizing radiation.* In:"The early effects of radiation on DNA", eds. Fielden E M and O'Neill P, Springer-Verlag, London, pp 85-101.
Terrissol M, Beaudre A, 1990. *Simulation of space and time evolution of radiolytic species induced by electrons in water*, Radiat. Protec. Dos., 31, pp 175-177.
Turner J E, Paretzke H G, Hamm R N, Wright H A, Ritchie R H, 1982. *Comparative study of electron energy deposition and yields in water in the liquid and vapour phases.* Radiat. Res., 92, 47-60.
Umrania Y, Nikjoo H, Goodfellow J M, 1994. *A knowledge-based model of DNA hydration.* Int. J. Radiat. Biol., (in press).
Ward J F 1991, *Mechanisms of radiation action on DNA in model systems - their relevance to cellular DNA. In: The early effects of radiation on DNA.* eds. by E M Fielden and P O'Neill. Springer-Verlag, London, 1-16.
Ward J F 1988. *DNA damage produced by ionizing radiation in mammalian cells: identities, mechanisms of formation and repairability. Progress in Nucleic Acid* Research and Molecular Biology, **35**, 95-125.
Wilson W E, Miller J H, Nikjoo H, 1994. *PITS: A code system for positive ion track simulation, in: Computational approaches in molecular radiation biology*, eds. M Varma and A Chatterjee. Plenum Press (in press).
Zaider M, 1993. *A mathematical formalism describing the yield of radiation-induced single- and double-strand breaks, and its dependency on radiation quality.* Radiat. Res. **134**, 1-8.

Chapter 2

Ionizing Radiation Induces Multiply Damaged Sites in DNA and Damage Can Migrate

MULTIPLY DAMAGED SITES

J.F. Ward and J.R. Milligan
Department of Radiology
U.C. San Diego, 0610
La Jolla, CA 92093

G.D.D. Jones
Radiobiology Program
Cross Cancer Inst.
Edmonton AB T6G 1Z2

R.C. Fahey
Department of Chemistry
U.C. San Diego, 0506
La Jolla, CA 92093

Introduction

With the ultimate goal of developing a continuous mechanistic link between the physical descriptions of energy deposition and the biological consequences of ionizing radiation the chemical stages of the process are being investigated. Such developments allow the definition of the chemical structure of the lesions which face the cellular repair enzymes. This is a step forward from the earlier terms, lesions and sublesions which were used in physical models of biological actions. The early investigations of the radiation chemistry of water provided a framework to describe what occurs in the short times post-energy deposition with the intracellular milieu. In 1953 Samuel and Magee refined the observations of Wilson (1923) who had observed clusters of ionizations in water vapor in his cloud-chamber. They calculated for low LET radiation the spacings and sizes of the energy deposition events to liquid water. Thus the distribution of radical content of "clusters of ions" as described by Lea (1947) were defined for liquid water. These calculations for low LET in liquid water have recently been refined by Pimblott and Mozumder (1991). The latter calculations have been extrapolated to determine the distribution of radical content per moiety. This confirms that the majority of events give rise to a single radical pair (58%), 28% have two radical pairs, 8% have three, 3.5% have four and 0.4% five, etc..

Multiplicity Of Damaged Sites

The question therefore arises can all of these events give rise to a biologically significant lesion. If we consider the DNA DSB as the simplest biologically significant lesion (see below) then we should consider the minimal requirements for the formation of such a lesion. Under normal conditions a single ion pair can cause such a lesion only if either each ion (radical) can cause a break, or one radical can cause two breaks. The former mechanism is achieved when the DNA is substituted with halouracil (see below) and the latter by the "swing over" mechanism proposed by Siddiqi and Bothe. To explain an observed yield of DNA DSB in irradiated dilute aqueous solution Siddiqi and Bothe proposed a single radical mechanism of DSB production, whereby a radical induced break on one strand transfers the radical to the other strand causing a second SSB.

It should be pointed out that the result of Siddiqi and Bothe is not unique in finding a yield of DSB resulting from single radical attack. Many others (e.g. Krisch et al. (1991), Roots et al. (1985), Hempel and Mildenberger (1987)) have observed a low yield of DSB upon irradiating DNA in dilute aqueous solution. The DSB yield measured is generally about 1% of the yield of SSB. It might be expected that the swing over mechanism be constrained on DNA packaged within a cell, however we have shown (Jones et al. 1993) that in SV40 chromatin the DSB yield is 1% of the SSB yield under modest scavenging conditions. Thus if the swing over mechanism does occur to the same extent that it occurs in SV40 chromatin it would account for ~25% of the DSB yield.

Obviously an energy deposition event giving rise to two or more radical pairs (two oxidizing radicals) within a nanometer sized volume can cause a DNA DSB. The yield will be determined by the probability that both of these radicals react with the DNA and that the sites at which they react on opposite strands both cause strand breaks.

The question arises as to whether a simple DSB is a sufficient amount of damage to be able to have a biological consequence. This is clearly so, since the production of double strand breaks by restriction enzymes leads to cell death and mutation (Bryant, 1988). The correlation between the number of DSB produced by these agents and their biological consequence is not possible, since these DSB are being repaired at the same time as they are produced. However, these clean, readily ligatable DSB do lead to biological consequences. More recently Phillips and Morgan (1994) have shown that the mutations caused by restriction enzymes occur precisely at the site within the chromosomal DNA at which the enzymes cut. This finding rules out any scenario in which double strand cutting of DNA causes an instability in the genome which gives rise to a more general rearrangement phenomena.

If events giving rise to two ion pairs can cause a DSB, then events with more ion pairs can also give rise to DSBs. The probability of a DSB per event increases as the number of ion pairs per event increase; i.e. a greater probability of two attacks on opposite strands at the correct sites. This point has been covered in a previous publication (Ward, 1994).

Complexity Of LMDS

This topic has been discussed previously (Ward, 1988) and is presented here for the sake of completeness and further discussion. There are three variables in the chemical structure of LMDS:

1. The ratio of base damages to strand breaks.
2. The size of the site - over how many base pairs.
3. The numbers of damages per site.

Each of these parameters will be important in defining whether the lesion can be repaired or not. The consideration of multiplicity of radical attack on DNA is the increasing complexity of the MDS (multiply damaged site). In the original concept of locally multiply damaged sites the preponderance of base damage over strand breakage was emphasized (Ward, 1981), and can be inferred from the review of Preston (1993). If we assume that the yields of base damage relative to SSB in the LMDS is the same as that in the total DNA, then literature values can be used to determine this ratio. Mori et al. (1993) have measured the yields of base damage in hepatic chromatin irradiated in the mouse. Using their values from Table 1 a yield of base damage which is 10 times that of SSB is calculated. This number differs from the 2.7 predicted from damage measurements in irradiated DNA (Ward, 1988). A site containing a base damage and a strand break on opposite strands may potentially be as damaging as a DSB (Ward, 1988).

Cellular Evidence For MDS

a. Halouracils

While the normal rule is that the oxidizing species ·OH radicals (or DNA^+ ion-radicals) cause strand breaks, when the DNA has the normal deoxynucleoside thymidine substituted by 5-bromodeoxyuridine (BrdU) a strand break can be produced by the reducing species i.e. e^-_{aq}. BrdU reacts with the reducing species (hydrated electron) produced by radiation, and after dissociative electron attachment a uracilyl radical, the precursor of a strand break is formed (Zimbrick et al. 1969). Thus a single ion pair can cause a DSB in BrdU substituted DNA. Since the majority of the energy deposition events from low LET radiation cause one radical pair (58%) (Pimblott and Mozumder, 1991), the presence of BrdU greatly increases the potential for DSB formation by low LET radiation. By the same token, the reactions of the BrdU with reducing species to cause strand breaks increases the probability of DSB formation when an energy deposition event causes more than two ion pairs. Our experimental data using V79 Chinese hamster cells verify these predictions (Webb et al. 1993): The yield of DNA DSB (and cell killing) is markedly increased by the presence of BrdU in the cellular DNA and the increase in yield is dependent on the amount of BrdU present and does not necessitate the presence of BrdU in both strands.

b. Thiols

If we consider the mechanisms by which thiols can reduce the yield of DNA DSB and SSB we can derive equations which show the dependence of such reactions on thiol concentration. In an earlier publication (Ward, 1983) it was predicted that the doubly positively charged thiol compound, WR1065 (WRSH), radioprotects by both scavenging of radiation induced hydroxyl radicals and by chemically repairing radicals produced on DNA by hydrogen atom transfer. Recently (Zheng et al. 1992) this has been experimentally confirmed. To account for the protection afforded by way of hydroxyl radical scavenging is straightforward from a knowledge of the rate constants for reaction of ·OH with the substrates. The fitting of repair reactions to a kinetic mechanisms is more complex and their consequences have been examined in a pulse radiolysis mode (Prise et al. 1993). In that work it was found that the DSB precursors had a shorter half life than the SSB precursors and that the oxygen enhancement ratio was for DSB is 2.5 that for SSB.

We have investigated the effect of WRSH concentration on the radiation induced induction of SSB and DSB in plasmid DNA in the presence of varying concentrations of the hydroxyl radical scavenger dimethyl sulfoxide (DMSO). To prevent the radiation induced consumption of WRSH the irradiations were carried out in the presence of 10 mM mercaptosuccinic acid (MSA) as a sacrificial thiol. (Since MSA has a net negative charge of 2 it does not participate in repair of DNA radicals (Zheng et al. 1992).

If we consider the chemical repair reactions occurring, e.g. in the presence of high radical scavenging concentrations of DMSO where the WRSH does not markedly affect the scavenging capacity) the reactions of interest are:

For SSB:

$$RH + \text{"radiation"} \text{ -----> } R\cdot \qquad (1)$$

Where RH represents the deoxyribose moieties of DNA and "radiation" represents both direct ionization and ·OH radical induction of the deoxyribose radicals. Then the reactions which compete in the chemical repair reactions are:

$$R\cdot + O_2 \text{ -----> } RO_2\cdot \text{ ---> SSB} \qquad (2)$$

$$R\cdot + RSH \text{ ----> } RH \text{ (No SSB)} \qquad (3)$$

From these three reactions the yields of SSB as a function of the relative concentrations of oxygen and thiol has been (Ward, 1991) calculated to be:

$$\frac{SSB_0}{SSB_{WRSH}} = 1 + \frac{k_3[WRSH]}{k_2[O_2]} \qquad (4)$$

Where SSB_0 is the yield of SSB in the absence of WRSH and SSB_{WRSH} the yield in its presence. Thus if irradiations are carried out at a fixed concentration of DMSO in the presence of a constant concentration of oxygen the reciprocal of the yield of single strand breaks should be linearly dependent on the thiol concentration. Our data are shown in Figure 1 plotted in this fashion for different concentrations of DMSO as an ·OH scavenger. When plotted in this manner the linear relationships are clear; even at low DMSO concentrations. However, the change in slope with DMSO concentration is less than that which would be predicted if it were to be accounted for by bulk scavenging of ·OH by WRSH. If we interpret these changes in slope as due to ·OH scavenging by WRSH condensed on the DNA the concentration is calculated to be a factor of 15-30 which is in agreement with the condensation measured by more standard techniques (Smoluk et al. 1988).

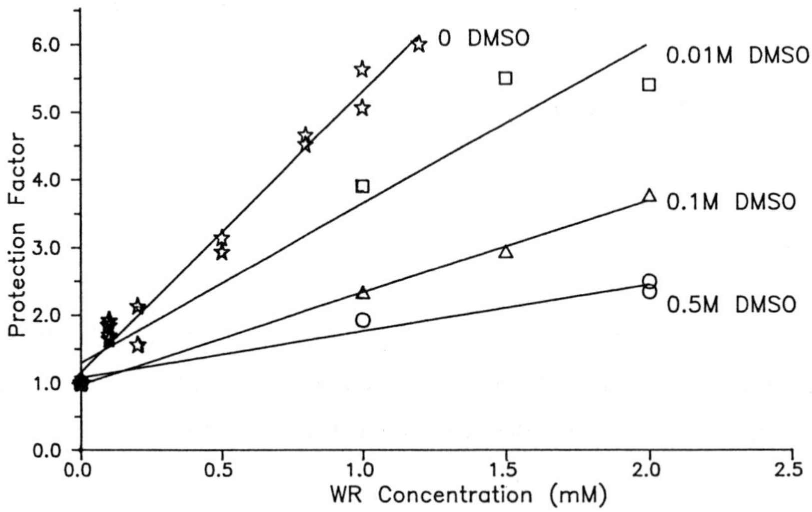

Figure 1. Protection of DNA against ionizing radiation induced SSB by WRSH at different DMSO concentrations.

If we consider the reactions involved in the production of DSB assuming that they are formed from a cluster of radicals causing two strand break precursors on opposite strands we must take into account both strand break precursors:

HR:RH + "radiation" -------> ·R:R· (5)

Where HR:RH represents the sites on both strands from which hydrogen atoms are abstracted to initiate the formation of the DSB. The succeeding reactions representing those giving rise to DSB and hydrogen atom donation repair by thiols are as follows:

$\cdot R{:}R\cdot + O_2$	------->	$\cdot R{:}RO_2\cdot$	(6)
$\cdot R{:}RO_2\cdot + O_2$	------->	$\cdot O_2R{:}RO_2\cdot$ (DSB)	(7)
$\cdot R{:}R\cdot$ + WRSH	------->	$\cdot R{:}RH$ (No DSB)	(8)
$\cdot R{:}RO_2\cdot$ + WRSH	------->	$HR{:}RO_2\cdot$ (No DSB)	(9)

If we make the reasonable assumptions that to a first approximation that $k_6 = k_7 = 2k_2$, and, $k_8 = k_9 = 2k_3$, then again as has been shown previously (Ward, 1991):

$$\frac{DSB_0}{DSB_{WRSH}} = \frac{(k_2[O_2] + k_3[WRSH])^2}{(k_2)^2\,[O_2]^2} \qquad (10)$$

Whence:

$$\left\{\frac{DSB_0}{DSB_{WRSH}}\right\}^{1/2} = 1 + \frac{k_3[WRSH])}{k_2\,[O_2]} \qquad (11)$$

This indicates that a plot of the square root of the ratio of DSB yields plotted against thiol concentration should give a line with the same slope as that for SSB dependence. Our data plotted in this fashion are shown in Figure 2.

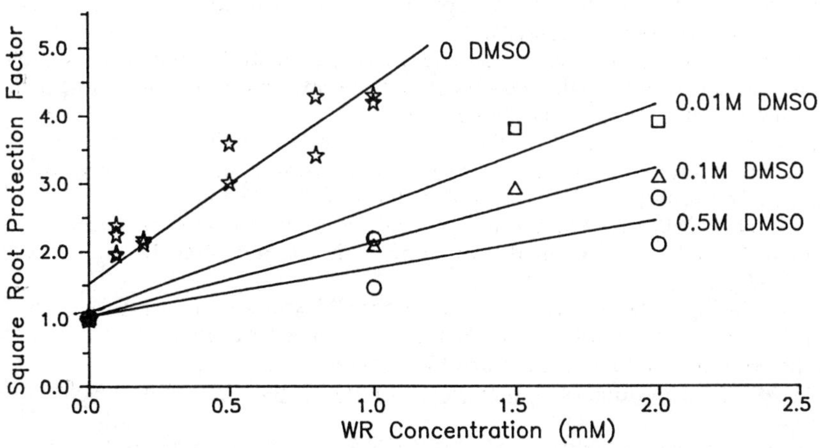

Figure 2. Effect of WRSH concentration on the square root of the protection factor for DSB formation at various concentrations of DMSO.

Unfortunately it is not possible to subtract out the reduction in yield caused by the ·OH radical scavenging properties of WRSH since the yield of DSB is not linearly dependent on scavenger concentration.

Equations (4) and (11) indicate that the slopes of the lines as plotted in Figures 1 and 2 should be identical. However, as mentioned it is difficult to interpret the DSB protection in the situations where the WRSH contributes significantly to ·OH radical scavenging. However, the slopes of 0.69 (SSB) and 0.72 (DSB) from the 0.5 M DMSO data are in good agreement. Such agreement, while not conclusive, is supportive of the proposed mechanism for DSB production by a two radical mechanism.

Acknowledgments

We acknowledge the following financial support of the work: Grant CA 46295 from NIH and Grant DEFG0393ER61592 from DOE (JFW) and Grant CA 39582 from NIH (RCF).

References

Aguilera, J, Newton, GL, Fahey, RC and Ward, JF (1992) Thiol uptake by Chinese hamstre V79 cells and aerobic radioprotection as a function of the net charge on the thiol. Radiat Res 130: 194-204.

Bryant PE (1988) The use of restriction endonucleases to study relationships between DNA double strand-breaks, chromosomal aberrations and other end-points in mammalian cells. Int J Radiat Biol 54: 869.-890

Hempel K and Mildenberger E (1987) Determination of G-values and for single and double strand beak induction in plasmid DNA using agarose gel electrophoresis and a curve fitting procedure. Int J of Radiat Biol 52: 125-138.

Krisch RE, Flick MB and Trumbore CN (1991) Radiation chemical mechanisms of single-and double-strand break formation in irradiated SV40 DNA. Radiat Res 126: 251-249.

Jones GDD, Milligan JR, Ward JF Calabro-Jones PM and Aguilera J. (1993) Yield of Strand Breaks as a Function of Scavenger Concentration and LET for SV40 Irradiated with ^{4}He Ions. Radiat Res 136: 190-196.

Lea, DH (1947) In: Actions of Radiation on Living Cells. MacMillan (Cambridge) p.27.

Mori, T, Hori, Y and Dizdaroglu, M (1993) DNA base damage generated *in vivo* in hepatic chromatin of mice upon whole body γ -irradiation. Int J Radiat Biol 64: 645-650.

Phillips JW and Morgan WF (1994) Illegitimate recombination induced by DNA double strand breaks in a mammalian chromosome. Molec Cell Biol (In Press).

Pimblott SM and Mozumder A. (1991) Structure of electron tracks in water. 2. Distribution of primary ionizations and excitations in water radiolysis. J Chem Phys 95: 7291-7300 (1991)

Preston RJ (1993) A consideration of the mechanism of induction of mutations in mammalian cells by low doses and dose rates of ionizing radiation. Adv Radiat Biol 16: 125-135.

Prise K, Davies S and Michael BD (1993) Evidence for induction of DNA double strand breaks at paired radical sites. Radiat Res 134: 102-106.

Roots R, Kraft G and Gosschalk E (1985) The formation of radiation induced DNA breaks: The ratio of double-strand to single-strand breaks. Int J Radiat Onc Biol and Phys 11: 259-265.

Samuel AH and Magee JL (1953) Theory of radiation chemistry. II. Track effects in radiolysis of water. J Chem Phys 21: 1080-87.

Siddiqi MA and Bothe E. (1987) Single- and double-strand break formation in DNA irradiated in aqueous solution: dependence on dose and scavenger concentration. Radiat Res 112: 449-467.

Smoluk GS, Fahey RC and Ward JF (1988). Interaction of glutathione and other low molecular weight thiols with DNA: Evidence for counterion condensation and coion depletion near DNA. Radiat Res 114: 3-10.

Ward JF (1981) Some biochemical consequences of the spatial distribution of ionizing radiation produced free radicals. Radiat Res 86: 185-195.

Ward JF (1983) Chemical aspects of radioprotectors. Radioprotectors and Anticarcinogens (O.F. Nygaard and M.G. Simic, eds.) Acad Press, New York, pp. 73-85

Ward JF (1988) DNA damage produced by ionizing radiation in mammalian cells: Identities, mechanisms of formation, and repairability. Prog Nucleic Acids Res Molec Biol 35: 96-128.

Ward JF (1991) Mechanisms of radiation action on DNA in model systems - Their relevance to cellular DNA. The Early Effects of Radiation on DNA (E.M. Fielden and P. O'Neill eds.) NATO ASI Series, Pages 1-14.

Ward JF (1994) The Complexity of DNA Damage - Relevance to Biological Consequences. Int J Radiat Biol (In Press).

Webb CF, Jones GDD, Ward JF, Moyer DJ, Aguilera JA and Ling LL (1993) Mechanisms of radiosensitization in bromodeoxyuridine substituted cells. Int J Radiat Biol 64: 695-705.

Wilson CTR (1923) Investigation on X-rays and β-rays by the cloud chamber method. Part II — β-rays. Proc Roy Soc (London) Ser A 104: 192-212.

Zheng S, Newton GL, Ward JF and Fahey RC (1992) Aerobic radioprotection of pBR322 by thiols: Effect of thiol net charge upon scavenging of hydroxyl radicals and repair of DNA radicals. Radiat Res 130: 183-193.

Zimbrick JD, Ward JF and Myers LS, Jr. (1969) Studies on the chemical basis of cellular radiosensitization by 5-bromouracil substitution of DNA. I. Pulse and steady-state radiolysis of 5-bromouracil and thymine. Int J Radiat Biol 16: 502-523.

REGIONALLY MULTIPLY DAMAGED SITES: THEORETICAL AND EXPERIMENTAL EVIDENCE

B. Rydberg, W. R. Holley and A. Chatterjee
Life Sciences Division
Lawrence Berkeley Laboratory
University of California
Berkeley, CA 94720
USA

Introduction

The DNA in the mammalian cell is organized in nucleosomes which in turn are arranged into the 30 nm chromatin fiber. This basic fiber consists of a helical arrangement of 6-7 nucleosomes per turn, comprising 11 nm length of fiber. The way the linker DNA connects between the nucleosomes is not known, however. Current models suggest either a solenoid arrangement, a twisted ribbon with the DNA following a zigzag path between the nucleosomes, or a crossed-linked arrangement (Widom 1989). With an average of 1300 bp per 11 nm length of fiber, the compaction factor is about 40 relative to naked B form DNA.

During irradiation of a mammalian cell, ionizing particles will intersect segments of the 30 nm chromatin fiber, with the resulting possibility of generating clusters of lesions over kilobasepair distances (Fig 1). Our Monte-Carlo calculations have suggested that multiple DNA double strand breaks may occur in these heavily damaged regions, generating small (kb-sized) DNA fragments after cell lysis. The calculations predict that the yield of such fragments will increase as the LET increases. In this presentation a brief summary of the theoretical calculations is given and experiments are described that detect such small fragments in irradiated human cells and correlate them with LET and the conformation of the chromatin fiber. We propose the term RMDSs, or Regionally Multiply Damaged Sites[1], to refer to such clusters of lesions.

[1] This term was first suggested to us by Ed Blazek.

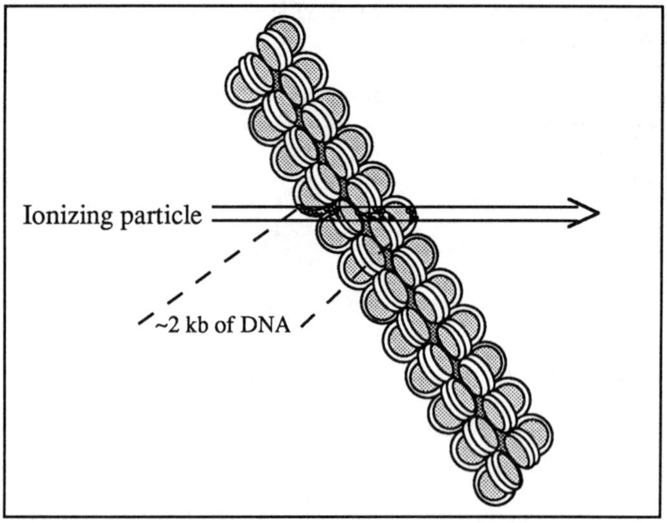

Figure 1. An ionizing particle traverses the 30 nm chromatin fiber and induces a Regionally Multiply Damaged Site containing multiple lesions spread out over a 2 kb region of DNA.

Theory

We are curently in the process of developing a general theory of radiation-induced DNA damage. The theoretical model uses Monte-Carlo techniques to calculate DNA strand break production by both direct and indirect effects of ionizing radiation, and is based on general features of track structure, stopping power theory and the detailed geometry of DNA structures (Holley *et al.* 1990, Chatterjee and Holley 1991, 1993). We have recently incorporated into our calculations a computer model of the 30 nm chromatin fiber (Schmidt 1993, Chatterjee *et al.* 1994). The chromatin model takes the form of an idealized solenoid consisting of twenty turns of a right handed helix with 6 nucleosomes per turn. Each nucleosome-linker unit contains 171 base pairs and the whole fiber contains more than 20,000 base pairs. The computer codes implementing the computational scheme outlined above provide detailed pictures of the extent of damage induced by each randomly generated charged particle which passes through or near the DNA fiber.

Our calculated results predict the existence of strong spatial correlations in damage sites associated with the symmetries in the soleoidal model. We have calculated spectra of short fragments of single and double stranded DNA produced by multiple strand breaks induced by both high and low LET radiation. The spectra exhibit peaks at multiples of ~80 basepairs (the nucleosome periodicity) and ~1000 basepairs (the solenoid periodicity).

Experimental Methods

Cell culture and preparation of agarose plugs

Cultures of primary human skin fibroblasts GM38 were supplemented with 1 μCi/ml (37 kBq/ml) of [methyl-^{3}H]thymidine (Amersham, 3.1 TBq/mmole) 4 days before experiments and changed to fresh non-labeled medium 24 hrs before experiments. The cells incorporated about 0.5 dpm/cell in DNA and were close to confluency at the time of experiments.

The ^{3}H-labeled cells were embedded in agarose plugs before or after irradiation. Trypsinized cells were suspended at a concentration of ~2×10^7 cells per ml in 1% low gelling temperature agarose (Sigma Type VII) in PBS at 37°C, and plugs of size 3 mm x4.5 mm x10 mm (X-ray experiments) or size 1.2 mm x 9 mm x 20 mm (N-ion experiments) were cast and allowed to set at 4°C.

Treatments to alter chromatin conformation

Agarose plugs with ^{3}HTdR-labeled cells were immersed in 0.1% Triton X-100, 10 mM Tris-HCl pH 8.0, 2 mM EDTA, 150 mM NaCl in order to isolate nuclei, or with the same buffer lacking 150 mM NaCl (low salt condition) in order to induce a change in chromatin conformation. The treatments were for 1.5 hr at 0°C with one change of buffer before irradiations.

Irradiation and dosimetry

X-irradiations were performed using a Philips Medical Systems X-ray unit (RT250) operated at 225 kVp, 15 mA with 0.35 mm Cu filtration at a dose rate of 6.5 Gy/min. Dosimetry was performed with a Victoreen dosimeter. Cells were embedded in agarose plugs in ice-cold buffers during and after irradiations.

N-ion irradiations at an initial energy of 32.4 MeV/n were carried out at the 88 inch cyclotron at Lawrence Berkeley Laboratory at a dose rate of ~100 Gy/min. Dosimetry was performed using an ion chamber. At the targets (agarose plugs of 1.2 mm thickness) the mean energy was calculated to be 22 MeV/n and the mean LET

120 keV/μm. When dishes with cell monolayers were irradiated the energy at the cells was 29 MeV/n with an LET of 97 keV/mm.

Cell lysis and agarose gel electrophoresis

Plugs with about $3x10^6$ irradiated cells were immersed in 1 ml ice-cold lysis solution (50 mM EDTA, 1% sarkosyl, 1 mg/ml proteinase K, pH 8) for 30 min, followed by 60 min lysis at 37°C. The plugs were then washed 4 times for 20 min each in 1 ml TE (10 mM Tris-HCl, 1 mM EDTA, pH 8) at room temperature, with 0.1 mg/ml RNase A present in the last two washes. The plugs were then inserted in 3 x 10 mm wells in a 1.5% agarose gel made up in 0.5x TBE (1x TBE is 89 mM Tris base, 89 mM boric acid, 2 mM Na_2EDTA, pH 8.0) and run in 0.5x TBE at 4 V/cm for 2-3 hrs. DNA size standards (1 Kb DNA Ladder, BRL) were run in separate lanes, or were included in some of the plugs containing cells. After the run, the gel was stained for one hr in 0.5 mg/ml of ethidium bromide in water and photographed under UV illumination. Each lane in the gel was then reproducibly sliced up in 3.44 mm wide slices using an array of razor blades. Special care was taken not to contaminate the part of the gel with low levels of radioactivity with material from the part near the top of the gel with high levels of radioactivity. Each slice was mixed with 0.2 ml 1 M HCl in a scintillation vial and melted in a boiling water bath for 10 min, followed by addition of 10 ml Opti-Fluor scintillation fluid (Packard). Samples that had low levels of radioactivity in the gel were counted in a liquid scintillation counter for at least 2 x 10 min per sample and corrected for background counts obtained from multiple (10-20) samples prepared from non-radioactive parts of the gel.

Results

Induction of double-stranded small DNA fragments.

^3H-labeled human primary fibroblasts were N-ion irradiated (97 keV/mm, 325 Gy) and then embedded in agarose blocks, lysed and subjected to 1.5% agarose gel electrophoresis. The gel was sliced and the radioactivity in each slice determined. As shown in Figure 2A, an increase in radioactivity as a result of radiation is seen at the location in the gel corresponding to DNA of sizes 0.1-2.0 kb. The fractions of DNA

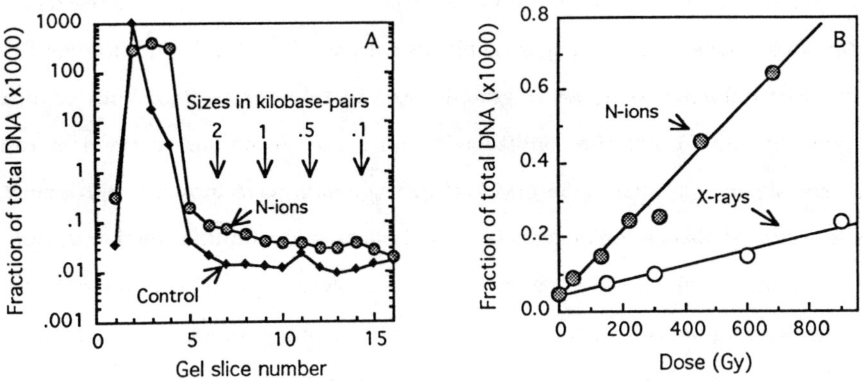

Figure 2. Small DNA double-stranded fragments induced by N-ions and X-rays. (A): Human primary fibroblasts were non-irradiated or irradiated with 325 Gy of N-ions. ^{3}H-labeled DNA (~$6x10^5$ cpm per lane) from lysed cells was subjected to agarose gel electrophoresis, after which the gel was sliced and radioactivity measured in each slice. Fraction number 2 contains the gel plug with lysed cells. Migration was from left to right. Arrows point to sizes in kilobase-pairs, as obtained from size standards. (B): The fraction of total DNA in the form of fragments of sizes 0.1-2 kb, as obtained by adding the counts in gel slice 7 to 15, is plotted as a function of dose. Also shown is the result of a similar experiment using X-rays.

released as small fragments in this size range are higher than expected from random breakage of DNA, and increase linearly with dose, as seen in Figure 2B. High LET radiation in the form of accelerated N ions was more efficient than X-rays in induction of small DNA fragments.

Irradiation of chromatin isolated from agarose-embedded cells.

^{3}H-TdR-labeled cells in agarose blocks were treated with 0.1% Triton X-100, 10 mM Tris-HCl pH 8.0, 2 mM EDTA (low salt condition) or with the same buffer also containing 150 mM NaCl in order to isolate nuclei. It has been shown previously that chromatin in isolated nuclei will adopt an extended "bead on a string" configuration under low salt conditions, while at 150 mM NaCl the 30 nm chromatin fiber will be maintained (Widom 1989). The treated samples were irradiated with X-rays or N-ions (120 keV/μm). The induction of total dsbs was determined for both salt conditions by pulsed field gel electrophoresis (Rydberg *et al.*

1994), and found to be 2-4 times higher than for intact cells, with slightly more breaks induced at the low salt condition (data not shown). This is consistent with other studies where nuclei in agarose blocks were irradiated (Elia and Bradley 1992).

The induction of DNA fragments was also determined for both conditions (Figure 3A and B). For the condition in which the 30 nm chromatin fiber can be expected to remain intact (150 mM NaCl), fragments were induced with about the same yield per dsb as for intact cells. For the low salt condition, the background in non-irradiated cells was higher but there was very little, if any, increase in the number of fragments with dose. These results strongly support the idea that the small fragments are induced as a result of ionizing particles interacting with the 30 nm chromatin fiber.

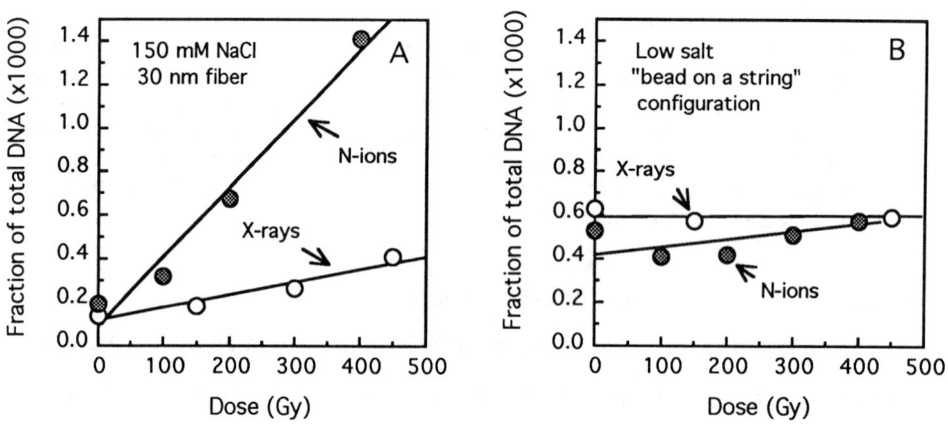

Figure 3. Induction of DNA fragments in the size range of 0.1 - 2 kb was determined in chromatin from GM38 human fibroblasts embedded in agarose blocks at two salt concentrations. A. Physiological salt concentration that preserves the 30 nm chromatin fiber. B. Low salt concentration that induces a transition to a "bead on a string" conformation.

Further experiments (to be published elsewhere) show that the kb-sized fragments are reduced in number when cells are incubated at 37°C after irradiation, suggesting a repair activity. This result also provides evidence that the fragments are not enzymatically induced in the first place. Further increase in LET has also demonstrated an increase in yield of fragments with a saturation at several hundred keV/mm, when the majority of DNA double strand breaks may occur within RMDSs. We are currently developing ^{32}P end-labeling methods to increase the sensitivity by which very small fragments can be detected.

Discussion

It is shown here that ionizing radiations induce small DNA fragments linearly with dose, with an LET-dependent yield. It is suggested that these fragments arise as a consequence of the organization of DNA in the mammalian cell in the form of the 30 nm chromatin fiber, where traversals of ionizing particles through the fiber may result in multiple breaks over kb-sized DNA segments. The data presented here directly implicates the 30 nm chromatin fiber as the target for generating such fragments. The yield and size-distribution of the fragments are in good agreement with our calculations using a Monte-Carlo approach. The yields obtained theoretically agree within a factor of two with the measured values, both for X-rays and for the N ions.

The suggested clustering of lesions may have biological implications. Crowding of multiple lesions within one kb could impose difficulties for repair enzymes which probably could not operate without interference with each other at such locations. Thus, in addition to the previously hypothesized interaction of lesions over several hundred nm (dual radiation action, Kellerer and Rossi 1972) and over a few nm (locally multiply damaged sites, Ward 1985), an intermediate level of clustering over 30 nm distances may be important. These three levels correspond to chromosomal dimensions, DNA dimensions and chromatin dimensions respectively. It is proposed here that the intermediate level of clustering over kb-sized DNA regions be called RMDSs, for Regionally Multiply Damaged Sites, in analogy with the concept introduced by Ward (Ward 1985) of LMDSs, for Locally Multiply Damaged Sites. The former results from consideration of the 30 nm chromatin fiber and the latter from consideration of the DNA molecule in the interaction with ionizing radiations.

Each kb-sized cluster will be detected only as a single break by standard techniques of measuring dsbs and ssbs. Thus the total number of breaks will be underestimated for high LET radiations. This may account for a low apparent yield of ssbs measured at high LET (Rydberg 1985, Aufderheide *et al.* 1987), and explain RBEs of less than one for dsb production by heavy ions (Heilmann *et al.* 1993, Rydberg *et al.* 1994, Löbrich *et al.* 1994). Correct rejoining of the large DNA on each side of the cluster will require joining of all breaks within the cluster. As an alternative mechanism, rejoining could proceed with loss of DNA segments between the breaks, which would result in small deletion mutations. The requirement for multiple rejoining events could contribute to inefficient repair of total breaks (ssbs and dsbs) observed after high LET radiations (Ahnström and Edvardsson 1974, Ritter *et al.* 1977, Roots *et al.* 1979, Van Der Schans *et al.* 1983, Rydberg 1985), and inefficient repair of dsbs at high LET (Blöcher 1988, Peak *et al.* 1991, Rydberg *et al.* 1994, Löbrich *et al.* 1994).

Acknowledgements

This work was supported by Order No. W-18,265 from the U. S. National Aeronautics and Space Administration to the NSCORT in Radiation Health at the Lawrence Berkeley Laboratory and by the Office of Health and Environmental Research, Office of Energy Research, U. S. Department of Energy, under Contract No. DE-AC03-76SF00098.

References

Ahnström, G. and Edvardsson, K.-A., 1974, Radiation-induced single-strand breaks in DNA determined by rate of alkaline strand separation and hydroxylapatite chromatography: an alternative to velocity sedimentation. International Journal of Radiation Biology, 26, 493-497.

Aufderheide, E., Rink, H., Hieber, L. and Kraft, G., 1987, Heavy ion effects on cellular DNA: strand break induction and repair in cultured diploid lens epithelial cells. International Journal of Radiation Biology, 51, 779-790.

Blöcher, D., 1988, DNA double-strand break repair determines the RBE of a-particles. International Journal of Radiation Biology, 54, 761-771.

Chatterjee, A. and Holley, W.R., 1991, Energy deposition mechanisms and biochemical aspects of DNA strand breaks by ionizing radiation. International Journal of Quantum Chemistry, 39, 709-727.

Chatterjee, A. and Holley, W.R., 1993, Computer simulation of initial events in the biochemical mechanisms of DNA damage. Advances in Radiation Biology, 17, 181-226.

Chatterjee, A., Schmidt, J.B. and Holley, W.R., 1994, Monte Carlo approach in assessing damage in higher order structures of DNA. Computational Approaches in Molecular Radiation Biology, in press.

Elia, M.C. and Bradley, M.O., 1992, Influence of chromatin structure on the induction of DNA double strand breaks by ionizing radiation. Cancer Research, 52, 1580-1586.

Heilmann, J., Rink, H., Taucher-Scholz, G. and Kraft, G., 1993, DNA strand break induction and rejoining and cellular recovery in mammalian cells after heavy-ion irradiation. Radiation Research, 135, 46-55.

Holley, W.R., Chatterjee, A. and Magee, J.L., 1990, Production of DNA strand breaks by direct effects of heavy charged particles. Radiation Research, 121, 161-168.

Kellerer, A. M. and Rossi, H. H., 1972, The theory of dual radiation action. Current Topics in Radiation Research Quarterly, 8, 85-158.

Löbrich, M., Rydberg, B. and Cooper, P.K., 1994, DNA double-strand breaks induced by high-energy neon and iron ions in human fibroblasts. II. Probing individual NotI fragments by hybridization. Radiation Research, 139, 142-151.

Peak, M. J., Wang, L., Hills, C. K. and Peak, J. G., 1991, Comparison of repair of DNA double-strand breaks caused by neutron or gamma radiation in cultured human cells. International Journal of Radiation Biology, 60, 891-898.

Ritter, M. A., Cleaver, J. E. and Tobias, C. A., 1977, High-LET radiations induce a large proportion of non-rejoining DNA breaks. Nature, 266, 653-655.

Roots, R., Yang, T. C. H., Craise, L., Blakely, E. A. and Tobias, C. A., 1979, Impaired repair capacity of DNA breaks induced in mammalian cellular DNA by accelerated heavy ions. Radiation Research, 78, 38-49.

Rydberg, B., 1985, DNA strand breaks induced by low-energy heavy ions. International Journal of Radiation Biology, 47, 57-61.

Rydberg, B., Löbrich, M. and Cooper, P.K., 1994, DNA double-strand breaks induced by high-energy neon and iron ions in human fibroblasts. I. Pulsed-field gel electrophoresis method. Radiation Research, 139, 133-141.

Schmidt, J.B., 1993, Heavy ion induced lesions in DNA: a theoretical model for the initial induction of DNA strand breaks and chromatin breaks. Ph.D. Thesis, University of California at Berkeley.

Van Der Schans, G. P., Paterson, M. C. and Cross, W. G., 1983, DNA strand break and rejoining in cultured human fibroblasts exposed to fast neutrons or gamma rays. International Journal of Radiation Biology, 44, 75-85.

Ward, J. F., 1985, Biochemistry of DNA lesions. Radiation Research, 104, S103-S111.

Widom, J., 1989, Toward a unified model of chromatin folding. Annual Review of Biophysics and Biophysical Chemistry, 18, 363-395.

MIGRATION OF RADIATION DAMAGE IN DNA

P. O'Neill, T. Melvin and S.W. Botchway
MRC Radiobiology Unit
Chilton
Didcot
Oxon OX11 ORD
UK

Introduction

The various biological effects of ionising radiation are generally thought to be a consequence of chemical modifications to DNA (O'Neill and Fielden, 1993). The initial chemical events occurring in DNA following exposure to ionising radiation arise through interactions of water radicals or through the direct deposition of energy within DNA (direct effect). The majority of the information on the various types of DNA radicals formed from the direct effects of radiation comes from epr studies on DNA -irradiated at 77 K (Becker and Sevilla 1993; Bernard *et al.*, 1994; Close, 1993). Exposure of DNA to γ-radiation results in ionisation of DNA with the formation of a radical cation and an electron. It is estimated that ~75% of these ion pairs undergo ion recombination (Bernhard *et al.*, 1994). From luminescence studies (Al-Kazwini *et al.*, 1988), these recombination processes occur at times less than a few nanoseconds at room temperature. Within the time scale of these geminate ion recombination processes, the positive hole does not migrate significantly (Al-Kazwini *et al.*, 1988). For those ions which do not undergo recombination it is inferred from the epr studies that the radical cation (or electron loss centre) ultimately becomes localised on the purines and the electron becomes localised upon cytosine and to a lesser extent on the thymine base moiety (Becker and Sevilla, 1993; Bernhard *et al.*, 1994; Boon *et al.*, 1984). It has been estimated that the energetic electrons which escape recombination may migrate ~10-200 base pairs depending on the conditions (Van Lith *et al* 1986; Al-Kazwini *et al.*, 1988; 1990; 1994). In contrast, the thermalised, hydrated electrons on addition to DNA only migrate a few base pairs (Anderson *et al.*, 1991; Fuciarelli *et al.*, 1994). Since the initial ionisation events are expected to be random, it is inferred from the esr studies that the electron loss centre may migrate at 77 K from the base radical cations and become localised predominantly upon guanine (>90%), the most readily oxidised base (Becker and Sevilla, 1993; Bernhard *et al.*, 1994; Boon *et al.*, 1984; Yan *et al.*, 1992).

Up to date, the majority of the studies on radiation-induced migration of damage in DNA at 295 K have concentrated on the ejected electron. It is equally important to know whether hole migration occurs in DNA at room temperature and the distance over which it occurs. It is difficult however to produce and investigate the chemistry of radical cations of DNA in aqueous solution using pulse radiolysis. One approach is to use the sulphate radical anion as a powerful one electron oxidant of DNA (O'Neill and Davies, 1987; Wolf *et al.*, 1993; Candeias and Steenken, 1993). These studies on DNA or model systems in aqueous solution have indicated, in a very qualitative manner, that hole migration occurs in DNA. The oxidative damage becomes localised on guanine (~70-85%) (Wolf *et al.*, 1993) although it is necessary to consider the other bases as sites for localisation of oxidative damage. Although migration of damage occurs to guanine in single stranded DNA, the oxidised guanine sites initiated by the $SO_4^{\cdot-}$ radical are not thought to be major precursors to prompt strand breakage (t < 10s) (Wolf *et al.*, 1993).

More recently we have used the approach based upon the ability of 193 nm light to ionise DNA and associated model compounds in an aqueous environment (Candeias and Steenken, 1992; Candeias *et al.*, 1993; Gorner, 1994) to assess whether hole migration occurs at 295 K. In this short review we will present some of our preliminary indications on hole migration induced by 193 nm light in specifically designed oligodeoxynucleotides (see Table 1). The time resolved technique of flash photolysis of these oligonucleotides with 193 nm light has been described in detail (Melvin *et al.*, 1994). The solutions are generally saturated with oxygen in order to remove the ejected electrons (Buxton *et al.*, 1988). The resulting superoxide anion is essentially transparent at > 300 nm (Bielski, 1978) and is assumed to be inert at times < 100 s. At present, the oligodeoxynucleotides contain mixtures of adenine, guanine and cytosine, due to potential complications with thymine. The quantum yield for photoionisation of the oligodeoxynucleotides (Table 1) are generally in the range 0.04-0.08, depending on the base composition of the samples.

Table 1. Estimated photon distribution in a series of oligodeoxynucleotides and the expected hole migration interaction.

Oligonucleotides	Predicted hole migration	% photo ejected electrons from bases[a)]	
dA9		benchmark	spectrum
dG9		benchmark	spectrum
dC9		benchmark	spectrum
AGAAGAAGA	$A^{\cdot+} \rightarrow G$	40%G	60%A
CGCCGCCGC	$C^{\cdot+} \rightarrow G$	50%G	50%C
CACCACCAC	$C^{\cdot+} \rightarrow A$	55%C	45%A

a) based up extinction coefficients at 193 nm (Candeias and Steenken, 1992) and quantum yield for photoionisation

Evidence for hole transfer between neighbouring bases

Table 1 shows the various oligodeoxynucleotides that have been studied to assess whether photo-induced transfer of the hole occurs between neighbouring base moieties. The compounds, $(dA)_9$ $(dG)_9$ and $(dC)_9$ have been used to obtain benchmark spectra of the one electron oxidised radicals, A, G and C. In Table 1 the initial distribution of ionised base moieties is equivalent to that for photo-ejected electrons. Therefore, if hole migration is important the spectral characteristics of the optical absorption will not reflect this initial distribution.

Adenine/Guanine

The optical absorption spectrum of the transient species produced upon photoionisation of $(AGA)_3$ by 193 nm light have been compared with the benchmark spectra from photoionisation of $(dG)_9$ and $(dA)_9$ to assess transfer from $A^{\cdot+}$ to guanine and are shown in Figure 1. It is assumed that the predominant

A

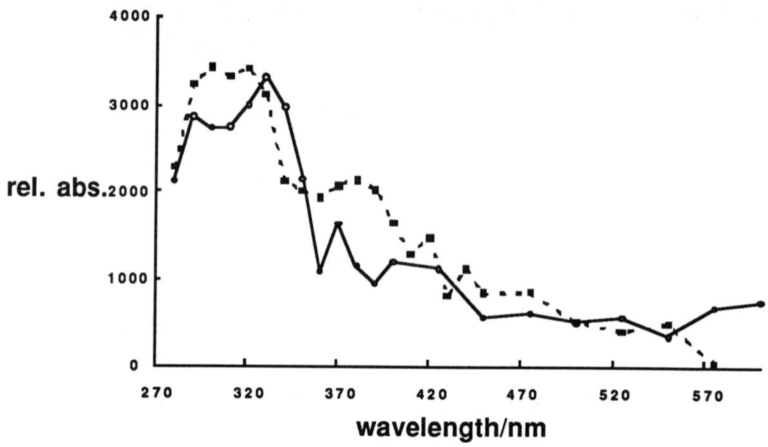

B

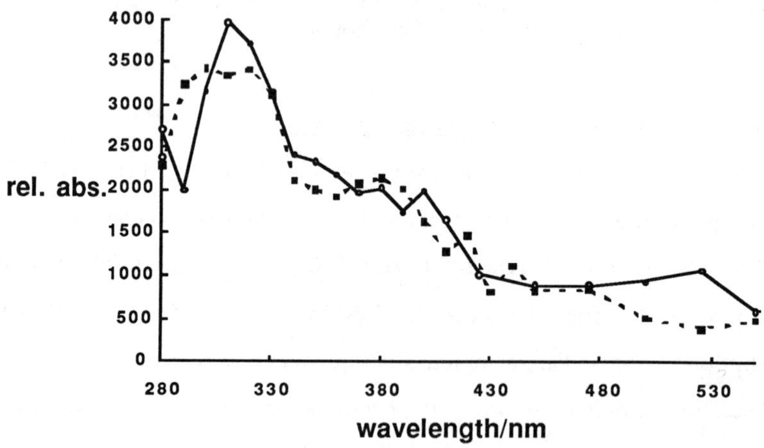

Figure 1 Optical absorption spectra determined 20 μs after 193 nm photolysis of (A) $(dG)_9$ (- - - -) and $(dA)_9$ (———) and (B) $(dG)_9$ (- - - -) and $(AGA)_3$ (——) at pH 7.4 - 7.8

species remaining 20 s after the light pulse are the one electron oxidised radicals of the bases. From the similarity of the transient optical absorption spectrum of $(AGA)_3$ with that from $(dG)_9$ it is inferred that efficient transfer occurs from a neighbouring $A^{\cdot+}$ to guanine. This transfer is consistent with the redox properties of the two bases (Jovanovic and Simic, 1986; 1989) and the known order of their ionisation potentials, eg A > G.

Cytosine/Guanine

The optical absorption spectrum determined upon photoionisation of $(CGC)_3$ is similar to the benchmark spectra of $(dG)_9$ although it is apparent that a minor component due to the cytosine radical is present. Therefore transfer of the hole from a neighbouring $C^{\cdot+}$ to guanine does occur but the efficiency is significantly < 100%. Although this process would be expected to be thermodynamically favoured, it is apparent that the radical site may remain on cytosine.

Cytosine/Adenine

Although it is expected on thermodynamic grounds that $C^{\cdot+}$ should transfer to adenine, the optical absorption spectrum following photoionisation of $(CAC)_3$ is a composite of the benchmark spectra from $(dC)_9$ and $(dA)_9$. It is inferred that the efficiency of transfer from neighbouring $C^{\cdot+}$ to adenine is low. This lower efficiency for transfer from $C^{\cdot+}$ to adenine emphasises the observation with $(CGC)_3$, whereby the cytosine radical is still present at 20 μs.

From comparison of these different oligodeoxynucleotides, efficient transfer occurs from a neighbouring $A^{\cdot+}$ to guanine but this efficiency decreases in the following order:

$$A^{\cdot+} \rightarrow G > C^{\cdot+} \rightarrow G > C^{\cdot+} \rightarrow A$$

Although thermodynamically favoured, we do not know, at present, why the transfer between $C^{\cdot+}$ and these purines is not 100% efficient. One possibility is that the cytosine radical cation may oxidise a neighbouring purine base in competition with its conversion to a species which may not oxidise purines. With deoxycytidine the lifetime of $C^{\cdot+}$ is estimated to be very short, << 200 ns. From the time dependencies of the optical changes, it is inferred that the hole transfer in these samples is complete within times <1μs ($k > 10^6$ s^{-1}). This rapid transfer is not inconsistent with that previously estimated for $A^{\cdot+}$ to guanine (Candeias and Steenken, 1993).

Migration over 2 base units

Two oligodeoxynucleotides were specifically synthesised in order to address the question whether the hole could migrate up to two base units. With these oligonucleotides, CAACGCAACGAAC and CAACCGCCAAC, it is estimated that <20% of the photons are absorbed by guanine but ~50% of the photons are absorbed by adenine. The optical absorption spectra of these two samples following photolysis with 193 nm light are similar and reflect a composite spectrum of all the bases present in these samples, namely A, G and C.

It is apparent that the transfer from $A^{\cdot+}$ to guanine via one or two cytosine base units is not very efficient, if it occurs at all. This transfer as detailed below

$$- A^{\cdot+} - C - G - \rightarrow \quad A - C - G^{\cdot+}$$

$$A^{\cdot+} - C - C - G - \rightarrow \quad A - C - C - G^{\cdot+}$$

requires processes which are considered to be thermodynamically unfavourable, leaving aside the difficulties of transfer involving cytosine discussed above.

From this limited series of oligodeoxynucleotides so far investigated, it is evident that the hole migration process may only occur over a limited range of up to 2-3 base units. From esr studies, it has been inferred that hole migration may occur over a limited range in DNA (Yan *et al.*, 1992). Similar studies with thymine containing oligodeoxynucleotides are being undertaken in order to assess the generality of this limited migration. From a knowledge of the migration distances, it will be possible to comment upon whether migration influences the fingerprint of the radiation track in cellular systems.

Summary

At present, we have only investigated hole migration in single stranded oligodeoxynucleotides. Photoionisation of these compounds results predominately in the formation of the radical cations of the bases. Subsequent hole migration occurs from neighbouring base radicals to guanine and preliminary indications are that this migration occurs over a limited range of up to 2-3 base units. These migration processes also reflect a sequence specificity whereby cytosine interferes with migration. The implications of limited migration of the hole produced in DNA following radiolysis are that, for a sample containing equal proportions of the four bases randomly distributed and assuming no interstrand transfer, the yield of $G^{\cdot +}$ would represent ~70% of the total yield of radical cations for an intrastrand migration distance of two base units. If migration is limited, the yield of $G^{\cdot +}$ would be predicted to reflect the base composition of the DNA.

References

Al-Kazwini AT, O'Neill P, Fielden EM, Adams GE (1988) Radiation-induced luminescence from 'dry' and hydrated DNA and related macromolecules. Radiat Phys Chem 2:385-389.

Al-Kazwini AT, O'Neill P, Adams GE, Fielden EM (1990) Radiation-induced energy migration within solid DNA: The role of misonidazole as an electron trap. Radiat Res 121:149-153.

Al-Kazwini AT, Adams GE, O'Neill P, Naylor MA, Fielden EM (1994) Radiation-induced luminescence in DNA: Evidence for long-range electron migration. Radiat Res 138:307-311.

Anderson RF, Patel KB, Wilson WR (1991) Pulse radiolysis studies of electron migration in DNA from DNA base radical anions to nitroacridine intercalators in aqueous solution. J Chem Soc Faraday Trans 1. 87:3739-3746.

Becker D, Sevilla MD (1993) The chemical consequences of radiation damage to DNA. Adv Radiat Biol 17:121-180.

Bernhard WA, Mroczka N, Barnes J (1994) Recombination is the dominant free radical process initiated in DNA by ionising radiation: An overview based on solid state epr studies. Int J Radiat Biol *(in press)*

Bielski BHJ (1978) Re-evaluation of the spectral and kinetic properties of HO_2 and $O_2^{\cdot -}$ free radicals. Photochem Photobiol 28:645-649.

Boon PJ, Cullis PM, Symons MCR, Wren BW (1984) Effects of ionising radiation in DNA; the role of oxygen. J Chem Soc Perkin Trans 2 1393-1399.

Buxton GB, Greenstock CL, Helman WP, Ross AB (1988) Critical review of rate constants for reaction of hydrated electrons, hydrogen atoms and hydroxyl radicals $(OH/O_2^{\cdot -})$ in aqueous solution. J Phys Chem Ref Data 17:513-886.

Candeias LP, Steenken S (1992) Ionization of purine nucleosides and nucleotides and their components by 193nm laser photolysis in aqueous solution. J Amer Chem Soc 114:699-704.

Candeias LP, Steenken S (1993) Electron transfer in di(deoxy)nucleoside phosphates in aqueous solution: rapid migration of oxidative damage (via adenine) to guanine. J Amer Chem Soc 115:2437-2440.

Candeias LP, O'Neill P, Jones GDD, Steenken S (1993) Ionization by polynucleotides and DNA in aqueous solution by 193nm pulsed laser light: identification of base-derived radicals. Int J Radiat Biol 61:15-20.

Close DM (1993) Radical ions and their reactions in DNA constituents: Esr/ENDOR studies of radiation damage in the solid state. Radiat Res 135:1-15.

Fuciarelli AF, Sisk EC, Zimbrick JD (1994) Electron migration in oligonucleotides upon γ-irradiation in solution. Int J Radiat Biol 65:409-418.

Görner H (1994) Photochemistry of DNA and related biomolecules: quantum yields and consequences of photo ionization. J Photochem Photobiol B: Biol *(in press)*

Jovanovic SV, Simic MG (1986) One-electron redox potentials of purines and pyrimidines. J Phys Chem 90:974-978.

Jovanovic SV, Simic MG (1989) The DNA guanyl radical-kinetics and mechanisms of generation and repair. Biochim Biophys Acta 1008:39-44.

Melvin T, Plumb MA, Botchway SW, O'Neill P, Parker AW 193 nm light induced single strand breakage of DNA occurs predominantly at guanine. Photochem Photobiol *(to be submitted)*

O'Neill P, Davies SE (1987) Pulse radiolytic study of the interaction of $SO_4^{\cdot -}$ with deoxynucleosides. Possible implications for direct energy deposition. Int J Radiat Biol 52:577-587.

O'Neill P, Fielden EM (1993) Primary free radical processes in DNA. Adv Radiat Biol 17:53-120.

Steenken S (1992) Electron-transfer-induced acidity/basicity and reactivity changes of purine and pyrimidine bases. Consequences of redox processes for DNA base pairs. Free Rad Res Comm 16:349-379.

Van Lith D, Warman JM, de Haas MP, Hummel A (1986) Electron migration in hydrated DNA and collagen at low temperature. Part I Effect of water concentration. J Chem Soc Faraday Trans 1. 82:2933-2943.

Wolf P, Jones GDD, Candeias LP, O'Neill P (1993) Induction of strand breaks in polyribonucleotides and DNA by the sulphate radical anion: role of electron loss centres as precursors of strand breakage. Int J Radiat Biol 64:7-18.

Yan M, Becker D, Summerfield S, Renke P, Sevilla MD (1992) Relative abundance and reactivity of primary ion radicals in γ-irradiated DNA at low temperatures. 2. Single - vs - double stranded DNA. J Phys Chem 96:1983-1989.

Chapter 3

Early Radical Events Following Deposition of Energy

FREE RADICAL FORMATION IN DNA: SOME NEW ASPECTS OF AN OLD PROBLEM

Jürgen Hüttermann, Werner Gatzweiler, Michael Lange and Barbara Weiland
Fachrichtung Biophysik und Physikalische Grundlagen der Medizin
Universität des Saarlandes
Klinikum Bau 76
66421 Homburg/Saar, Germany

Introduction

Despite extensive efforts over the past three decades or more, there is a continuing debate about the number of different free radicals and their detailed chemical structure obtained from DNA irradiated under various conditions concerning, among others, hydration state, temperature of irradiation or measurement, base composition, and radiation dose or quality. The only proposal accepted unanimously is the 5-yl radical on the base thymine involving net H-gain at carbon C_6 denoted **T(C_6 + H)$^\cdot$**. Its octet spectrum which usually appears, after irradiation at e.g. 77 K, at elevated temperatures (ca. 200 K) provides for an unmistakable fingerprint in Electron Spin Resonance (ESR) spectroscopy, the method of choice for investigating free radicals in general. All other proposals of radical structures, nothing to say about details of their mechanism of formation or their subsequent reactions, have been challenged and are under debate.

This apparently disturbing situation can, in our view, be traced back to several fundamental reasons which result from inherent limitations of both the sample itself and of the method of investigation. Some major reasons which have become apparent through the efforts mentioned can be formulated as

- multitude of possible radical sites together with lack of spectral differentiation
- variability of spectra from DNA subunits and DNA depending on environment (matrix-effects)
- lack of matrices for DNA with clearly defined contributions of "direct" and "indirect" radiation action combined with the possibility that radicals with similar or identical ESR-features may be formed by different mechanisms.

From this incomplete list it becomes clear that "progress" in the field has, by necessity, often meant dealing with conflicting assignments and mechanisms rather than with development of a coherent picture. This situation still prevails and applies also for the present discussion. Nevertheless, some of the results we shall present may provide a new insight into two of the various systems which are established for investigation of radical formation in DNA. These are oriented fibers and frozen aqueous solutions. For each system we delineate shortly the development of results and ideas together with the main points of controversy existing at present.

1. Oriented DNA fibers

Oriented DNA fibers equilibrated at 76% relative humidity were first used by Ehrenberg and colleagues (1-3) after irradiation at 77 K. Specifically, equilibration in D_2O vapor provided for sufficient resolution to discern spectral features obtained with the magnetic field direction parallel vs. perpendicular to the helix axis of double-stranded A- and/or B-DNA. The results obtained were eventually interpreted in terms of ionic radicals formed at 77 K, cations residing at the guanine base, **($G^{+\cdot}$)**, and anions at the thymine base, **($T^{-\cdot}$)**. The latter were shown in a kinetic study to convert, upon annealing, into the above mentioned thymine 5-yl radical. Subsequent work from our own laboratory appeared to support that model; in addition, we proposed a N_1 deprotonated guanine species, **($GN^{\cdot}$)**, as cation successor so that only four radical components were to be considered in the temperature range 77 K to about 260 K, at which temperature all radicals are lost (4-6). Upon replacement of thymine by fully carbon-deuterated thymine in the DNA it became clear, however, that a third component was involved at 77 K (7); this was tentatively ascribed to a cytosine anion, **($C^{-\cdot}$)**, mainly since other radical candidates were expected to give spectra differing largely from those found for the component in question, for which a successor radical upon annealing could not be discerned.

The main points of controversy which developed over the years and which are debated today involve both radical assignment and number of radical components. In short, there is no sufficient match between the spectra parameters of the guanine cation or its deprotonated successor between DNA and model compounds (8-10) nor have the highly resolved substructures of the putative cytosine anion spectra in DNA ever been found in a model compound (11). Thus, if there are uncertainties in component separation and assignments, there could be more and different radicals (12).

In a very recent study, we have applied the new method of pulsed ESR to unravel the composition of D_2O-equilibrated DNA fiber spectra after irradiation at 77 K. Through extensive variation of pulse distance times in two-pulse field sweep electron spin echo experiments altogether nine different radical components could be separated as being stabilized at 77 K (13). This finding has broadened the picture of primary free radical formation extensively and necessitates a renewed effort in securing structural assignment of known and new components. The present report deals with fibers equilibrated in H_2O and with those from fully deuterated DNA

2. Frozen aqueous solutions of DNA and model compounds

Over the last decade, the system of frozen aqueous solutions as matrix for stabilizing free radicals was thought to offer the possibility of studying "direct" radiation effects i.e. oxidation and reduction of substrates upon irradiation at low temperatures, typically 77 K. The influence of the $OH^{\cdot}$ radical from ice radiolysis on substrate radical formation was considered to be negligible, its reaction upon annealing being confined to the ice compartment. The role of water was suggested to consist of delivering H^+ and OH^- for neutralization of substrate anions and cations at elevated temperatures (14-23). For DNA, the picture emerging from a series of studies by Gregoli and coworkers involving single nucleotides and nucleotide mixtures (14-16) was about the same as for the oriented fibers. Primary species were thought to be **($G^{+\cdot}$)** and **($T^{-\cdot}$)**, the only secondary species being the thymine 5-yl radical (17). Subsequent studies by Symons and coworkers as well as by Sevilla and coworkers followed the same line of arguments. While the former group used additives and connected the radical balance thus changed to strandbreak induction (18-20), the latter concentrated on the delineation of DNA spectra in terms of relative contributions of different primary and secondary radicals using "benchmark" spectra of model compounds for computer manipulation (21-23). One of the main results was addition of a cytosine anion, ($T^{-\cdot}$), to the list of primary species together with a large reduction of the ($T^{-\cdot}$) contribution for double-stranded

DNA as well as an about even distribution of anions over the pyrimidines and cations over the purines in single stranded DNA.

In a recent study of the nucleotide thymidine-5'-monophosphate (dTMP) we found that the picture should, at least for single nucleotides, be revised in that oxidation of the substrate at 77 K does not take place to any significant extent and that $OH^{\cdot}$ radicals from the ice do react with the substrate upon annealing. We proposed, that the border between the ice phase and the solute compartment was transparent for electrons which, moreover, have an efficient sink in form of the nucleobase at which they produce electron adduct radicals. On the other hand, oxidation appeared to be largely confined to the ice phase in which $OH^{\cdot}$ radicals stabilize the hole (24). In this report we address the contribution of oxidation in DNA and the delineation of spectra after irradiation and measurement at 77 K.

Methods and Materials

All nucleic acid constituents and lyophilized DNA were obtained from Sigma in the highest available purity and were used as purchased. Oriented fibers prepared by the wet spinning method were obtained from A. Rupprecht (Stockholm). Fully deuterated DNA was made available through A. Gräslund (Stockholm). Sample preparation, irradiation, ESR measurements and data evaluation were performed as described previously (13, 24).

Results

1. Oriented DNA fibers

A. Equilibration in H_2O vs. D_2O

Fig. 1 (top half) shows the known ESR- spectra from D_2O equilibrated fibers in first (left) and second derivative (right) display for the two orientations parallel and perpendicular to the magnetic field direction. The sharp features marked 1-10 as well as the broader lines 11 and 12 in the top right spectrum were shown recently to be mainly due to two spectra components, denoted **C** and **G**, respectively (13).

The corresponding spectra of the fiber from H_2O equilibration (Fig. 1, bottom half) seem to have lost, at first glance, mainly the sharp lines of **C**. The dominating

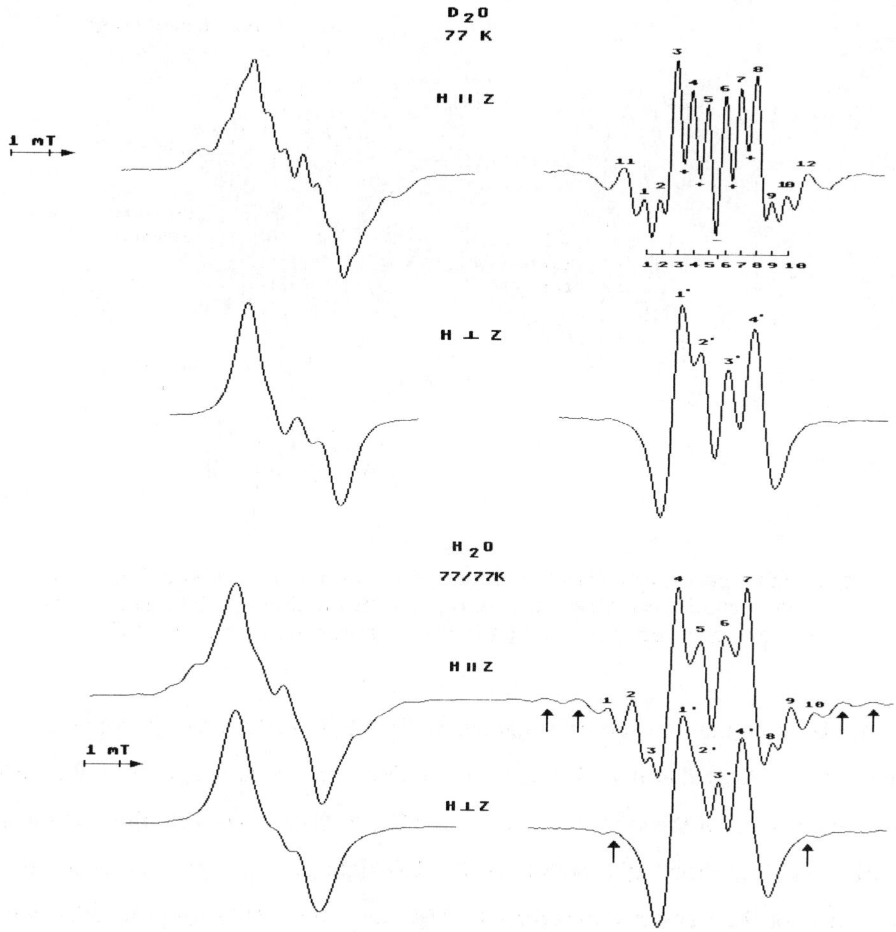

Fig. 1: ESR-spectra from D_2O (top half) and H_2O (bottom half) equilibrated DNA fibers in first (left) and second derivative (right) display for the two orientations parallel and perpendicular to the magnetic field direction.

pattern in second derivative display for the direction H| |z consists of, again, ten lines which are indicated by numbers together with flanking lines marked by arrows. In the perpendicular direction, four lines marked l'-4' are discernible, flanked by smaller, outer lines (arrows).

Considering first the ten lines (H| |z) of the H_2O sample we compare, in Fig. 2, their features in second derivative display (top left) with those previously reported (13) from D_2O after removal of component **C** by annealing to 175 K (bottom left). These were shown, in conventional ESR, to be dominated by **G.** The close resem-

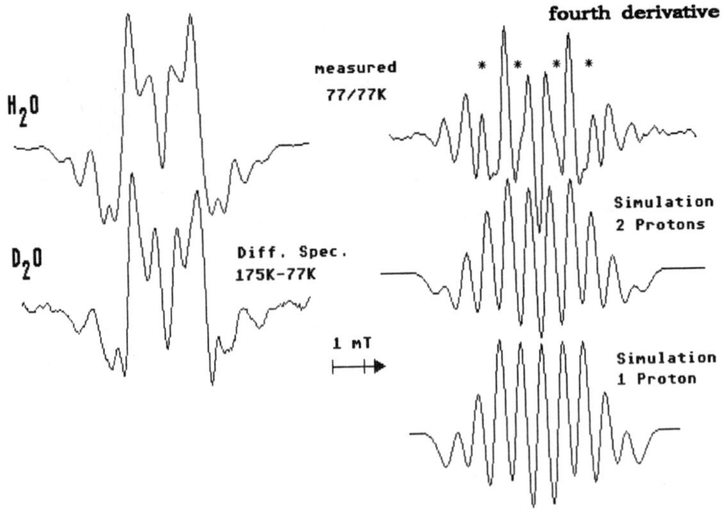

Fig. 2: ESR-spectra from H_2O equilibrated fibers for the orientation H| |z in second derivative (top left). Spectrum (bottom left) shows the previously reported D_2O pattern after removal of component **C**.

blance between the two spectra suggests, that the H_2O pattern mostly contains that component, except that line 11 and 12 of the D_2O case are now split into two each, yielding, instead of the eight line pattern for G (D_2O) now one with those ten lines marked in Fig. 1 (top right, bottom half). The right part of Fig. 2 shows a fourth derivative of the respective experimental H_2O spectrum (H| |z) (top) together with two simulated spectra. The simulations use the same parameters as given previously for component **G** in the D_2O case (13) but add either one or two proton couplings to account for labile protons. It is apparent that the simulation with two labile protons, neglecting the positions marked (*) reproduce well the experimental situation. We therefore assign the ten-line group of the H_2O sample dominating at 77 K (cf. Fig. 1) to **G.** This component is discussed again below in connection with results from fully deuterated DNA fibers.

Turning now to the four lines marked (*) in Fig. 2 we note that they approximately occupy positions which coincide with those of component **C** in the D_2O fiber which are marked in Fig. 1 (top half, top right) with numbers 1-10. These are known, in the D_2O case, to vanish upon annealing to about 175 K, as mentioned already above. Application of the same procedure for the H_2O fibers yields a comparable behaviour. Fig. 3 displays the respective experimental spectra for the parallel (top half) and perpendicular directions (bottom half) at 77 K and after annealing to 185 K in fourth and eighth derivative display. Clearly, the changes induced by annealing can be rationalized by assuming loss of one component for both field directions, four of its lines are marked again (*) for the direction Hllz.

Indications that a component with comparable features to **C** of the D_2O case are present in the H_2O fibers also derive from field sweep electron spin echo (FS ESE) experiments.

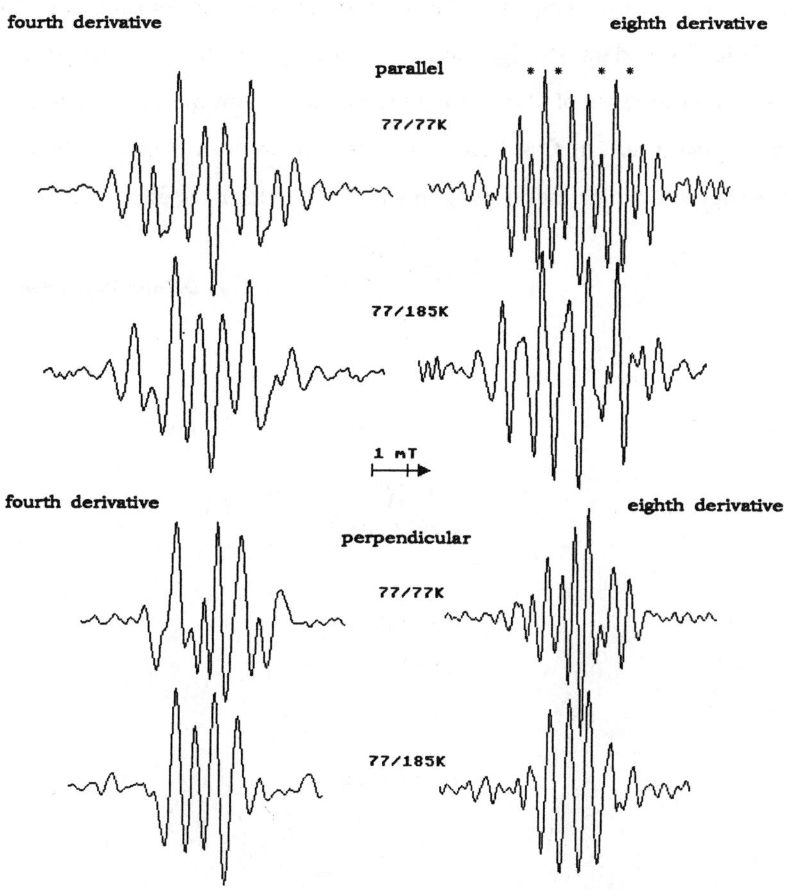

Fig. 3: ESR spectra for H_2O equilibrated fibers for the parallel (top half) and perpendicular directions (bottom half) at 77 K and 185 K in fourth and eighth derivative.

Fig. 4 (top half) shows the respective spectra obtained at 18 K after irradiation at 77 K for a pulse distance of 136 ns. The direction H| |z yields, apart from a dominating doublet group, some inner smaller lines which have the same spacing, position and relaxation behaviour as those of **C** in the D_2O fibers. The second derivative display (top right) enhances their contribution but, at the same time shows, that there should be more lines than the ten of **C** (D_2O). Comparison with the respective second derivative of the CW difference spectrum between 77 K and 165 K which enhances **C** contributions (third from top, right) shows coincidence for six of the inner lines. Roughly the same conclusion applies to the FS ESE and CW difference spectra for the perpendicular direction (second and fourth group of Fig. 4).

While these data strongly argue for the presence of a component with features related to those of **C** in the D_2O fibers, it is more difficult to extract its full spectral appearance. In CW ESR, the lines can be seen clearly only in higher derivative spectra, which makes assignment of the outer lines difficult due to

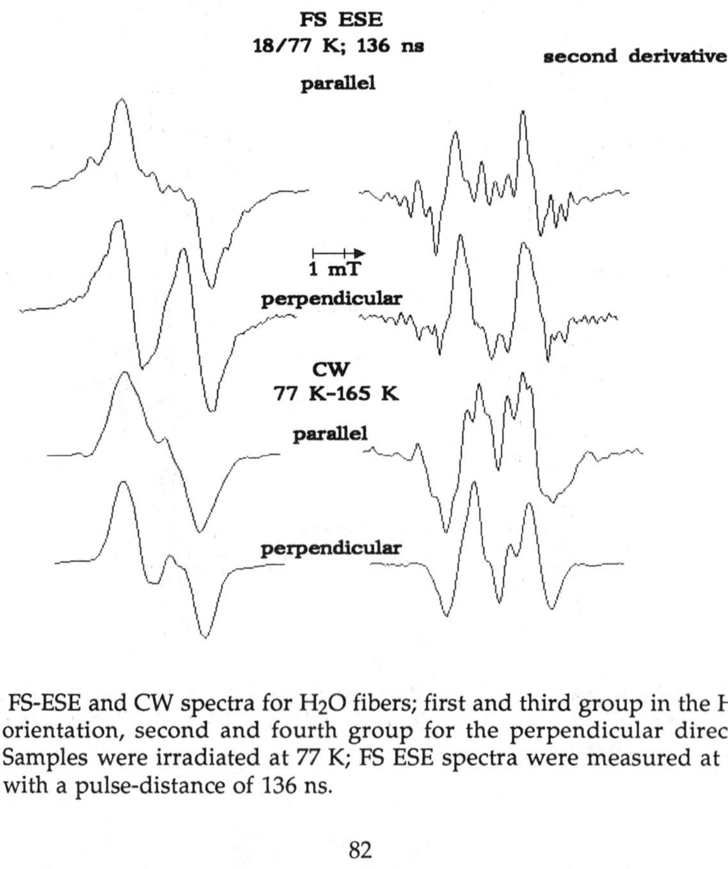

Fig. 4: FS-ESE and CW spectra for H_2O fibers; first and third group in the H| |z orientation, second and fourth group for the perpendicular direction. Samples were irradiated at 77 K; FS ESE spectra were measured at 18 K with a pulse-distance of 136 ns.

signal-to-noise problems. A full extraction was not possible for either field direction. Certainly, there are more lines than the ten for the parallel and the four for the perpendicular direction of **C** in D_2O fibers. Fig. 5 (top tracing) shows a proposed pattern for H| |z in fourth derivative gained by simulation using the parameters for **C** in D_2O and adding two labile proton couplings of 0.3 mT each. This can be compared with a summed spectrum of difference spectra (CW: 77 K - 165 K; FS ESE 18 K/136 ns -18 K/296 ns) which enhance the contributions of this component. The summation was performed to gain a better S/N ratio.

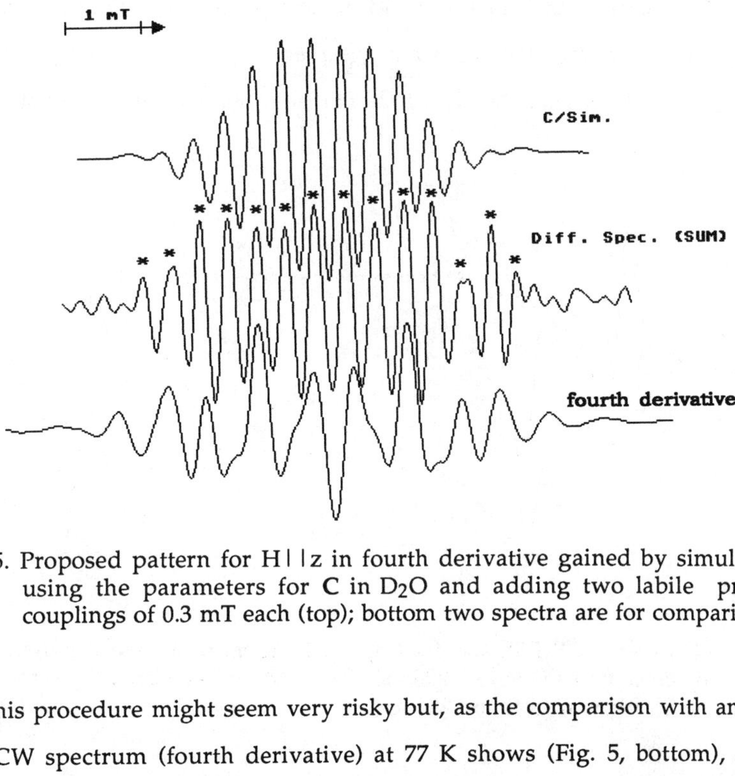

Fig. 5. Proposed pattern for H| |z in fourth derivative gained by simulation using the parameters for **C** in D_2O and adding two labile proton couplings of 0.3 mT each (top); bottom two spectra are for comparison.

This procedure might seem very risky but, as the comparison with an experimental CW spectrum (fourth derivative) at 77 K shows (Fig. 5, bottom), all lines produced by both simulation and summation of difference spectra can be found in the latter spectrum in places were the dominating component **G** does not contribute.

The FS ESE spectrum in D_2O fibers were dominated by doublet denoted **DA** since component **C** was suppressed in this type of measurement. The use of DNA containing carbon deuterated thymine allowed us to ascribe it to the thymine anion (13). A doublet is also seen to prevail in the FS ESE spectra of H_2O fibers in both field directions (cf. Fig 3, top half, left). It is reasonable to assume that it is comparable to **DA.** Fig. 6 exemplifies this relation.

The top two spectra are from the D_2O fibers in parallel (left) and perpendicular direction (right). They were gained from the difference of FS ESE spectra at 77 K taken at two different pulse distance times (136 ns - 296 ns) in order to enhance the **DA** contribution by further elimination of **C**. The bottom two spectra are from H_2O fibers taken at 18 K and 296 ns. Clearly, although not representing pure components, the dominating doublet has the same splittings as **DA.** Further conformation for relating it to **DA**, and thus to the thymine anion, is found in the D_2O case, in which this doublet is not observed in DNA containing 2 deuterated thymine (data not shown). We conclude that no apparent labile proton couplings exist for the doublet.

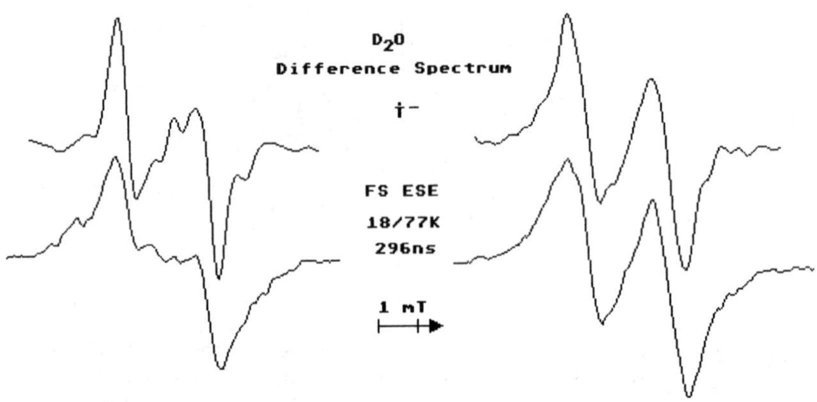

Fig. 6: Difference FS ESE spectra (top) at 77 K taken at two different pulse distance times (136 ns - 296 ns). The bottom two spectra are from H_2O fibers taken at 18 K and 296 ns.

The same holds for two other components which can be derived from H_20 fibers. One, denoted **M** in the D_2O case, was ascribed to the thymine allyl radical. The other remained unassigned since only its perpendicular component was well discernible as doublet **DB** broadening **DA** at 77 K in FS ESE spectra but dominating at very low temperatures and large pulse distance times. Fig. 7 shows the corresponding patterns from the D_2O fibers for **M** (top, left) and **DB** (top, right).

parallel

perpendicular

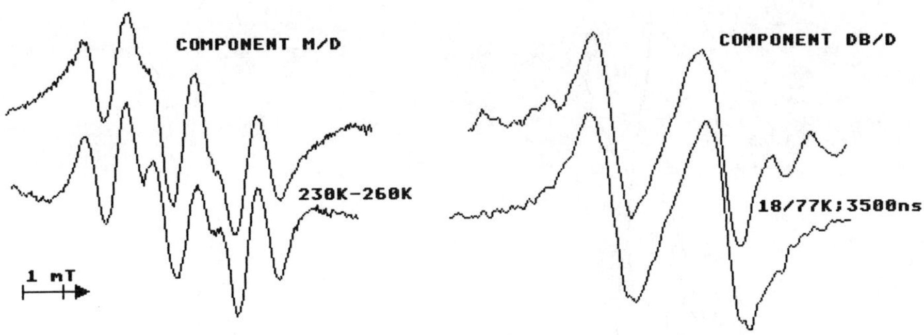

Fig. 7: Patterns isolated from the D_2O fibers denoted as components **M** (top, left) and **DB** (top, right). The bottom two spectra are from the H_2O fibers.

The bottom two spectra are from the H_2O fibers. The pattern of **M** can be produced by difference between two CW spectra (230-260 K) but is present, as in the D_2O fibers, already at 77 K. Its contribution at this temperature is indicated in Fig. 1 by arrows. Doublet **DB** is accessible, again, only in FS ESE spectra for the perpendicular direction (bottom, right).

The components **DC, S** and **U** of the D_2O fiber for which no structural assignment is available at present (13) can also be shown to exist in the H_2O sample and to display the same spectral features (25). So far, the corollary of component **A** was not detected. Thus, out of the nine different components from the D_2O fiber, eight are so far established for the H_2O specimen. Of these, two show the existence of labile proton interactions which should aid in securing their assignment to the specific radical structures. For a third component, **DA,** a stringent assignment to the thymine anion comes from DNA containing deuterated thymine.

B. Fully deuterated DNA in D_2O

Fig. 8 gives CW ESR-spectra for the direction H| |z (top left) and the perpendicular direction (top right) after irradiation at 77 K and measurement at 100

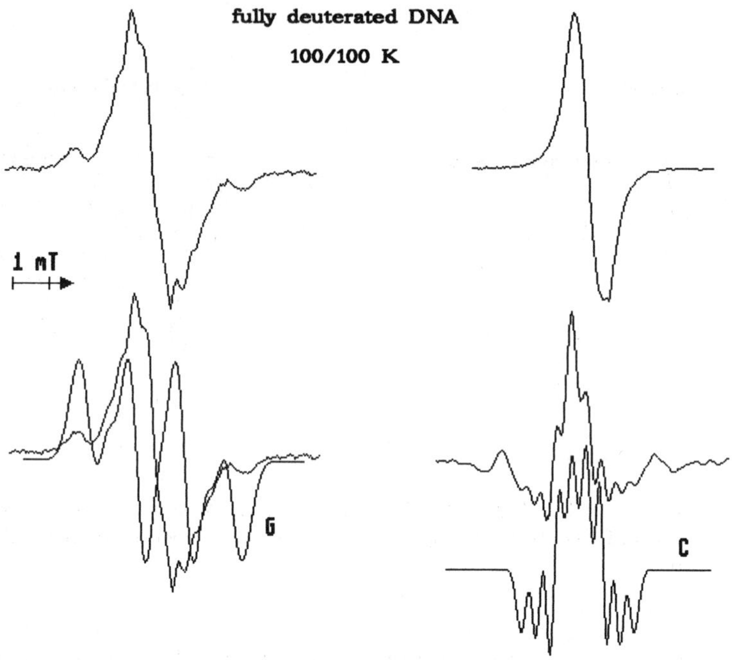

Fig. 8: Top half. ESR-spectra from fully deuterated DNA in parallel (left) and perpendicular direction (right). Bottom: Overlay of the experimental spectrum for H | | z with a simulation of component **G** using the nitrogen interaction parameters as given in (13) left). Experimental and simulated spectra of component **C** in H | | z in second derivative (right).

K. As expected for this sample, the spectral width and resolution are largely decreased due to loss of all proton couplings. Nevertheless, some structures can be discerned in the spectra for H | | z which are completely lost in the perpendicular direction. These should be ascribed to nitrogen interactions in DNA-base π-radicals for which the symmetry is such as to expect the variation between the two directions. The bottom left part in Fig. 8 shows an overlay of the experimental spectrumfor H | | z with a simulation of component **G** using only the nitrogen interaction parameters as given in (13). Clearly, the outer, unoverlapped lines are well reproduced in the simulation, as are those parts in the inner section at which the simulation requires a contribution of **G**. The bottom right recordings give the experimental H | | z feature in second derivative together with a simulation for component **C**. Again, using only the nitrogen interactions assigned to **C** (13) there is good match. These findings can be taken as strong additional proof for the correctness of the assignment of **G** and **C**.

2. DNA in frozen aqueous solution

A. The use of "benchmark"-spectra in computer reconstruction

One of the techniques which has been frequently applied in the delineation of "powder-type" multicomponent ESR spectra is the reconstruction by weighed addition of simulated or experimental spectra of single components by computer. Provided one finds the right component spectra and the S/N ratio of both component and multicomponent spectra are satisfactory, this is a straightforward matter when normalization with respect to the g-factor and the spin concentration is taken care of.

This approach has been applied recently by Sevilla and coworkers in order to unravel the low temperature spectra obtained from DNA in frozen D_2O solutions (21, 22). The results for double-stranded DNA are shown in the top right spectrum of Fig. 9. The fit between reconstructed and experimental spectra which are superimposed is very convincing. We have applied the same procedure and obtained the results given in the top left part of Fig. 9. Apparently, apart from small

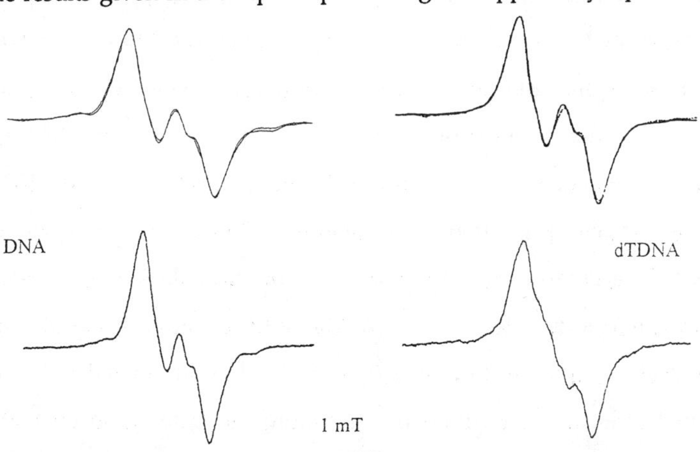

Fig. 9: Top half: Overlay of an experimental ESR-spectrum of DNA in frozen D_2O solution with a deconvolution fit reported by Sevilla et al. 1991 (top right) and a fit using "anion" spectra only (top left). Bottom half: ESR spectrum of DNA in frozen D_2O solution at 140 K (left) and ESR spectrum from DNA containing carbon deuterated thymine in frozen D_2O solution.

flanking outer lines the experimental spectra in both cases are identical. Likewise, our fit is of the same quality. Nevertheless, we have used a completely different set of component spectra. This can be seen in Table 1 which lists seven possible primary components and shows their relative contribution used in reconstruction. In both cases there are three major components employed but only one of them, that of **($C^{-\cdot}$)** is common, albeit with drastically different concentrations.

DNA	A	A^{+}	C^{-}	T^{-}	G^{-}	G^{+}	Allyl	R
	-	-	74.9	-	14.1	-	11.0	0.052
Sevilla et al. '91	-	-	40.1	18.7	-	41.2	-	0.043

Table 1: List of seven possible primary components and their relative contribution used in reconstruction together with resulting R-values.

The problem connected with this comparison has at least two aspects. One is, of course, finding the suitable component spectra, the "benchmark" spectra. The other is finding algorithms and numerical criteria to distinguish between spectra components with bad differentiation. Having established a solution for the latter part it should be possible to judge the quality of the benchmark spectra.

With n components with weighted contribution c_j (j=l..,n) the problem of finding intensities K_{ij} of component j at magnetic field position i by solving $\Sigma c_j * K_{ij} = S_i$ with S being the intensity of the summed component intensities at magnetic field position i; the summation is over j. Matrix notation yields finding the solution of the equation $\mathbf{K} * \mathbf{c} = \mathbf{S}$. In our digitized spectra there are 1024 magnetic field values, much more than components. Therefore, the system of linear equations is overdetermined and there is a unique solution to a reduced set of equations written as $(\mathbf{K}^T * \mathbf{K}) * \mathbf{c} = \mathbf{K}^T * \mathbf{S}$. The solution vector **c** has the property to solve the original equation in a way that $(S_i - S_i`)^2$ is minimized for all i ($S_i`$ is the experimental intensity at position i). Summing the latter term over all magnetic field values i yield the R-value used in the literature as a measure for the quality of the procedure.

We have applied this procedure to weighed additions of our component spectra. Fig. 10 shows the spectra we used. There are two very much alike doublets

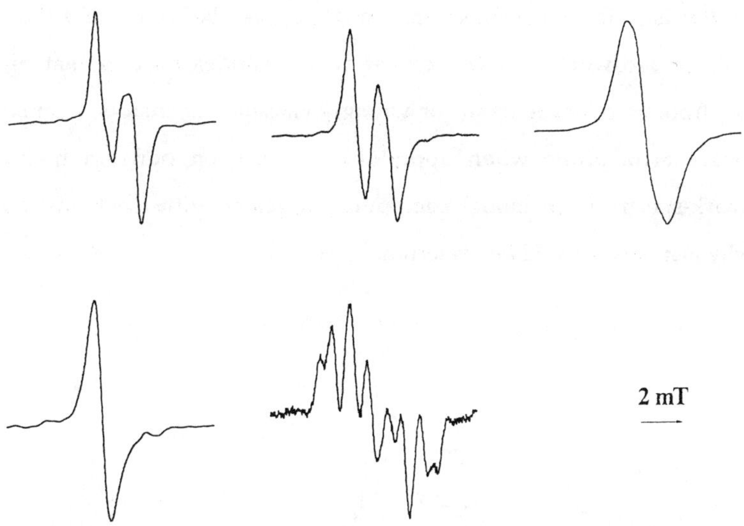

Fig. 10: "Benchmark-spectra used in the deconvolution. Spectrum top left: T⁻ top middle: C⁻, top right: A⁻, bottom left: Allyl, bottom right: G⁻.

(C⁻ and T⁻), two singlets (G⁻ and A⁻) plus a pattern denoted "Allyl". Fig. 11 gives some 1:1 additions of two or more components together with the result of the delineation. In all cases, the delineated spectra reproduced the given pattem with an R-value of zero and gave the correct weight. In many trials we found the algorithm to be very stable and reliable. In short, if the component spectra are present in the multicomponent spectra, the delineation will give the correct answer about their contributions.

Since we applied the same algorithm and the same benchmark spectra to the delineation of the DNA spectrum of Fig. 9, one is led to believe that only our component distribution should be correct. This would mean that, as an example, the pattern T⁻ which we associate somehow with the thymine anion, is not at all presented in the primary DNA spectra (cf. Table 1). On the other hand, we can directly show by employing DNA with carbon-deuterated thymine, that the thymine anion must be formed (cf. Fig. 9 bottom). As a serious consequence we have to postulate strictly, that one or indeed potentially any of the so-called benchmark spectra are not in the set really necessary for delineating the DNA spectra, despite the very impressive fit.

If this is so for our delineation, it might as well be the case for the delineation of Sevilla and coworkers. The caveat which applies here is that any R-value differing from zero leaves room for wrongly chosen benchmarks, a situation which is strongly error-prone when applying, as has been done in both examples, benchmark-spectra from model compounds together with mechanistic arguments as to why just these should be preferred.

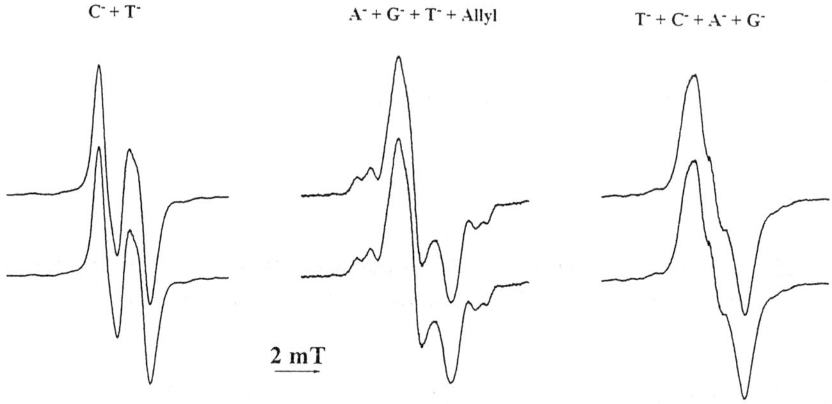

Fig. 11: 1:1 additions of two or more components together with a result of the deconvolution.

B. DNA in frozen aqueous solution

If the qualitative picture we presented recently for the nucleotide TMP is correct in that substrate oxidation in frozen aqueous solutions irradiated at 77 K is negligible and that most if not all of the radicals formed are due to reaction of electrons and OH$^{\cdot}$ radicals, one should expect to prove it quantitatively by use of scavengers. For TMP we found that more than 80% of all TMP radicals are due to electron reaction which can indeed be scavenged (e.g. by $FeCl_3$). The remaining fraction consists nearly exclusively of allyl radicals which, however, could not be scavenged by those substances known to be OH$^{\cdot}$ scavengers at 300 K in aqueous solution (26, 27). We ascribe this to lack of mobility of OH$^{\cdot}$ radicals and scavengers at 77 K in the frozen system.

Using electron scavengers in the system DNA (30 mg/ml) in frozen solutions we find the results shown in Fig. 12 for irradiation and measurement at 77 K. Without scavenger (O mM), about 65% of the radicals observed are OH˙, 35% are DNA radicals. This fraction reduces to about 10% at 10 mM scavenger concentration

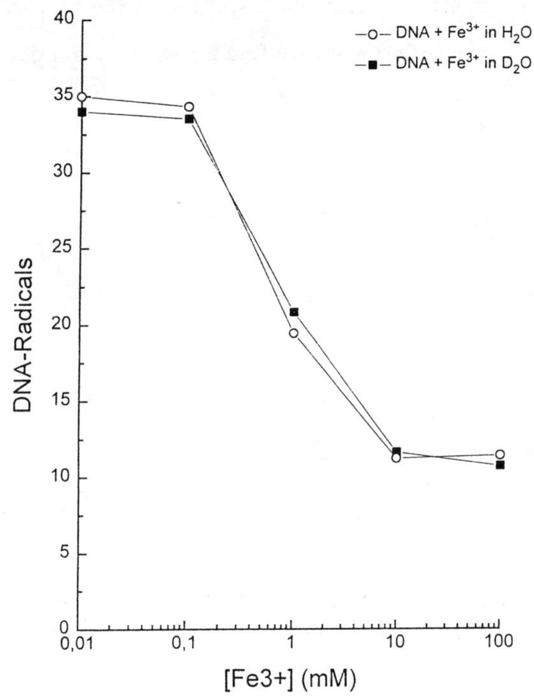

Fig. 12: Plot of the DNA and OH˙ radical concentration versus the scavenger concentration after irradiation and measurement at 77 K.

and appears to remain the unscavengable fraction also at 100 mM. Thus, about 28% of DNA radicals formed cannot be scavenged by external electron scavengers. If we assume an even distribution of these between oxidation and reduction, about 14% of the DNA radicals formed in this system are produced by DNA oxidation, a drastic difference to the prevailing numbers of 50%! It thus appears that the mechanistic picture applied for this system has to be revised.

It is of interest to inspect the structural features obtained under scavenger influence. Fig. 13 gives (left column), the experimental spectra obtained after annealing to 140 K (to remove OH˙ contributions) and measurement at 77 K for H_2O

(Fig. 13a) and D_2O (Fig. 13b). Consider first the H_2O case.

Apparently, the doublet-type feature for 0 M scavenger changes to a more singlet type and again to a doublet feature, which, however, differs in splitting from the initial doublet. These changes can be accounted for by three spectral components present which are delineated, by difference spectra, in the right column. Applying the same procedure to the D_2O-spectra we find again three components, one of

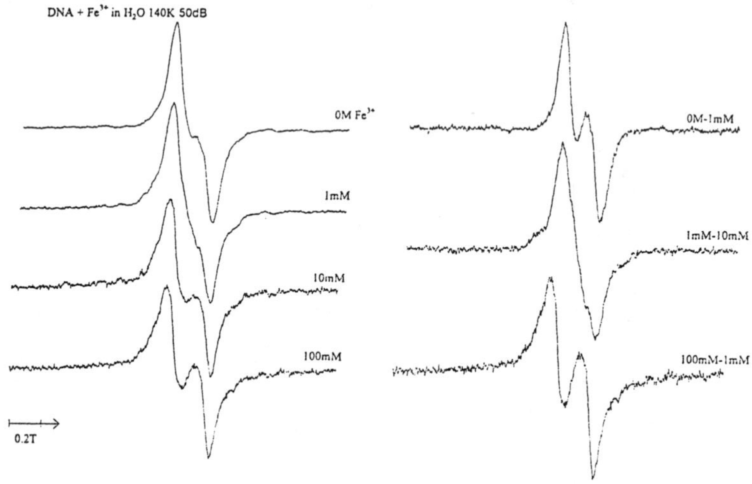

Fig. 13a: Experimental ESR-spectra of DNA in frozen H_2O solution with various scavenger concentration, annealed to 140 K and measured at 77 K.

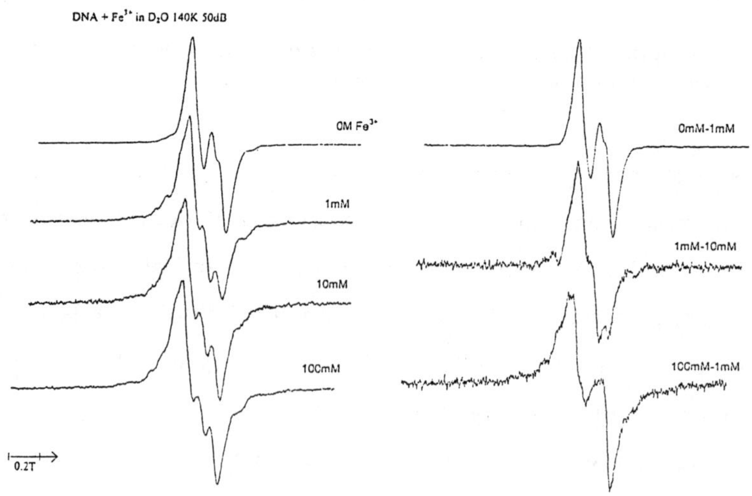

Fig. 13b: Same as a) but frozen D_2O solution.

which, that at high scavenger concentration, is unchanged in appearance.

It is reasonable to assume that the two components which vanish with increasing scavenger concentration should be anion radicals. Without implying them to be "benchmark" spectra we can compare the single nucleotide anion spectra with the doublet and the singlet. This comparison is shown in Fig. 14. The doublet **Dl** in H_2O seems to match closely with the **($T^{-\cdot}$)** pattern from TMP while the **($C^{-\cdot}$)** doublet from dCMP has different splitting and line-width. In D_2O, both **($T^{-\cdot}$)** and, perhaps somewhat more, **($C^{-\cdot}$)** appear to be good candidates. For the singlet **Sl,** we find that a mixture of **($G^{-\cdot}$)** and **($A^{-\cdot}$)** may represent both the H_2O and D_2O patterns.

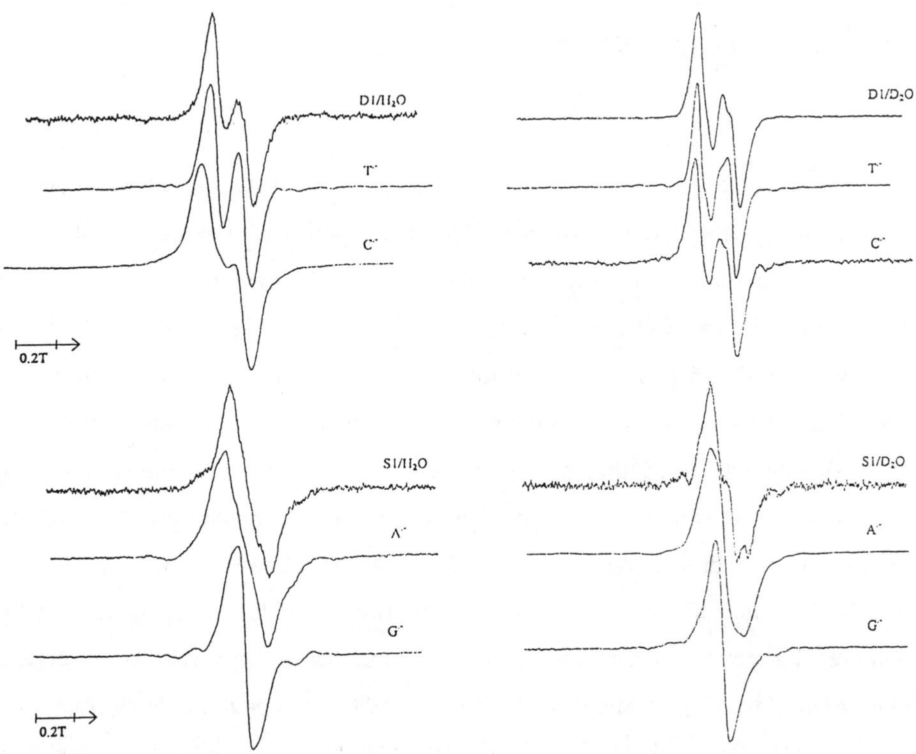

Fig. 14: Difference spectra yielded by subtraction of DNA spectra shown in Fig. 13a and 13b respectively.

In conclusion we find that since more than 70% of all DNA radicals are due to electron reactions which can be scavenged, the original DNA-spectrum at 77 K should be comprised mostly from anion patterns. It therefore becomes clear why we have the delineation of Table 1 only used those. (We note, that there are between 5 and 10% allyl radicals present in addition which are not apparent in the components of Fig. 14 due to the subtration procedure.)

Another conclusion which contrasts again with the prevailing discussion in the literature concerns the guanine cation. If we assume that it is formed then it should show up strongly in the pattern at high scavenger concentration. Certainly, all features discussed for this species do not give a large weight of **($G^{+\cdot}$)** to the doublet seen. Again, this implies that **($G^{+\cdot}$)**, does not play a dominant role and is among the small contributions of oxidation.

Discussion

In oriented DNA fiber, the new method of pulsed EPR has enhanced largely the knowledge about the number of radical components. Of the nine different patterns described most recently in D_2O equilibrated specimen (13), eight were found to be present under H_2O conditions (25) of which five are discussed here in detail. Two of these, **G** and **C**, show proton interactions which are exchanged under D_2O condition. These two can also be found in fully deuterated DNA on account of their nitrogen interactions so that a coherent set of couplings applying for all three systems of DNA can be presented. The quantitative contribution of **C** in H_2O appears to be reduced compared to D_2O if we normalize to component **G**. The doublet **DA,** unambiguously assignable to the thymine anion by using specifically deuterated thymine, is also present in H_2O and has no apparent labile proton interaction. Likewise, component **M**, (the allyl radical) can be found in H_2O without a change in pattern, as holds for doublet **DB** which has a somewhat larger splitting than **DA.**

In assigning component **G** and **C** we still, despite a very consistent set of nitrogen and proton interactions, can only be tentative. The guanine cation **($G^{+\cdot}$)** with two exchangeable amino protons still appears to be a good candidate for **G** since, in addition, its nitrogen couplings match well with the values from model compounds. Assigning the cytosine anion **($C^{-\cdot}$)** to **C** still is unsupported by model compound studies but could be a reasonable assumption, particularly since the features of the other pyrimidine anion, **($T^{-\cdot}$)** (component **DA**) are now established.

More work is needed before the other four thus far unassigned components become amenable. One would then be able to gain a more complete structural picture about radical formation in that system.

While oxidation processes take place in the oriented fibers at 77 K (30% water by weight) the system of DNA (~30 mg/ml) in frozen solution has a very reduced contribution of oxidation derived radicals. We propose that all four base anions are formed in the absence of electron scavengers, together with a doublet which is unaffected by scavengers. Radicals from base oxidation could be formed in the range of about 15% or less. The change from H_2O to D_2O appears to enhance **($C^{-\cdot}$)** contributions like in the fiber. This would explain some of the discrepancies apparent in the literature.

We have shown that delineation of DNA spectra by benchmark patterns can be a very risky process. A numerical procedure should be applied allowing for transformation of initial component spectra in iterative deconvolution.

References

(1) Ehrenberg A., Rupprecht A., Ström G.; *Science* **157** (1967) 1317
(2) Gräslund A., Ehrenberg A., Rupprecht A., Ström G.; *Biochim. Biophys. Acta* **254** (1971) 172
(3) Gräslund A., Ehrenberg A., Rupprecht A., Ström G.; *Int. J. Radiat. BioL* **28** (1975) 379
(4) Hüttermann J., Volt K., Oloff H., Köhnlein W., Gräslund A., Rupprecht A.; *Faraday Discussion of the Chemical Society* **78** (1984) 135
(5) Hüttermann J., Röhrig M., Köhnlein W.; *Int. J. Radiat. Biol-* **61** (1992) 299
(6) Volt K., Hüttermann J.; *Electronic Magnetic Resonance of the Solid State,* edited by J. A. Weil (Canadian Society for Chemistry, Ottawa, Canada) p. 267
(7) Zell I., Hüttermann, J., Gräslund, A., Rupprecht, A., Köhnlein, W.; *Free Radical Res. Comms.* **6** (1989) 105
(8) Hole E. O., Nelson W. H., Close M. D., Sagstuen E.; *J. Chem. Phys.* **86** (1987) 5218
(9) Kim H., Budzinski E. E., Box H. C.; *J. Chem. Phys.* **90** (1989) 1448
(10) Close D. M., Sagstuen E., Nelson W. H.; *J. Chem. Phys.* **82** (1985) 4386
(11) Hüttermann J., Ohlmann J., Schaefer A., Gatzweiler W.; *Int. J. Radiat. Biol* **59** (1991) 1257
(12) Bernhard W. A., *Adv. Radiat. Biol.* **9** (1981) 199

(13) Gatzweiler W., Hüttermann J., Rupprecht A.; *Radiat. Res.* **138** (1994) 151
(14) Gregoli S., Olast M., Bertinchamps A.; *Radiat. Res.* **60** (1974) 388
(15) Gregoli S., Olast M., Bertinchamps A.; *Radiat. Res.* **70** (1977) 255
(16) Gregoli S., Olast M., Bertinchamps A.; *Radiat. Res.* **77** (1979) 417
(17) Gregoli S., Olast M., Bertinchamps A.; *Radiat. Res.* **89** (1982) 238
(18) Boon Ph. J., Cullis P. M., Symons M. C. R., Wren B. W.; *J. Chem. Soc. Perkin Trans.* **II** (1985) 1057
(19) Cullis P. M., Symons M. C. R., Wren B., Gregoli S.; *J. Chem. Soc. Perkin Trans. 11* (1985) 1819
(20) Cullis P. M., Symons M. C. R., Swenberg M. C., Jones G. D. D., McClymont J. D.; *J. Chem. Soc. Perkin Trans.* II (1986) 1671
(21) Sevilla M. D., Becker D., Yan M., Summerfield S. R.; *J. Phys. Chem.* **95** (1991) 3409
(22) Yan M., Becker D., Summerfield S., Renke P., Sevilla M. D.; *J. Phys. Chem.* **96** (1992) 1983
(23) Wang W., Sevilla M. D.; *Radiat. Res.* **138** (1994) 9-17
(24) Hüttermann J., Lange M., Ohlmann J.; *Radiat. Res.* **131** (1992) 18
(25) Gatzweiler W.; *PhD thesis,* University of Saarland, Germany (1993)
(26) Von Sonntag C.; *The Chemical Basis of Radiation Bioloby,* Taylor and Francis (1987)
(27) Weiland B.; *Diploma thesis,* University of Saarland, Germany (1994)

SOLID STATE RADIATION CHEMISTRY OF CO-CRYSTALLIZED DNA BASE PAIRS STUDIED WITH EPR AND ENDOR

W. H. Nelson and S. Nimmala
Department of Physics and Astronomy
Georgia State University
Atlanta, GA 30303 USA

D. M. Close
Department of Physics,
Eastern Tennessee State University
Johnson City, TN 37614 USA

E. O. Hole and E. Sagstuen
Department of Physics
University of Oslo
P.O. Box 1048
Blindern, 0316 Oslo 3, Norway

Introduction

DNA is assembled from a specific set of basic chemical components, *i.e.*,purine and pyrimidine bases, deoxyribose sugar, phosphate groups, and water. Nevertheless, ionization of DNA leads to a distribution and relative concentration of products from these components unlike that expected from a simple mixture. Evidently, therefore, product formation and stability in the DNA system not only depends on the specific components, but also depends on interactions arising from their spatial arrangement. Base stacking and base pairing by hydrogen bonds provide obvious possibilities for charge and spin transfer between the components following ionization. Other possibilities are provided by the many additional hydrogen bonds in the system, especially those to water in the inner hydration layer.

Thus, it is clear that understanding the initiation and progression of ionization effects in DNA requires understanding charge and spin transfer mechanisms in ***that system***. Of particular importance is learning the DNA-related factors which may promote or inhibit these mechanisms. For example, Steenken (1992) used pK_a considerations to predict the probability of proton transfers between hydrogen-bonded bases following ionization; in this discussion, he pointed out the importance of ***reversible*** vs. ***irreversible*** proton transfers. From a survey of solid-state work, Nelson *et al.* (1992) tabulated results showing proton transfer behavior involving specific hydrogen bond types. With extensive experimental and quantum chemical work, Sevilla's group has investigated electron and proton transfer between bases in DNA and DNA-like systems (Colson 1992; Sevilla *et al.*,

1991). Very recently, Bernhard *et al.* (1994) discussed the importance of molecular packing on the stabilization of primary ionic pyrimidine radicals.

For a number of years, our group has focused on identification of radicals formed from x-irradiation of DNA components by application of EPR and ENDOR spectroscopic techniques to samples in the form of single crystals. With single crystals as samples, it is possible to use the detailed packing and structural information available from x-ray or neutron diffraction reports. This report summarizes results from two crystal systems in which DNA bases are paired by hydrogen bonding. Extensive results are available from one of these, 1-methylthymine:9-methyladenine (**MTMA**), in which the base pairing is the Hoogsteen configuration (Hoogsteen 1963). Although this configuration is different from that found by Watson-Crick in DNA, nonetheless the hydrogen bond between T(O4) and $A(NH_2)$ is present, as shown in the schematic below. Although **MTMA** crystals have been studied previously (Schmidt and Snipes 1967), our objective was to apply the high-resolution technique of ENDOR to crystals irradiated and studied at temperatures of 10 K or lower in the effort to obtain direct evidence for specific proton transfers. The second system, from which the results are only preliminary, is 9-ethylguanine:1-methyl-5-fluorocytosine (**GFC**) in which the G:C bases pair is in the Watson Crick configuration as shown in the structure (O'Brien 1967). Both crystal systems are anhydrous, so the results include no possible effects from water interactions.

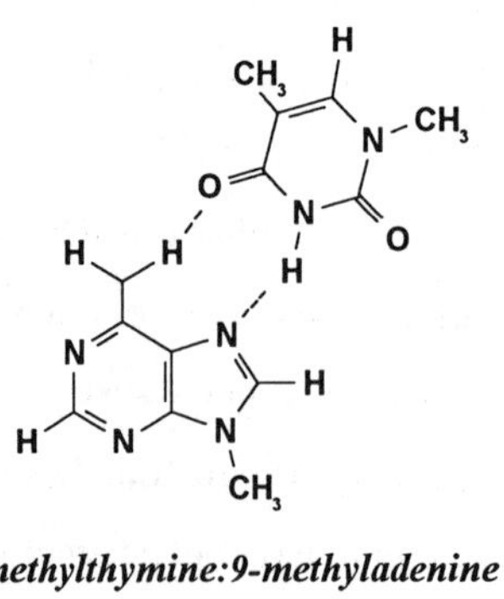

1-methylthymine:9-methyladenine

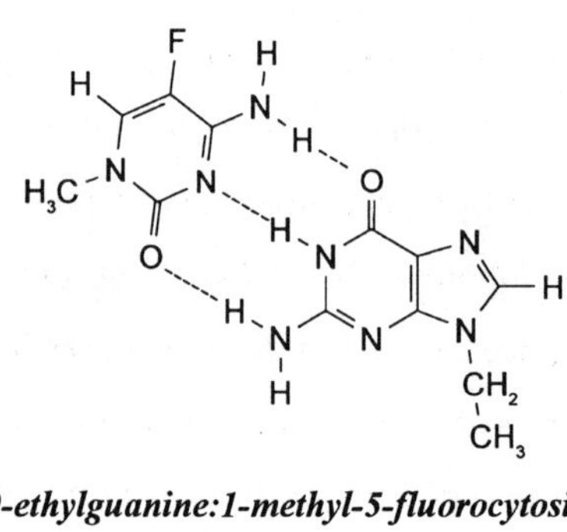

9-ethylguanine:1-methyl-5-fluorocytosine

Results

Using techniques described at length previously (Nelson *et al.*, 1988), EPR and ENDOR were obtained following irradiation at 10 K or 65 K. By means of field-swept ENDOR (FSE), and by extensive analysis of the ENDOR spectra, it was possible to detect and identify 6 radicals (labelled **MTMA1 - 6** in the following discussion) from the **MTMA** system.

Radical **MTMA1** was identified as the product of net H-abstraction from the C5 methyl of the MT unit. This radical, shown below and initially identified by Hüttermann (1970), has been detected in all thymine derivatives studied to date as crystals. The coupling constants listed in the **Table** were derived from the data and are consistent with results from other systems. The **MTMA1** radical species is known to form by deprotonation of the thymine cation (Sevilla *et al.*, 1972), although an alternative mechanism is dissociation of excited species. Radical **MTMA2**, characterized by coupling constants as listed in the **Table,** was identified as the product of net H-abstraction from the N1 methyl. In contrast to **MTMA1**, the **MTMA2** species was not previously detected in methylthymine; however, abstraction from the N1 methyl is well-known from other N1-methyl pyrimidine derivatives, *e.g.* 1-methyl uracil (Flossmann *et al.*, 1973). This product is thought to be the result of dissociative excitation (Bernhard *et al.*, 1994). Radical **MTMA3**, identified as the product of net hydrogen addition to O4 of MT, was characterized by the major coupling listed in the **Table,** and at least one minor coupling yielding an ENDOR line at 40.6 Mhz when the magnetic field was oriented along the crystallographic axis. This minor coupling is as expected from the rotating C5 methyl group as discussed previously (Hole *et al.*, 1991), and is the basis for inferring the presence of the proton at O4 as shown in the **MTMA3** structure. This radical is

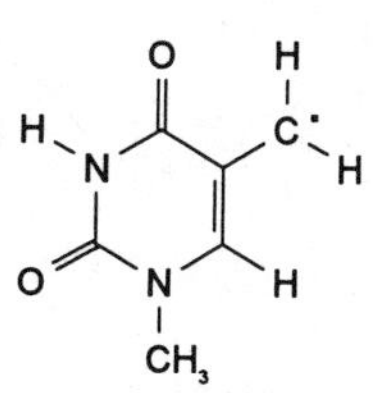

MTMA1

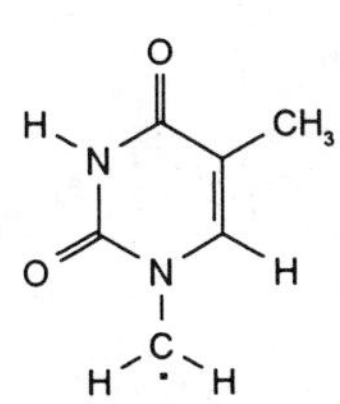

MTMA2

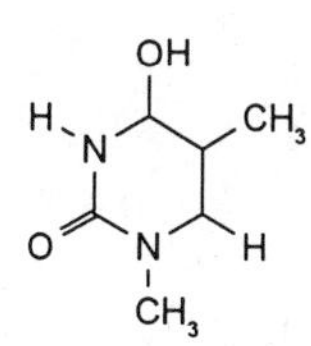

MTMA3

evidently the result of proton transfer to MT(O4) from MA(NH_2) following electron capture (anion formation) by MT. Unlike previous cases of thymidine (Hole *et al.*, 1991) and anhydrous thymine (Sagstuen *et al.*, 1992), the molecular arrangement in MTMA crystals, with the MT and MA units rigorously co-planar, constrains the proton at O4 to the MT molecular plane; consequently the isotropic component of the •COH ß-coupling will be very small. Evidence for radical **MTMA4**, the product of net H-addition to C5 of MT, comes from an axially symmetric ENDOR line appearing in the 63-67 Mhz range. FSE from this line gave the characteristic 8-line pattern of the MT 5-yl radical, and the coupling tensor is like that from the rotating methyl group of the similar radical in other thymine derivatives (Box *et al.*, 1974). It is interesting that the methyl group rotates freely in this case even at 10 K while the onset of free rotation occurs near 75 K in thymidine. Also, it is interesting that the MT 5-yl radical was absent after low temperature irradiation in methylthymine crystals, and formed only after prolonged warming above 250 K. Features from this radical were only weakly observable in the EPR from MTMA, however. Radicals **MTMA5** and **MTMA6** were identified as the well-known C2 and C8 H-addition radicals of the MA unit. Both were present immediately following irradiation at 10 K, although the C2 H-addition radical decayed on warming the crystal beyond 170 K. It is particularly notable that no evidence was found in the MTMA system for the product of net H-addition to N3 of the MA unit, or for the product of net H-abstraction from the 9-methyl. The N3 adduct, evidently the product of reduction followed by protonation was found in crystals of the MA unit (Hole *et al.*, 1995), and previously in adenosine:bromouridine crystals (Kar and Bernhard 1983), in adenosine crystals (Nelson and Close 1989), and in deoxyadenosine monohydrate

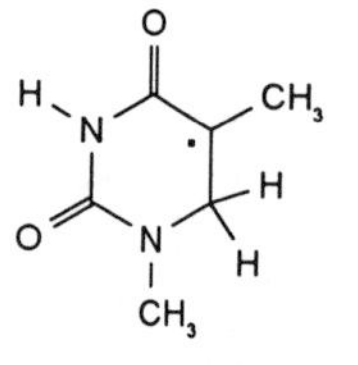

MTMA4

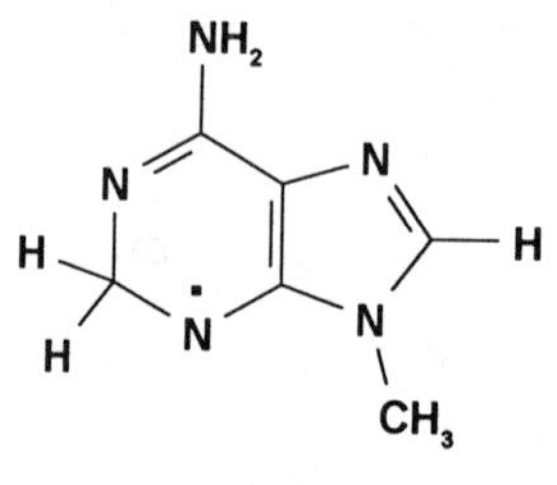

MTMA5

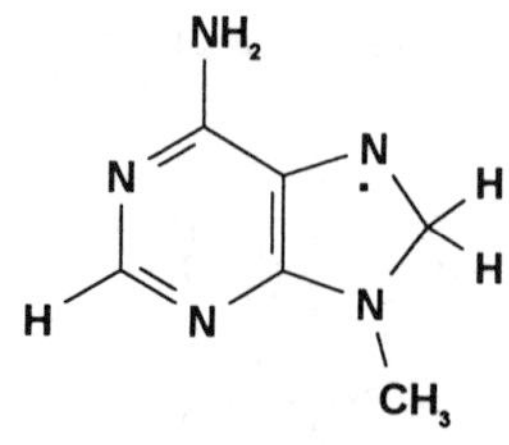

MTMA6

crystals (Close *et al.*, 1994). Previous study of 9-methyl adenine found a major product to be the 9-methyl abstraction radical, thought to be the result of oxidation followed by proton loss (Schmidt and Snipes 1969).

Results from GFC crystals are very preliminary, and come from analysis of only one coupling. This coupling, listed in the **Table**, is like that previously found for the guanine base in 5'-dGMP (Hole *et al.*, 1992a) and 3',5'-cGMP (Hole, *et al.*, 1992b) following net H-abstraction from the amino group. If further investigation confirms this preliminary assignment, this will be direct evidence for proton transfer from the guanine amino group to cytosine presumably following guanine oxidation.

Table. Coupling constants derived from the data for MTMA and GFC.

Coupling / Assignment	Coupling / Assignment
MTMA1:	**MTMA4:**
(-38.8, -28.1, -13.8) MHz / HC6 (MT)	(57.8, 58.2, 52.6) MHz / MeC5 (MT)
(-62.7, -41.0, -19.9) MHz / H_1C7 (MT)	**MTMA5:**
(-68.6, -43.0, -23.2) MHz / H_2C7 (MT)	(-26.5, -17.5, -8.7) MHz / HC8 (MA)
MTMA2:	**MTMA6:**
(-88.6, -54.7, -30.7) MHz / H_1C8 (MT)	(-19.7, -17.5, -8.7) MHz / HC2 (MA)
(-83.8, -52.7, -24.4) MHz / H_2C8 (MT)	**GFC:**
MTMA3:	(-48.0, -22.3, -10.8) MHz / *
(-61.4, -35.8, -19.0) MHz / HC6 (MT)	

* For the amino-deprotonated guanine cation, the constants are *ca.* (-46, -29, -5) MHz. (Hole, *et al.*, 1992a; Hole, *et al.*, 1992b)

Summary

The set of radicals found in the MTMA system can be associated with basic mechanisms as follows: MTMA1, oxidation (or excitation); MTMA2, excitation; MTMA3, reduction; MTMA4, reduction (or •H attack); MTMA5, reduction (or •H attack); MTMA6, reduction (or •H attack). In this system, therefore, radicals from oxidation, reduction and excitation of the pyrimidine component were stabilized under the experimental conditions employed. The purine radicals were products of reduction or •H attack. Evidently, the base-pairing in this system does not permit the ready charge and spin transfer necessary for the pyrimidine anion / purine cation behavior discussed for DNA. Further work will focus on measurement of relative concentrations of the species reported above to obtain additional information which may permit distinguishing between the mechanisms. A more complete description of these results will be published elsewhere.

Acknowledgements

This work was supported by NIH grant CA36810, and by NATO Travel Grant RG 0426/88. Additional travel support came from Norges Allmenvitenskapelige Forskningsråd (NAVF).

References

BERNHARD WA, BARNES J, MERCER K, MROCZKA H (1994) The influence of packing on free radical yields in crystalline nucleic acids: the pyrimidine bases. Radiat Res, in press

BOX HC, BUDZINSKI HC, POTTER WR (1974) ENDOR characterization of the free radical in irradiated thymidine. J Chem Phys 61: 1136-1139

CLOSE DM, NELSON WH, SAGSTUEN E, AND HOLE EO (1994) ESR and ENDOR study of single crystals of deoxyadenosine monohydrate x-irradiated at 10 K. Radiat Res 137: 300-309

COLSON A-O, BESLER B, CLOSE DM, SEVILLA MD (1992) *Ab initio* molecular orbital calculations of DNA bases and their radical ions in various protonation states: evidence for proton transfer in GC base pair radical anions. J Phys Chem 96: 661-668

FLOSSMANN W, HÜTTERMANN J, MÜLLER A, WESTHOF E (1973) Electron spin resonance of free radicals and radical pairs in irradiated single crystals of 1-methyluracil. Z Naturforsch 28C: 523-532

HOLE EO, SAGSTUEN E, NELSON WH, CLOSE DM (1991) Primary reduction and oxidation of thymine derivatives. ESR/ENDOR of thymidine and 1-methylthymine x-irradiated at 10K. J Phys Chem 95: 1494-1503

HOLE EO, NELSON WH, SAGSTUEN E, CLOSE DM (1992a) Free radical formation in single crystals of 2'-deoxyguanosine 5'-monophosphate tetrahydrate disodium salt: and EPR/ENDOR study. Radiat Res 129:119-138

HOLE EO, SAGSTUEN E, NELSON WH, CLOSE DM (1992b) The structure of the guanine cation: ESR/ENDOR of cyclic guanosine monophosphate single crystals after x-irradiation at 10 K. Radiat Res 129: 1-10

HOLE EO, SAGSTUEN E, NELSON WH, CLOSE DM (1995) Radiation damsge to DNA base pairs. EPR and ENDOR study of single crystals of 9-methyladenine and of the complex 1-methylthymine•9-methyladenine x-irradiated at 10K. Radiat Res, to be published

HOOGSTEEN K (1963) The crystal and molecular structure of a hydrogen-bonded complex between 1-methylthymine and 9-methyladenine. Acta Crystallogr 12: 907-916.

HÜTTERMANN J (1970) Electron-spin-resonance spectroscopy of radiation-induced free radicals in irradiated single crystals of thymine monohydrate. Int J Radiat Biol 17: 249-259

KAR L, BERNHARD WA (1983) Electron gain and electron loss radicals stabilized on the purine and pyrimidine of a co-crystal exhibiting base-base interstacking: ESR-ENDOR of x-irradiated adenosine:5-bromouracil. Radiat Res 93: 232-253

NELSON WH, HOLE EO, SAGSTUEN E, CLOSE DM (1988) EPR/ENDOR study of guanine•HCl•2H2O x-irradiated at 20 K. Int J Radiat Biol 54: 963-986

NELSON WH, CLOSE DM (1989) ESR and ENDOR study of adenosine single crystals x-irradiated at 10 K. Radiat Res 117: 367-378

NELSON WH, SAGSTUEN E, HOLE EO, CLOSE DM (1992) On the proton transfer behavior of the primary oxidation product in irradiated DNA. Radiat Res 131: 10-17

O'BRIEN EJ (1967) Crystal structures of two complexes containing guanine and cytosine derivatives. Acta Cryst 23:92-106

SAGSTUEN E, HOLE EO, NELSON WH, CLOSE DM (1992) Radiation-induced free-radical formation in anhydrous thymine single crystals x-irradiated at 10 K. J Phys Chem 96: 1121-1126

SCHMIDT J, SNIPES W (1967) Free radical formation in a gamma-irradiated pyrimidine-purine co-crystal complex. Int J Radiat Biol 13: 101-109

SCHMIDT J, SNIPES W (1969) Electron spin resonance of gamma-irradiated single crystals of 9-methyladenine. Radiat Res 38: 274-284

SEVILLA MD, VAN PAEMEL C, ZORMAN G (1972) Reactions of the cation and anion radicals of several DNA bases. J Phys Chem 76: 3577-3582

SEVILLA MD, BECKER D, YAN M, SUMMERFIELD SR (1991) Relative abundances of primary ion radicals in γ-irradiated DNA: cytosine vs thymine anions and guanine vs. adenine cations. J Phys Chem 95: 3409-3415

STEENKEN S (1992) Electron-transfer induced acidity / basicity and reactivity changes of purine and pyrimidine bases. Consequences of redox processes for DNA base pairs. Free Radical Res Commun 53: 349-379

PROTONATION/DEPROTONATION PROCESSES OF PRIMARY PRODUCTS IN X-IRRADIATED CYTOSINE DERIVATIVES: EPR AND ENDOR STUDIES AT 10 K

Eli O. Hole and E. Sagstuen
Department of Physics
University of Oslo, Box 1048
Blindern, 0316 Oslo, Norway

D. M. Close
Department of Physics
East Tennessee State University
Johnson City, TN, USA

W. H. Nelson
Department of Physics and Astronomy
Georgia State University, Atlanta, GA, USA

Introduction

Recently, evidence has accumulated that cytosine is the initial electron trap in irradiated DNA (1-4). It has previously been observed that the protonation/ deprotonation mechanisms and sites of the pristine purine/pyrimidine electron gain and loss centers depend upon the matrix. Local variations as e.g. water of crystallization, hydrogen bond pattern, initial protonation state of the cytosine bases, and associated counter ions may be important in the selection of reaction pathways following the initial ionization event (5). In response to this increased interest in the primary radical production in irradiated cytosine, five different crystalline cytosine derivatives have been investigated at 10 K following X-irradiation exposure to doses between 0.8 and 30 kGy. The techniques used are K-band EPR, ENDOR, and ENDOR-induced EPR (FSE) spectroscopy. The systems studied are: cytosine monohydrate (Cm) (6); cytosine-HCI (C:HCl); cytidine (CR); deoxycytidine:HCI (CdR:HCl); and 2'-deoxycytidine 5'-monophosphate (5'dCMP) (7). A general structure is shown below. In three of the systems the cytosine base is initially protonated at the N3 position (X = H^+).

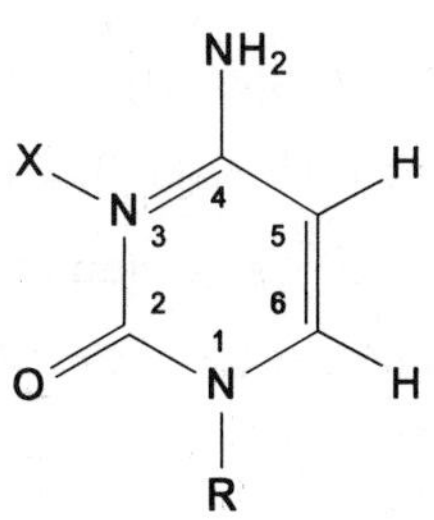

Cm	*X = blank*	*R = H*
C:HCl	$X = H^+$	*R = H*
CR:	*X = blank*	*R = ribose*
CdR:HCl	$X = H^+$	*R = deoxyribose*
5'dCMP	$X = H^+$	*R = deoxyribose monophosphate*

From previous studies of purine and pyrimidine derivatives in the solid state it is well known that most reduction and oxidation products exhibit a strong tendency to regain their original charge state as quickly as possible following ionization (5, 8). This appears to be the case with cytosine derivatives as well, even though the protonation state for some of the reduction products is unknown.

Table 1 : Radical formation in Cytosine derivatives [1]

System	Base reduction	Base oxidation	3α radical
Cm	N3 protonated anion	N1 deprotonated cation	no
C:HCl [2]	amino protonated anion[3]	amino deprotonation cation [5] + N1 deprot. cation [4] + N3 deprot. cation [4]	no
CdR:HCl [2]	unknown protonation state[3]	C1' H-abstraction [6]	yes [7]
CR	unknown protonation state[4]	no deprotonated cation	yes [8]
5'-dCMP [2]	amino protonated anion	N3 deprotonated cation	yes [9]

[1] In addition to the radicals summarized in this table, base centered H-addition radicals, sugar centered H-abstraction radicals, and alkoxy radicals were formed in several of the systems.
[2] The cytosine base is initially protonated at N3.
[3] Several spectroscopically similar products were observed.
[4] Minority radical.
[5] Dominant oxidation product.
[6] A sugar centered C1' H-abstraction radical observed at 10 K may result from base oxidation.
[7] Observed when irradiated with low X-irradiation doses at 10 K and annealed to 100 K.
[8] Dominant in the temperature interval between 8 K and 295 K.
[9] Observed following X-irradiation at 295 K.

Results

A preliminary analysis of the five cytosine derivatives shows that both the qualitative and the quantitative radical productions differ considerably in the various systems. Products linked to the reduction of the cytosine base is stabilized at 10 K in all five systems, whereas base centered deprotonated cations are found in only three of the systems at 10 K. The well-known but unidentified 3α-radical (9) was observed in the three systems containing sugar, but under considerably different experimental conditions. In CR this radical dominates the spectra at temperatures between 10 K and room temperature. In addition, base centered H-addition radicals, sugar centered H-abstraction radicals, as well as alkoxy radicals were observed in several of the systems studied. The results linked to base oxidation and reduction, as well as the 3α-radical are summarized in Table 1.

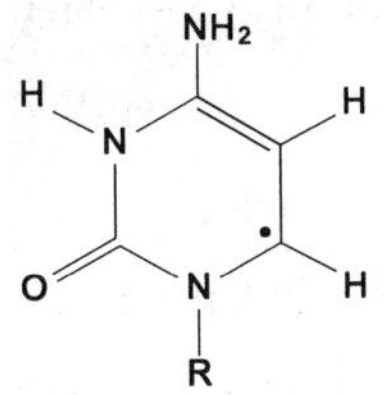

N3 protonated anion
(initially nutral base)

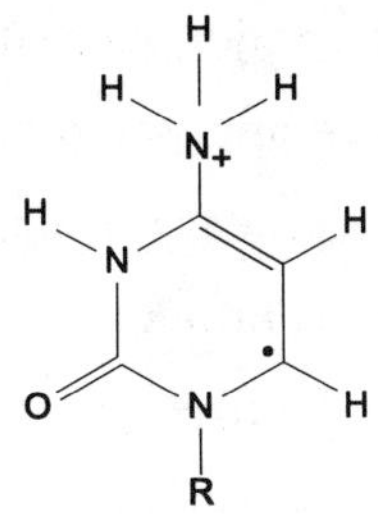

Amino protonated anion
(initially N3 protonated base)

Reduction: Base reduction products were observed at 10 K in all five systems, and were dominant radicals in all but one. In several systems, various spectroscopically similar forms of the base reduction products were stabilized at 10 K. It is not clear, however, whether these are due to slightly different geometric conformations of the same radical, whether the cytosine anions protonate at several different positions, or whether both pristine and protonated anions are stabilized at 10 K.

Two of the systems, Cm and CR, contain normal bases (i.e., not protonated at N3). In Cm the reduction product appears to be protonated at N3 (6). In CR, however, only traces of a reduction product is observed and the protonation state remains unknown.

The last three systems contain N3-protonated bases. Apparently, several different basereduction products are trapped in the two systems containing polar ions (C:HCl and CdR:HCl). It is not clear, however, whether these are actually different protonation states, or just different conformations of the same radical. The protonation state of the reduction products observed in CdR:HCl remains unknown, whereas the major reduction product formed in C:HCl is assigned to the amino protonated anion. The same reduction product (assigned to the amino protonated anion) is also observed in 5'dCMP.

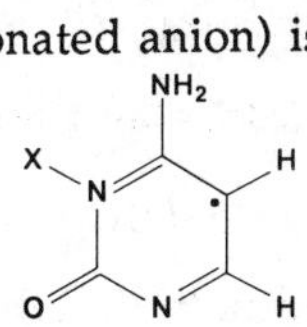

N1 deprotonated cation

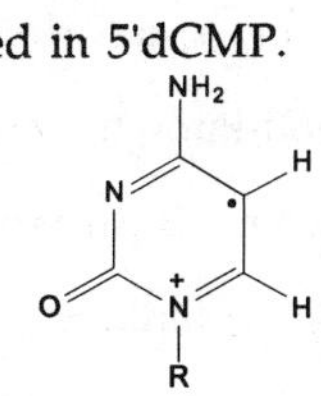

N3 deprotonated cation
(initially N3 protonated base)

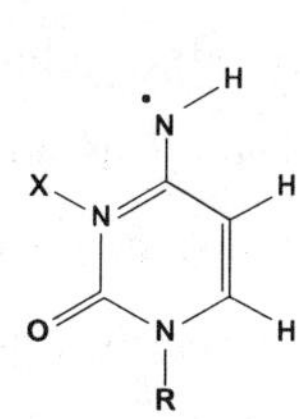

Amino deprotonated cation

Oxidation: Like reduction products, oxidation products exhibit a strong tendency to regain the original charge of the molecule immediately upon ionization. The major mechanism for this in recent and present nucleobase derivatives studied appears to be deprotonation of the primary electron loss center. Evidence for the primary oxidation product has not been observed in any of the systems, not even at 8-10 K.

No unique pattern for the pathway by which the pristine cytosine cation restores the charge balance has been discovered. In Cm the cation is found to deprotonate at N1, whereas no regular base oxidation product has been identified in CR. In the N3 protonated cytosine base derivatives there are also very few similarities. It appears that the electron loss centeres formed in C:HCl may deprotonate at different positions, N1, N3, or the amino group, but that the major deprotonation site is the amino group. In contrast, no base centered oxidation product is identified in CdR:HCl. It should be mentioned, however, that a sugar centered Cl' H-abstraction radical present at 8 K may be a successor to the primary base oxidation product. In the third system containing an N3-protonated cytosine base, 5'dCMP, the oxidation product observed is most probably the N3 deprotonated species.

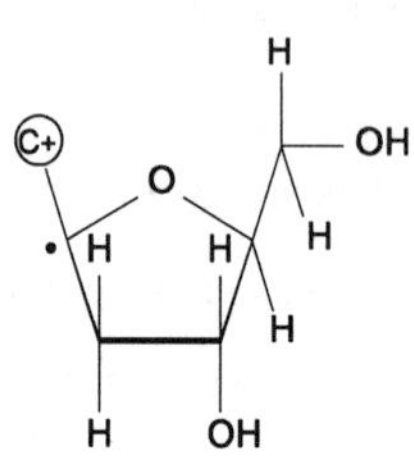

C1' H-abstraction

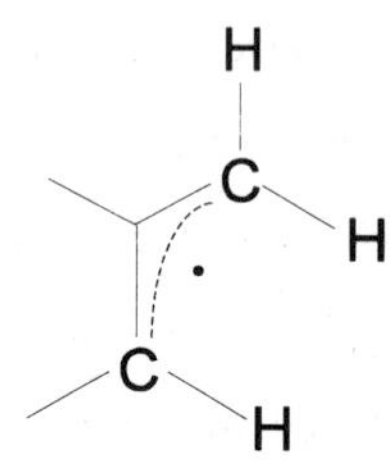

3α-radical

The 3α-radical: Radicals associated with three anisotropic proton hyperfine couplings were observed in the cytosine derivatives containing ribose or deoxyribose. This 3α-radical is a well-known product in nucleosides and nucleotides of cytosine, even though the exact radical structure remains unknown (9). In fact, it is uncertain whether it is a base centered or a sugar centered radical. Whereas the hyperfine coupling tensors fit with a base centered radical, the type of radical fragment as well as the fact that it is only observed when ribose or deoxyribose is present, points towards a sugar centered radical. It has been suggested that this unknown 3α-radical species may represent the missing link in the radical transfer from the base to the sugar.

In the three present sugar-containing systems, the 3α-radicals were observed under considerably different conditions. In CR it was the dominant radical at 10 K, as well as at room temperature. In 5'dCMP, however, it was only observed when the crystals were irradiated at room temperature (7). In CdR:HCl the 3α-radical was only observed in crystals which had been irradiated with low radiation doses, and then it appeared only after annealing to about 100 K. Unfortunately, more than one of the primary radicals observed at 10 K decayed upon annealing to 100 K; thus it has so far not been possible with certainty to identify the precursor to the 3α-radical formed in CdR:HCl.

Summary

In the present study of five different cytosine derivatives no unique pattern by which the primary reduction and oxidation products restore the charge balance has been found. For two of the systems the protonation states of the observed reduction products remain unknown. The reduction products observed in the remaining systems, indicate that the primary electron gain center prefers to protonate at N3, but protonates at the amino group if the base is originally protonated at N3. The present results on the base oxidation products, indicate that the charge balance is restored almost immediately upon oxidation and that the electron loss centers may deprotonate at all hetroatom positions in the cytosine base.

The well-known, but unidentified 3α-radical has been observed in the three systems containing ribose or deoxyribose. Considerable structural changes must be involved to form a base centered radical with the observed 3α-radical characteristics. Nevertheless, based on the hyperfine coupling tensors, the conditions under which the radicals are observed, as well as the anisotropic behaviour of the associated EPR spectra, it is suggested that this radical must be located to the cytosine base. Since the presence of a sugar group apparently is requested for the formation of the 3α-radical it is likely that the sugar group is an active part in the formation of a base centered 3α-radical. A possible mechanism may involve a sugar-to-base radical transfer. Upon ionization sugar radicals are formed. If their formation or decay involve rupture of intermolecular interactions, e.g., hydrogen bonds, the sugar group may reorient to a position which permits a close contact with the C5 region in the base, and thereby leading to the formation of a 3α-radical in the C5-C6 region of the cytosine base. The role of the 3α-radical in the radiation chemistry of cytosine nucleosides and nucleotides will continue to be investigated in our laboratories.

Acknowledgment

This work has been supported by PHS Grant CA36810 by DHHS, and by NATO Travel Grant 0426/88

References

1. W. Wang, M. Yan, D. Becker, and M.D. Sevilla, The influence of hydration on the absolute yields of primary ionic radicals in gamma-irradiated DNA at 77 K: II. Individual radical yields. *Radiat. Res.* **137,** 2-10 (1994).
2. J. Bames and W.A. Bernhard, The protonation state of one-electron-reduced cytosine and adenine. 1. Initial protonation sites at low temperatures in glassy solids. *J. Phys. Chem.* **97,** 3401-3408 (1993).
3. M. Yan, D. Becker, S. Summerfield, P. Renke, and M.D. Sevilla, Relative abundance and reactivity of primary ion radicals in g-irradiated DNA at low temperatures. 2. Single vs double-strand DNA. *J. Phys. Chem.* **96,** 1983-1989 (1992).
4. J. Barnes, W.A. Bernhard, and K.R. Mercer, Distribution of electron trapping in DNA: protonation of one-electron-reduced cytosine. *Radiat. Res.* **126,** 104-107, (1991).
5. W. H. Nelson, E. Sagstuen, E.O. Hole, and D.M. Close, On the proton transfer behavior of the primary oxidation product in irradiated DNA, *Radiat. Res.* **131,** 10-17 (1992).
6. E. Sagstuen, E.O.Hole, W.H. Nelson, and D.M. Close, Protonation state of radiationproduced cytosine anions and cations in the solid state. EPR/ENDOR of cytosine monohydrate single crystals X-irradiated at 1 0 K. *J. Phys. Chem.* **96,** 8269-8276 (1992).
7. D.M. Close and W.A. Bernhard, ESR and ENDOR study of 5'dCMP at 6 K. *J. Chem. Phys.* **70,** 210-215 (1979).
8. E.O. Hole, Model studies of radiation-induced oxidation and reduction processes in DNA, Dr. Philos. thesis, University of Oslo, 1992. NEI-NO-297, 1992.
9. W. A. Bernhard, J. Hüttermann, A. Muller, D.M. Close, and G.W. Fouse, A stable radical in the ribose moiety of X-irradiated 3'-cytidylic acid and its relation to a similar radical in cytidine: An ESR-ENDOR study. *Radiat. Res.* **68,** 390-413 (1976).

REACTIONS OF THIOLS WITH INDIVIDUAL DNA BASE RADICALS IN DNA AND DNA MODEL SYSTEMS: THE EFFECT OF OXYGEN ON RADIOPROTECTION

Michael D. Sevilla, David Becker and Wending Wang
Department of Chemistry
Oakland University
Rochester, MI 48309-4401 USA

Introduction

Thiol radioprotection of DNA in the living cell has been suggested to be a result of hydrogen or electron donation to free radicals generated on DNA resulting in a repair of the original radical lesion (Becker and Sevilla 1993) (reactions 1 and 2).

$$DNA^{\dot{+}} + RS^{-} \rightarrow RS\bullet + DNA \qquad (1)$$

$$DNA(\text{-}H) + RSH \rightarrow RS\bullet + DNA \qquad (2)$$

Several studies (Willson 1974, Prütz 1989) (Cullis *et al.* 1987, Sevilla *et al.* 1991a, Prütz *et al.* 1987, O'Neill 1983) have shown that the guanine cation radical $G^{+}\bullet$ as well as the hydroxyl adduct G(C4)OH•, can be repaired by electron transfer from the thiolate anion (RS^{-}) or the disulfide anion ($RSSR^{\dot{-}}$). Product analyses of irradiated solutions of thymidine with thiols (Cadet *et al.* 1988) suggest that all thymine radicals produced by OH• attack, $TCH_2\bullet$, T(C5)OH• and T(C6)OH•, undergo reduction by hydrogen (or electron) transfer.

Thus far there are no reports of reactions of DNA anion radicals with thiols. Several studies (Holian and Garrison 1969, Cadet *et al.* 1988, Nucifora *et al.* 1972, Becker *et al.* 1993) suggest that under anoxic conditions the pyrimidine protonated anion radicals, T(C6)H• and C(N3)H•, abstract hydrogen from thiols, resulting in 5,6-dihydro-pyrimidine products TH_2 and CH_2, as well as, a thiyl radical RS•. The thiyl radical RS• has been shown in ESR studies (Becker *et al.* 1988, Grierson *et al.* 1992) to abstract hydrogen from weak C-H bonds forming carbon centered radicals.

Oxygen involvement in thiol free radical reactions has been implicated in oxygen radiosensitization of cells to the lethal effects of radiation (Alper 1979, Quintiliani 1986, Ewing 1987). A biological oxygen enhancement effect and a chemical oxygen enhancement effect have both been recognized (Greenstock 1984).

One hypothesis is that thiyl radical reactions with oxygen are the source of the chemical oxygen enhancement effect. For example, thiol peroxyl radicals (RSOO•) formed in an equilibrium (reaction 3) (Tamba *et al.* 1986), have been suggested to lead to DNA inactivation (Prütz and Mönig 1987).

$$RS\bullet + O_2 \rightleftarrows RSOO\bullet \qquad (3)$$

ESR work shows that RSOO• undergoes a cyclic reaction sequence (reaction 4) which results in several potentially damaging intermediates, i.e., $RSO_2\bullet$ and $RSO_2OO\bullet$ (Sevilla *et al.* 1988, Sevilla *et al.* 1989, Sevilla *et al.* 1990a, Sevilla *et al.* 1990b, Becker *et al.* 1994).

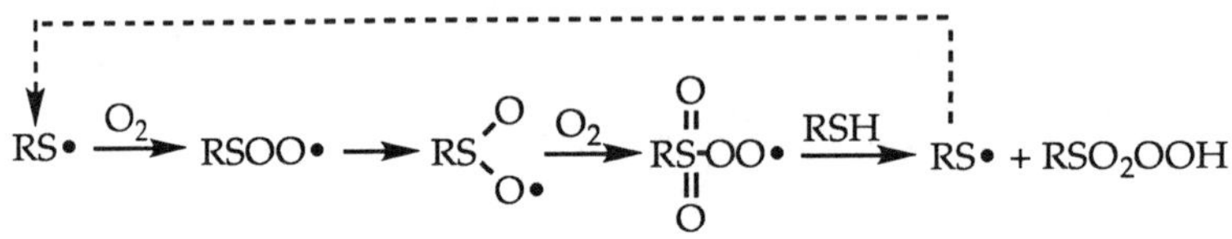

(4)

A side reaction of RSOO• with thiol (reaction 5) competes with its conversion to $RSO_2\bullet$ (Sevilla *et al.* 1990a).

$$RSOO\bullet + RSH \rightarrow RSO\bullet + RSOH \qquad (5)$$

The isomerization of RSOO• to form $RSO_2\bullet$ at low temperatures requires visible light; however at room temperature it is spontaneous (Zhang *et al.* 1994).

Recent Results

Nucleotide-Thiol Reactions: In this section we summarize results of an ESR investigation of the individual DNA base radicals produced by γ-irradiation of frozen solutions of the nucleotides, TMP, dCMP, dGMP and dAMP and their reactions with cysteamine upon annealing (Wang and Sevilla 1994b). The results found in this work on nucleotide-thiol systems (see Table I) provide a detailed understanding of the likely reactions between DNA components and the thiol, cysteamine. Reactions of the nucleotide radicals with cysteamine initially give rise to RS• which subsequently reacts with RS^- to form $RSSR^{\bar{\bullet}}$.

$$RS\bullet + RS^- \rightleftarrows RSSR^{\bar{\bullet}} \qquad (6)$$

Very little RS• (<5%) results from direct ionization of cysteamine in the frozen samples owing to its low concentration. Thus under our conditions, production of cysteamine radicals RS•/RSSR$\bar{•}$ mainly results from reaction of nucleotide radicals with cysteamine. It is found that the fraction of conversion of total nucleotide radicals (not including OH•) to RSSR$\bar{•}$ is: TMP (74%) > dCMP (53%) > dAMP ≈ dGMP (34%). All the pyrimidine nucleotide radicals, including those produced by both oxidation and reduction, react with cysteamine to form RS•/RSSR$\bar{•}$; whereas, only radicals from the oxidation pathway can do so for purine nucleotides. As a consequence larger amounts of RS•/RSSR$\bar{•}$ are formed in the pyrimidine nucleotides than in the purine nucleotides. Water radicals in bulk ice are not found to lead to the formation of DNA or cysteamine radicals.

Table I Reactions of DNA base and sugar radicals with cysteamine[a]

	Reaction[b]	Certainty[c]	Temperature (K)[d]
TMP	1. TH• + RSH → RS• + TH_2	1	190
	2. TCH_2• + RSH → RS• + T	2	
	3. T_{di}• + RSH → $T_{di}H$ + RS•	1[e]	210
dGMP	4. π-G^{+}• + RS^{-} → G + RS•	1[f]	100
	5. GOH• + RSH → GHOH + RS•	1	190
	6. GH• + RSH -/-> GH_2 + RS•	3	
dAMP	7. π-A^{+}• + RS^{-} → A + RS•	2[f]	
	8. AOH• + RSH → AHOH + RS•	1	170
	9. Sgr• + RSH → SgrH + RS•	1	170
	10. AH• + RSH -/-> AH_2 + RS•	3	
dCMP	11. π-C^{+}• + RS^{-} → RS• + C	1[f]	100
	12. CH• + RSH → RS• + CH_2	2	160
DNA	13. DNA(π-G^{+}•+ RS^{-} → G + RS•)	1[g]	170
	14. DNA(TH• + RSH→RS• + TH_2)	2[g]	200

[a] Wang and Sevilla 1994b.
[b] RSH or RS^{-} may be the primary reactant. Reactions are as depicted for simplicity. Also note that in all systems RSSR$\bar{•}$ (a strong reductant) is produced by RS• addition to RS^{-}.
[c] 1 = clearly observed; 2 = likely but not clearly observed; 3 = does not occur under our conditions.
[d] Reactions occur at these temperatures or greater.
[e] TOH• is also a possible contributor.
[f] The symbol for the π-cations is employed since these species have undergone one electron loss from the π-system. Their nitrogen protonation state is uncertain.
[g] Becker *et al.* 1994.

Radicals from the oxidation pathway which include the DNA base one electron oxidized radicals and their successors, G(C8)OH•, A(C8)OH•, thymine dimer, •T_{di} (and/or T(C6)OH•), readily react with cysteamine to form RS• and ultimately RSSR$\bar{•}$. While the TMP quintet radical is reduced by cysteamine, there is still some uncertainty in the nature of the quintet in TMP spectra, as it likely arises from the dimer ($\bullet T_{di}$) and/or the hydroxyl adduct (TOH•) (Malone *et al.* 1993). Reactions of dGMP and dCMP radicals from the oxidation pathway (π-cation radicals) with cysteamine occur at lower temperatures than those of dAMP and TMP, suggesting hole migration. This process was previously found to occur in crystalline nucleic acids (Sankovic *et al.* 1988, Herak and Hüttermann 1991). It is likely that holes in G and C migrate either before or despite any deprotonation reactions, whereas, A and T holes likely undergo deprotonation reactions at low temperatures effectively preventing hole migration. From previous work the sites of proton loss are known to be the methyl group in $T^{+\bullet}$ (Malone *et al.* 1993) and the amine group in $A^{+\bullet}$(Close *et al.* 1994).

Our results show that pyrimidine protonated anion radicals, T(C6)H• and C(N3)H• react with cysteamine to form RS• and diamagnetic products, but that those from the purines, G(C8)H• and A(C8)H•, do not. The reaction of T(C6)H• with thiol has been suggested in DNA-thiol systems (Becker *et al.* 1994, Cullis *et al.* 1987). The reaction of C(N3)H• with thiol (reaction 12 in Table I) is supported by earlier product analysis studies in which the likely product, 5,6-dihydrocytosine, CH_2, was reported (Holian and Garrison 1969, Cadet *et al.* 1988). Somewhat less than half of the total nucleotide radicals found result from the oxidative pathway.

The only non-based centered radical identified is the sugar radical at the C1' site on the deoxyribose portion of dAMP. This species is found to react with cysteamine which confirms an important radioprotective mechanism (reaction 9 in Table I). As this has implications for radioprotection against strand breaks it will be important in future work to distinguish between the two possible reaction mechanisms. From previous single crystal work (Close 1993), it is likely that a small amount of sugar radical, Sgr(G)•, is also present in the dGMP spectra; however, the uncertainties in our analyses are to great for any firm conclusions.

It is generally assumed that γ-irradiation of frozen aqueous nucleotides produces equal numbers of reduced and oxidized radicals (Gregoli 1977, Yan *et al.* 1992). While this is strictly true for the primary ionization event, it by no means is a condition which extends to stabilized species (Wang *et al.* 1994a). Our present results indicate that typically in γ-irradiated frozen aqueous nucleotides ca. 40% of total radicals result from the oxidative pathway while ca. 60% result from the reductive pathway. This is in agreement with earlier quantitative work by Gregoli *et al.* (1977, 1982) and previous studies on DNA (Wang *et al.* 1994a, Boon *et al.* 1984), but is in conflict with Hüttermann *et al.*'s recent conclusion (1992) from investigation of thymine and its derivatives which suggests a complete dominance of the reductive pathway.

Thiol Reactions with DNA Intermediates and the Influence of Oxygen: In our investigations of the reactions of thiols with DNA primary radical intermediates formed after γ-irradiation of frozen (77 K) anoxic and oxic solutions of DNA/thiol (cysteamine or glutathione) mixtures, we find the primary DNA radical anions and cations ($DNA^{\overset{+}{\bullet}}$ and $DNA^{\overset{-}{\bullet}}$) are the predominant radicals at low temperatures (Becker *et al.* 1994). In anoxic cysteamine samples, TH• and $RSSR^{\overset{-}{\bullet}}$ are observed as the temperature is increased. Whereas, for glutathione samples, •GSH [γ-glu-NH$\dot{C}$(CH_2SH)CO-gly] radicals are observed along with TH•. The presence of oxygen efficiently suppresses the formation of both $RSSR^{\overset{-}{\bullet}}$ and •GSH; instead, in oxic samples, $O_2^{\overset{-}{\bullet}}$, DNAOO•, RSOO•, and RSO• are observed at higher temperatures in keeping with reactions 3 and 5. Photolytic conversion of RSOO• to RSO_2• in γ-irradiated frozen DNA/cysteamine solutions confirms the presence of RSOO• in these systems. From the relative concentrations of radicals present, it is clear that the radicals observed at higher temperatures originate from the radical reactions of the primary $DNA^{\overset{+}{\bullet}}$ and $DNA^{\overset{-}{\bullet}}$ radicals. The major source of thiyl radical (RS•) is from electron transfer from RS^- to $DNA^{\overset{+}{\bullet}}$ and hydrogen addition from RSH to TH•. The subsequent thiyl oxygen species ensue through reactions 4 and 5.

Scheme I combines our identification of radical species and their reaction mechanisms with known rate constants at ambient temperatures (Becker *et al.* 1994). All the radicals in this scheme are found in our work in DNA/thiol mixtures or in work with model systems. The rate constants are well known as they have been determined through pulse radiolysis by several investigative groups (see references in the similar scheme in Becker et al. 1994). Since the scheme represents a good understanding of the mechanisms of thiol autoxidation in aqueous environments, it is of interest to compute concentrations over time using this set of reaction mechanisms and biologically realistic initial concentrations. In Figure 1 we present

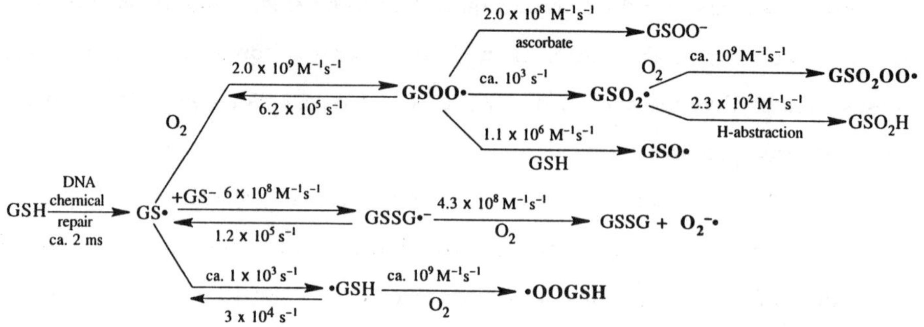

Scheme 1. Radical reaction mechanisms that result in sulphoxyl and peroxyl radicals from the glutathione thiyl radical and oxygen. The rate constants were found by pulse radiolyses investigations of various thiols at ambient temperature (see Becker *et al.* 1994 for specific references and Zhang *et al.* 1994 for the •GSH equilibrium).

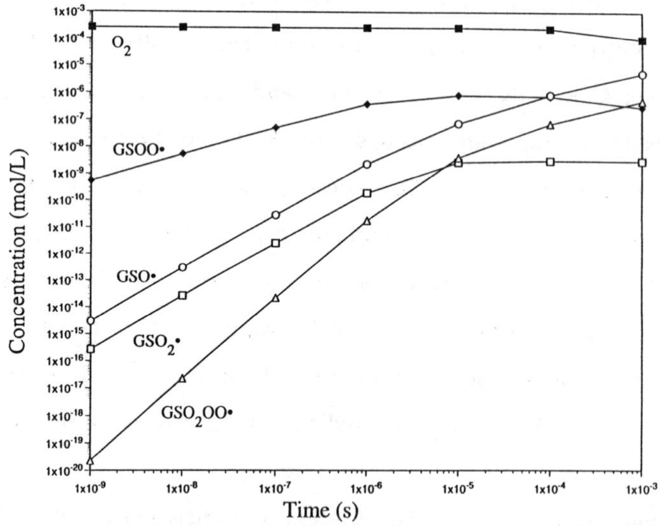

Figure 1. Sulfoxyl radical concentrations as a function of time resulting from a solution of the kinetic equations for the reactions in Scheme 1. The initial concentrations were: 10 mM GSH, 0.26 mM O_2, and 0.1 mM ascorbate. A constant concentration of 1×10^{-6} M GS• was assumed. GSO2OO• is found to be a major radical species at the msec time scale.

the results of a preliminary simulation which assumes homogeneous kinetics and uses small discrete steps (psec steps at the nsec time scale, nsec steps at the msec time scale, etc.) to solve the system of linear differential kinetic equations with time. The starting conditions are given in the legend and assumes a 1μM concentration of thiyl radical, an oxygen concentration of 0.26 mM, 10mM GSH, and an ascorbate concentration of 0.1 mM. As can be seen in the figure the level of GSOO• increases very rapidly with very low levels of GSO_2OO• found at the msec time scale. By the msec time scale GSO_2OO• is comparable to GSOO•. This suggests that this species likely has a biological role. As its oxygen-17 couplings suggest it to be a highly reactive peroxyl radical (Sevilla *et al.* 1990b), its involvement in the oxygen enhancement effect is suspected. The levels of the other sulfoxyl or peroxyl radicals, GSO•, GSO_2• and •OOGSH (not shown), reach significant levels at earlier time periods. Although less reactive these species are electron affinic and may act as radiosensitizers.

Acknowledgment

This work is supported by the National Cancer Institute of the National Institutes of Health (RO1CA45424) and the Office of Health and Environmental Research of the Department of Energy (DEFGO286ER60455).

References

Alper, T., 1979, Cellular Radiobiology. (Cambridge: Cambridge University Press).

Becker, D., Sevilla, M. D., 1993, The chemical consequences of radiation damage to DNA. Advances in Radiation Biology 17, 121-180.

Becker, D., Swarts, S., Champagne, M., Sevilla, M. D., 1988, An ESR investigation of the reactions of glutathione, cysteine and penicillamine thiyl radicals: competitive formation of RSO•, R•, $RSSR^{\bullet-}$, and RSS•. International Journal of Radiation Biology 53, 767-786.

Becker, D., Summerfield, S., Gillich, S., Sevilla, M. D., 1994, The influence of oxygen on the repair of direct radiation damage to DNA by thiols in model systems. International Journal of Radiation Biology,65, 537-548.

Boon, P. J., Cullis, P. M., Symons, M. C. R., Wren, B. W., 1984, Effects of ionizing radiation on deoxyribonucleic acid. Part 1. The role of oxygen. Journal of the Chemical Society, Perkin Transactions 2 1393-1399.

Cadet, J., Berger, M., Demonchaux, P., Lhomme, J., 1988, Modifying effects of cysteine and aromatic sulfhydryl and disulfide agents on the radiation-induced decomposition of thymidine. Radiation Physical chemistry 32, 197-202.

Close, D. M., 1993, Radical ions and their reactions in DNA constituents: ESR/ENDOR studies of radiation damage in the solid state, Radiation Research 135, 1-15.
Close, D. M., Nelson, W. H., Sagstuen, E., Hole, E. O., 1994, ESR and ENDOR study of single crystals of deoxyadenosine monohydrate X-irradiated at 10K, Radiation Research 137, 300-309.
Cullis, P. M. Jones, G. D. D., Lea, J., Symons, M. C. R., Sweeney, M., 1987, The effects of ionizing radiation on deoxyribonucleic acid. Part 5. The role of thiols in chemical repair. Journal of the Chemical Society, Perkin Transactions 2 1907-1914.
Ewing, D., 1987, Application of radiation chemistry to studies in the radiation biology of microorganisms. Radiation Chemistry, edited by Farhataziz and M. A. J. Rodgers. (VCH, New York), pp 501-526.
Gregoli, S., Olast, M., Bertinchamps, A., 1977, Free radical formation in deoxyguanosine-5'-monophosphate γ-irradiated in frozen solution. A computer-assisted analysis of temperature-dependent ESR spectra. Radiation Research 72, 201-217.
Gregoli, S., Olast, M., Bertinchamps, A., 1982, Radiolytic pathways in γ-irradiated DNA: Influence of chemical and conformational factors. Radiation Research 89, 238-254.
Greenstock, C. L., 1984, Oxy-radicals and the radiobiological oxygen effect. Israel Journal of Chemistry 24, 1-10.
Grierson, L., Hildenbrand, K., Bothe, E., 1992, Intramolecular transfer reaction of the glutathione thiyl radical into a non-sulphur-centered radical: a pulse radiolysis and EPR study. International Journal of Radiation Biology 62, 265-277.
Herak, J. N., Hüttermann, J., 1991, Long-range hole migration in irradiated crystals of nucleic acid bases. An EPR Study, International Journal of Radiation Biology 60, 423-432.
Holian, J., Garrison, W. M., 1969, Reconstitution mechanisms in the radiolysis of aqueous biochemical systems: Inhibitive effects of thiols, Nature 221, 57.
Hüttermann, J., Lange, M., Ohlmann, J., 1992a, Mechanistic aspects of radiation-induced free radical formation in frozen aqueous solutions of DNA constituents: Consequences for DNA? Radiation Research 131, 18-23.
Hüttermann, J., Röhrig, M., Köhnlein, W., 1992b, Free radicals from irradiated lyophilized DNA: influence of water of hydration. International Journal of Radiation Biology 61, 299-313
Malone, M. E., Symons, M. C. R., Parker, A. W., 1993, An EPR study of photoionised thymine and its derivatives at 77 K. Journal of the Chemical Society, Perkin Transactions 2 2067-2075.
Nucifora, G., Smaller, B., Remko, R., Avery, E. C., 1972, Transient radicals of DNA bases by pulse radiolysis. Effects of cysteine and cysteamine as radioprotectors. Radiation Research 49, 96-111.
O'Neill, P., 1983, Pulse radiolytic study of the interaction of thiols and ascorbate with OH adducts of dGMP and dG: Implications for DNA repair processes. Radiation Research 96, 198-210.
Prütz, W. A., Mönig, H., 1987, On the effect of oxygen or copper (II) in radiation-induced degradation of DNA in the presence of thiols. International Journal of Radiation Biology 52, 677-682.
Prütz, W. A., 1989, "Chemical Repair" in irradiated DNA solutions containing thiols and/or disulphides. Further evidence for disulphide radical anions acting as electrons donors. International Journal of Radiation Biology 56, 21-33.
Quintiliani, M., 1986, The oxygen effect in radiation inactivation of DNA and enzymes. International Journal of Radiation Biology 50, 573-594.

Sankovic, K., Herak, J. N., Krilov, D., 1988, ESR spectroscopy of the sulfur-centered radicals derived from thiocytosine, Journal of Molecular Structure 190, 277-286.

Sevilla, M. D., Yan, M., Becker, D., 1988, Thiol peroxyl radical formation from the reaction of cysteine thiyl radical with molecular oxygen: an ESR investigation. Biochemical and Biophysical Research Communications 155, 405-410.

Sevilla, M. D., Yan, M., Becker, D., Gillich, S., 1989, ESR investigations of the reactions of radiation-produced thiyl and DNA peroxyl radicals: Formation of sulfoxyl radicals. Free Radical Research Communications 6, 99-102.

Sevilla, M. D., Becker, D., Yan, M., 1990a, The formation and structure of the sulfoxyl radicals RSO•, RSOO•, RSO_2•, and RSO_2OO• from the reaction of cysteine, glutathione and penicillamine thiyl radicals with molecular oxygen. International Journal of Radiation Biology **57**, 65-81.

Sevilla, M. D., Becker, D., Yan, M., 1990b, Structure and reactivity of peroxyl and sulphoxyl radicals from measurement of oxygen–17 hyperfine couplings: Relationship with Taft substituent parameters. Journal of The Chemical Society Faraday Transactions 86, 3279-3286.

Sevilla, M. D., Yan, M., Becker, D., Gillich, S., 1991a, ESR investigations of the reactions of radiation-produced thiyl and DNA peroxyl radicals: Formation of sulfoxyl radicals. Free Radical Research Communications 6, 99-102.

Tamba, M., Simone, G., Quintiliani, M., 1986, Interactions of thiyl free radicals with oxygen: a pulse radiolysis study. International Journal of Radiation Biology 50, 595-600.

Wang, W., Yan, M., Becker, D., Sevilla, M. D., 1994a, The influence of hydration on the absolute yields of primary ionic free radicals in γ-irradiated DNA at 77 K, II. Individual radical yields. Radiation Research 137, 2-10.

Wang, W., Sevilla, M. D., 1994b, Reaction of cysteamine with individual DNA base radicals in g-irradiated nucleotides at low temperature. International Journal of Radiation Biology, In Press.

Willson, R. L., Wardman, P., Asmus, K. D., 1974, Interaction of dGMP radical with cysteamine and promethazine as possible model of DNA repair. Nature 252, 323-324.

Yan, M., Becker, D., Summerfield, S., Renke, P., Sevilla, M. D., 1992, Relative abundance and reactivity of primary ion radicals in γ-irradiated DNA at low temperatures. 2. Single-vs double-strand DNA. Journal of Physical Chemistry 96, 1983-1989.

Zhang, X., Zhang, N., Schuchmann, H. P., von Sonntag, C., 1994, Pulse radiolysis of 2-mercaptoethanol in oxygenated aqueous solution. Generation and reactions of thiylperoxyl radical. Journal of Physical Chemistry 98, 6541-6547.

ESR AND ENDOR STUDIES OF X-IRRADIATED SINGLE CRYSTALS OF DEOXY-CYTIDINE 5-PHOSPHATE MONOHYDRATE (5'dCMP) AT 10 K

D. M. Close
Physics Department
East Tennessee State University
Johnson City, TN

W. H. Nelson
Physics Department
Georgia State University
Atlanta, GA

E. O. Hole & E. Sagstuen
Institute of Physics
University of Oslo
Oslo, Norway

W. A. Bernhard
Department of Biophysics
University of Rochester
Rochester, NY

Introduction

Cytosine has received a great deal of attention recently. Results by Bernhard and co-workers (1989), and Sevilla and co-workers (1991) suggest that cytosine may be the primary reduction site in DNA. Also, a dissimilarity in the ESR spectrum from monomers and oligomers of cytosine has focused attention on the protonation state of the cytosine anion (Barnes *et al.* 1991). A recent review by Close (1993) covers most of the results on radical ions and their reactions in cytosine derivatives. The present work concerns ENDOR studies on single crystals of Deoxycytidine 5'phosphate monohydrate (5'dCMP).

Reduction of the Cytosine Base

ENDOR studies on 5'-dCMP have been performed by Close and Bernhard (1979). In the native molecule, the cytosine base is protonated at N3. Single crystals, x-irradiated at 10 K, exhibit a complicated EPR pattern composed of several radical species. The electron adduct exhibits the expected $>C^{\bullet}6\text{-}H_{\alpha}$ hyperfine coupling, with spin density of $\rho(C6) = 0.52$. However the protonation state of this radical was not determined in the original study. The reduction product was therefore considered to have one of the following possible structures (Structures **R1** and **R2**). This radical decays on warming to *ca.* 100 K.

Detailed ENDOR experiments have now been performed on many of the weak couplings in 5'dCMP. Using field-swept ENDOR techniques, a coupling with A_{iso} = 5 G has been linked to the reduction product. The eigenvectors for the minimum principal value of this hyperfine coupling do not correlate well with the

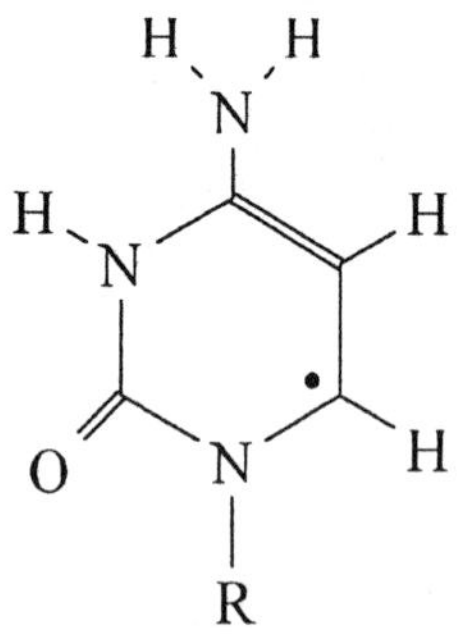

R1

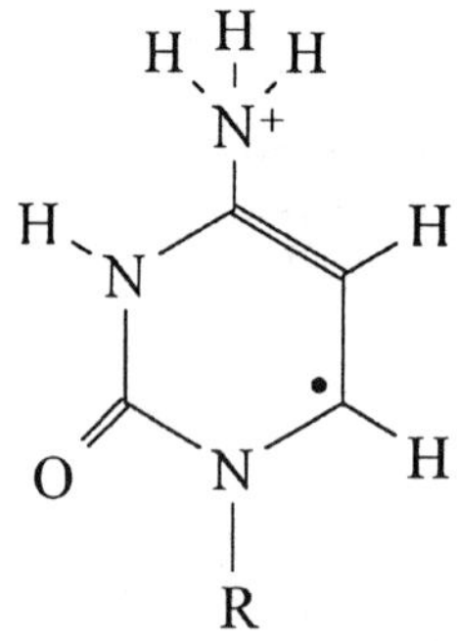

R2

direction of the proton on N3. The best correlation for this hyperfine coupling appears to be with a proton at N4 somewhat out of the molecular plane.

Barnes and Bernhard (1994) have described the protonation states of reduced cytosine before and after proton transfer in acidic glasses at low temperatures. They suggest that at 4 K the amino group has one N-H bond directed 30° out of the molecular plane due to the presence of a hydrogen bond. An intermediate state exists giving rise to a 5 G hyperfine coupling. Upon warming to 100 K the amino group protonates and reorients into an eclipsed position. These results suggest that a likely structure for the reduction product in 5'dCMP is structure **R2**. The precise arrangement of the protons attached to N4 remains to be determined.

Oxidation of the Cytosine Base

The base oxidation product observed in 5'dCMP exhibits couplings from >C•5-H_α and C1'-H_β, and significant nitrogen hyperfine couplings (Close and Bernhard 1979). The calculated spin density $\rho(C5) = 0.57$. Again, the protonation state of the oxidation product was not determined in the original study. Two possible structures for the oxidation product are shown as Structures **R3** and **R4**.

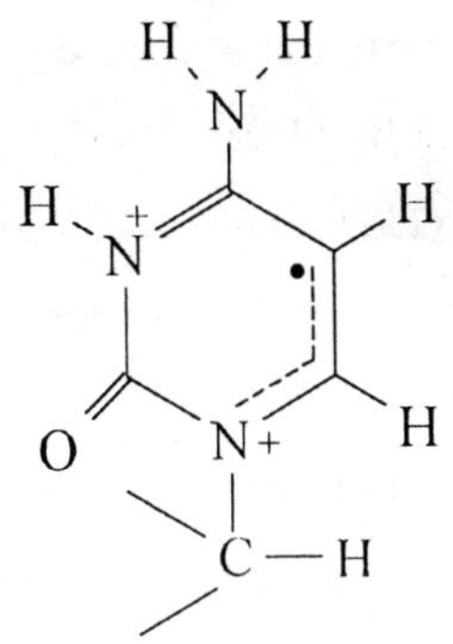

R3

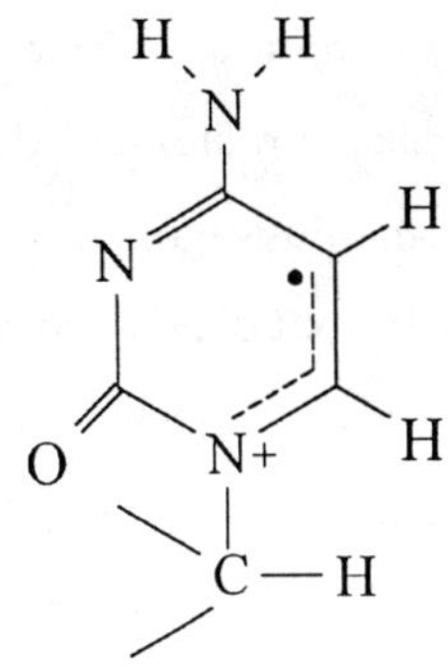

R4

Using field-swept ENDOR techniques, two weak ENDOR lines can be linked to the base oxidation product. The eigenvectors for the minimum principal values of these hyperfine couplings are close to the two N4-H proton directions. No coupling was observed that can be associated with the N3-H proton. Therefore it is most likely that the oxidation product observed at 10 K in 5'dCMP is Radical **R4**.

Oxidation of the Deoxyribose Moiety

Identification of an alkoxy radical in 5'dCMP initially led to several incorrect assignments. The problem is that one usually looks at the crystal structure to locate the C-O bonds (the expected direction of g_{max}). In the 5'dCMP crystal, both >C5'-OH and >C3'-OH are in nearly the same crystallographic direction. Therefore, when the alkoxy radical was first studied, it was assigned as a O•5'-C5'H_2- radical by Krilov *et al.* (1978). This model however couldn't explain the rather small β-proton coupling (actually two 10 G couplings, and perhaps several other smaller couplings). Therefore the alkoxy radical was reassigned as a O•3'-C3'< secondary alkoxy radical (Close and Bernhard (1979)).

There was one additional problem with the O•3'-C3'< assignment. There are actually several small hyperfine couplings to be explained. At the time it was not known that γ-proton hyperfine couplings could be of sufficient magnitude to explain these results. However, ENDOR results by Box *et al.* (1980) have accounted for one β-proton and three γ-proton couplings. Therefore, there is no doubt now that the alkoxy radical in 5'dCMP is a secondary alkoxy radical located at O3'.

Some time ago it was suggested by Hüttermann (1982) that in DNA there is a shift away from damage to the sugar. In a 3' nucleotide one often observes a O$^\bullet$-C5' alkoxy radical (Bernhard *et al.* (1976)). If this position is blocked by the attachment of a phosphate group as in a 5'-nucleotide, the site of oxidation may be the secondary alcohol group. Extrapolating to the polymers, one sees that all alcohol functions are used as phosphoester bonds. Therefore in DNA the site of oxidation is most likely the base.

The primary radiation induced products in 5'dCMP are therefore: (1) reduction of the base resulting in protonation at the amino group (**R2**), oxidation of the base resulting in the N3 deprotonated cation (**R4**), and oxidation of the deoxyribose moiety, resulting in a secondary alkoxy radical at O3'. A goal of this work is to relate the primary radiation induced radicals observed at 10 K with other radicals observed at elevated temperatures.

Radicals at 77 K

An interesting radical observed in 5'dCMP at 77 K by Krilov and Herak (1974) has the structure C$^\bullet$5'(H_2)-C4'(H)<. To form this radical it is necessary to cleave the phosphoester bond. This radical is therefore of interest as a possible intermediate in a strand break.

The most likely pathway for formation of this radical is *via* electron capture of the phosphate group or perhaps *via* deexcitation from a superexcited state. It is commonly assumed however that when bases are present electron capture by the phosphate is unlikely. An intriguing possibility exists that these radicals are formed by what Ward (1988) calls locally multiply damaged sites. Here one supposes that a base captures an electron. Then, when a second electron in this region cannot be scavenged by the base, it is captured by the phosphate.

Room Temperature Radicals

A radical that was first observed in cytidine at room temperature by Hampton and Alexander (1973) has the allylic features shown in Structure **R5**

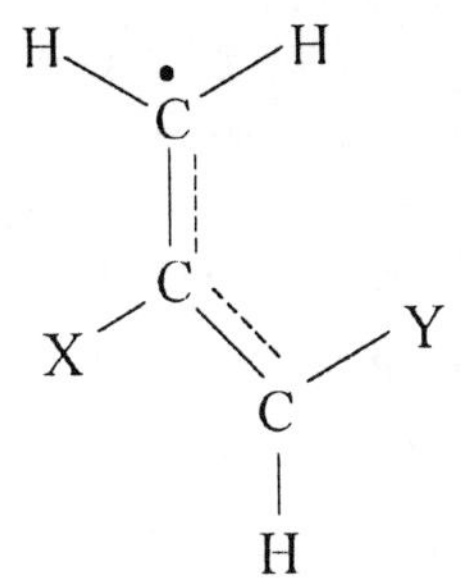

R5

and is referred to as the 3αH radical. This radical has also been observed in 3'CMP by Bernhard *et al.* (1976) and in 5'dCMP by Close *et al.* (1977). It is curious to note that the 3αH radical has only been observed in cytosine derivatives. The characteristics of this radical are similar to the allylic radical observed in thymine derivatives which is formed by net H-abstraction at the methyl group (Hüttermann (1970)). No possible mechanisms have been conceived to explain how an allylic radical could be formed on the cytosine base.

It has been proposed that Radical **R5** is formed in the ribose moiety, although its precise structure is unknown. To do so the ribose moiety must change to an unsaturated form. Although the reaction scheme for the formation of the 3αH radical has yet to be worked out, this radical may also be involved in the formation of a strand break.

In previous ESR studies it has commonly been assumed that the electron adduct is the precursor to the base H-addition radicals. One often sees the cytosine electron adduct signal decay on warming to 100 -150 K at which time the H-adduct signals appear to grow in. However detailed ENDOR studies have shown that when base electron adducts are formed they rapidly protonate (Close 1993). At low temperatures one often sees ENDOR signals from the H-adducts. Upon warming the electron adduct signal often decays with no apparent successor. Therefore in the crystalline state, at least, there doesn't seem to be a link between the electron adducts and the H-addition radicals.

Conclusions

This study indicates that a great deal is now known about the structure of the primary radiation induced defects in 5'dCMP. However not much is known about the reactions that these primary defects undergo. Therefore much remains to be learned about the pathways leading to the $C^{\bullet}5'(H_2)$-C4'(H)< radical and the 3αH radical.

Acknowledgements

This work is supported by PHS Grant CA 36810 awarded by the National Cancer Institute, DHHS.

References

Barnes JP, Bernhard WA and Mercer KR (1991) Distribution of electron trapping in DNA: Protonation of one-electron reduced cytosine. *Radiat. Res.* **126**, 104-107.

Barnes JP and Bernhard WA (1994) One-Electron Reduced Cytosine in Acidic Glasses: Conformational States before and after Proton Transfer. *J. Phys. Chem.* **98**, 887-893.

Bernhard WA, Close DM, Hüttermann J and Zehner H (1976) The alkoxy radical, $RCH_2O\bullet$, as a free radical product in x-irradiated single crystals of nucleosides and nucleotides. *J. Chem. Phys.* **67**, 1211-1219.

Bernhard WA, Hüttermann J, Müller A, Close DM and Fouse GA (1976) A stable radical in the ribose moiety of x-irradiated 3'-cytidylic acid at its relation to a similar radical in cytidine, An ESR-ENDOR study. *Radiat. Res.* **68**, 390-413.

Bernhard WA (1989) Sites of electron trapping in DNA as determined by ESR of one-electron-reduced oligonucleotides. *J. Phys. Chem.* **93**, 2187-2189.

Box HC, Budzinski EE and Potienko G (1980) Alkoxy radicals: γ proton hyperfine Interactions. *J. Chem. Phys.* **73**, 2052-2056.

Close DM, Fouse GW and Bernhard WA (1977) ENDOR Study of the 3αH radical in deoxycytidine 5'-phosphate H_2O. *J. Chem. Phys.* **66**, 4689-4693.

Close DM and Bernhard WA (1979) ESR and ENDOR study of 5'-dCMP at 6 K. *J. Chem. Phys.* **70**, 210-215.

Close DM (1993) Radical Ions and Their Reactions in DNA Constituents: ESR/ENDOR Studies of Radiation Damage in the Solid State. *Radiat. Res.* **135** 1-15.

Hampton DA and Alexander C (1973) ENDOR and ESR studies of irradiated cytidine. *J. Chem. Phys.* **58**, 4891-4898.
Hüttermann J (1970) Electron-spin-resonance spectroscopy of radiation-induced free radicals in irradiated single crystals of thymine monohydrate. *Int. J. Rad. Biol.* **17**, 249-259.
Hüttermann J (1977) Electron-spin-resonance spectroscopy of radiation-induced free radicals in irradiated single crystals of thymine monohydrate. *Int. J. Rad. Biol.* **17**, 249-259.
Hüttermann J (1982) Solid state radiation chemistry of DNA and its constituents. *Ultramicroscopy* **10**, 25-40.
Krilov D and Herak JN (1974) ESR evidence for the radiation-induced breakage of the sugar phosphate bonds in nucleotides: Single crystal of deoxycytidine 5'-monophosphate. *Biochim. et Biophys. Acta* **366**, 396-401.
Krilov D, Velenik A, and Herak JN (1978) ESR study of radiation-induced radicals in the sugar phosphate region of nucleotides. III. The alkoxy radical in deoxycytidine 5'-phosphate. *J. Chem. Phys.* **69**, 2420-2423.
Sevilla MD, Becker D, Yan M and Summerfield SR (1991) Relative abundances of primary ion radicals in γ-irradiated DNA: Cytosine vs thymine anions and guanine vs adenine cations. *J. Phys. Chem.* **95**, 3409-3415.
Ward JF (1988) DNA damage produced by ionizing radiation in mammalian cells: Identities, Mechanisms of formation, and reparability. In *Progress in Nucleic Acid Research and Molecular Biology* (W.E. Cohn and K. Moldave, Eds.) Vol. 35. pp. 95-125. Academic Press, San Diego.

Chapter 4

Effect of Hydration on DNA Radiation Damage

Radiation-Induced DNA Damage as a Function of DNA Hydration
Steven G. Swarts, Lan Miao, Kenneth T. Wheeler, Michael D. Sevilla and David Becker

Effects of Hydration on Free Radical Trapping in X-Irradiated Frozen Polynucleotides
Nicholas Mroczka and William A. Bernhard

γ-Irradiation of the Hydration Layer of DNA: ·OH Formation vs. Hole Transfer
David Becker, Therese La Vere and Michael D. Sevilla

Applicability of Experimental Models for Cellular Radiation Targets
John T. Lett

RADIATION-INDUCED DNA DAMAGE AS A FUNCTION OF DNA HYDRATION

S.G. Swarts, L. Miao, and K.T. Wheeler
Department of Radiology
Bowman Gray School of Medicine
Medical Center Boulevard
Winston-Salem, NC 27157

M.D. Sevilla and D. Becker
Department of Chemistry
Oakland University
Rochester, MI 48309

Introduction

Radiation-induced DNA damage is produced from the sum of the radicals generated by the direct ionization of the DNA (direct effect) and by the reactions of the DNA with free radicals formed in the surrounding environment (indirect effect). The indirect effect has been believed to be the predominant contributor to radiation-induced intracellular DNA damage, mainly as the result of reactions of bulk water radicals (e.g. OH•) with DNA. However, recent evidence suggests that DNA damage, derived from the irradiation of water molecules that are tightly bound in the hydration layer, may occur as the result of the transfer of electron-loss centers (e.g. holes) and electrons from these water molecules to the DNA (Becker et al 1992, Swarts et al 1992). Since this mechanism for damaging DNA more closely parallels that of the direct effect, the irradiation of these tightly bound water molecules may contribute to a "quasi-direct effect" (Becker et al 1992). These water molecules comprise a large fraction of the water surrounding intracellular DNA and could account for a significant proportion of intracellular radiation-induced DNA damage.

Consequently, we have attempted to characterize this "quasi-direct effect" to determine, 1) the extent of the DNA hydration layer that is involved with this effect, and 2) what influence this effect has on the types and quantities of radiation-induced DNA damage. The results from a study by Swarts et al (1992) on the release of unaltered DNA bases as a function of DNA hydration supported the hypothesis that radiation-induced damage in the tightly bound water molecules in the DNA hydration layer (12-15 water molecules per nucleotide [Γ]) can result in DNA damage that is indistinguishable from that caused by the direct ionization of DNA. In this extended work, we have determined the yields of a number of base damage products as a function of the degree of hydration in DNA that has been irradiated under nitrogen in order to examine the influence that the quasi-direct effect has on the formation of base damage products.

Methods

The analysis of base damage products was performed on DNA that had been hydrated, γ-irradiated to a dose of 90 kGy, and isolated as described previously (Swarts et al 1992). The isolated DNA was prepared for gas chromatography/mass spectrometry (GC/MS) analysis by first hydrolyzing the DNA (≈300 μg) in 300 μl of 88% formic acid at 140°C for either 90 or 30 min, then lyophilizing the sample and derivatizing the dried hydrozylate with 120 μl of BSTFA/acetonitrile (3:1). The derivatized base damage products were separated on a Hewlett-Packard 5980A GC using an HP Ultra-2 fused silicon capillary column (12.5 m x 0.2 mm, I.D.) coated with a crosslinked 5% phenyl methyl silicone gum phase (film thickness, 11 μm). Both the derivatized base damage products and two internal standards, 6-azathymine and 8-azaadenine, were measured using the selected-ion monitoring (SIM) mode of analysis.

Quantitation of the base damage products was performed using relative molar response factors that were obtained either from, 1) a multi-level calibration schemes which utilized authentic base damage products (Fuciarelli et al 1992, Dizdaroglu 1992) or 2) a plot of the area ratios of the base damage product and an int. std. versus varying amounts of hydrolyzed DNA that had been irradiated in aqueous solution under either oxygen or nitrogen (Dizdaroglu et al 1986). When using authentic compounds to calibrate the GC/MS-SIM measurements, the calibration samples containing both the base damage products and 300 μg of unirradiated DNA were treated under the same hydrolysis conditions as the hydrated DNA samples (Swarts et al). This calibration method corrects for the effects that formic acid hydrolysis has on the quantitation of base damage products, and has been shown to provide the most accurate estimation of the actual amounts of these base damage products in DNA (Swarts et al), unless the calibration can be performed using stable isotope dilution methods (Dizdaroglu 1993).

The yields for the base damage products were calculated based on the dry weight of DNA (Γ = 0) and expressed as nmoles/joule. Using yields that are based on the dry weight of DNA allows one to more readily observe the contribution of the radiation damage in the water molecules to the production of base damage in the hydrated DNA.

Results and Discussion

DNA Base Damage. In a study of the radiation-induced release of unaltered bases from hydrated DNA (Swarts et al 1992), it was found that the yields of the released bases were dependent on the extent to which the DNA was hydrated. However, the changes in the yields for base release were not proportional to the mass of water that was added to the DNA. Similarly, the yields for 14 base damage products that were analyzed in this study were usually not proportional to the mass of added water (Table 1). In addition, the dependence of these yields on Γ is different

Table 1. Yields of base damage products in hydrated DNA irradiated with 90 kGy under nitrogen

	G-values (nmoles/joule)				
Compound	$\Gamma = 2.5^a$	$\Gamma = 10.3^a$	$\Gamma = 13.2^a$	$\Gamma = 22.4^a$	$\Gamma = 32.8^a$
8-OH-Gua	120(20)[b]	130(30)	140(50)	160(20)	190(50)
FapyGua	13(1)	33(4)	23(5)	19(2)	45(12)
8-OH-Ade	1.6(0.2)	2.3(0.4)	3.1(1.0)	2.8(1.0)	3.8(1.3)
FapyAde	0.5(0.2)	1.2(0.2)	0.6(0.1)	2.6(0.5)	6.1(2.8)
2-OH-Ade	0.7(0.1)	0.6(0.2)	0.8(0.4)	0.3(0.1)	0.2(0.1)
5-OH-Ura	1.5(0.1)	1.1(0.3)	0.5(0.3)	2.5(0.5)	4.5(3.0)
5-OH-Cyt	0.5(0.2)	0.2(0.1)	0.4(0.2)	1.7(0.3)	2.3(0.8)
5,6-diOH-Ura	0.4(0.2)	0.5(0.4)	0.6(0.2)	1.3(1.0)	1.4(0.1)
5-OH-Hyd	2.2(1.5)	1.2(0.6)	0.7(0.6)	0.4(0.4)	0.7(0.2)
5,6-diHThy	30(12)	100(30)	87(17)	24(3)	26(4)
5-OHMe-Ura	7.0(0.9)	4.2(1.8)	3.8(1.8)	7.7(1.0)	8.2(1.3)
5-OH-5,6-diHThy	2.5(1.3)	2.4(0.7)	2.3(0.3)	0.9(0.1)	2.2(0.1)
Thy glycol	0.9(0.4)	0.5(0.2)	1.9(1.1)	1.5(0.3)	3.7(0.8)
5-OH-5-Me-Hyd	1.9(0.6)	0.9(0.8)	0.1(-)	0.6(0.6)	1.2(0.2)

a. Based on the dry weight (Γ=0) of DNA
b. Values represent the mean (±1 S.D.) from 3-5 independent experiments

for each of the base damage products. This can clearly be seen in Fig. 1 in which the yields of three of the base damage products, 8-OH-Gua, 5,6-diHThy, and FapyGua, have been plotted as a function of Γ. For 5,6-diHThy and FapyGua, the yields are maximized at approximately Γ = 10.3-13.2, then decrease with increasing hydration until Γ = 22.4. After Γ = 22.4, the yields either remain fairly constant (5,6-diHThy) or begin to increase (FapyGua) with increasing hydration. Conversely, the yield of 8-OH-Gua remains fairly constant between Γ = 2.5-13.2, and then increases as Γ is increased. Consequently, factors other than added water mass have a significant influence on the yields of the base damage products.

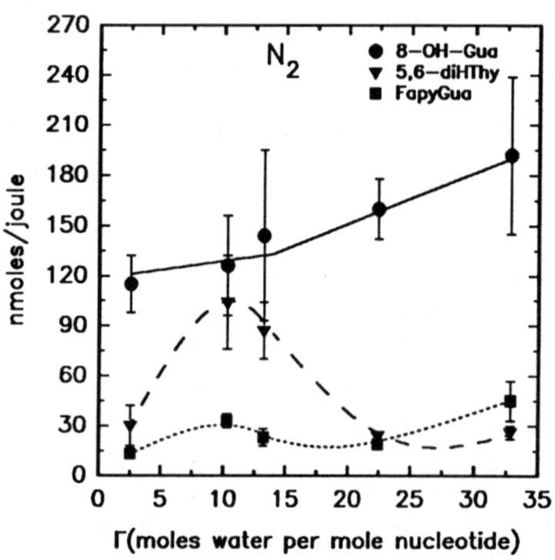

Figure 1. Yields of 5,6-diHThy (▾), 8-OH-Gua (•), and FapyGua (▪) as a function of the extent of hydration (Γ) in DNA irradiated under nitrogen. The values are expressed as nmoles/joule and are based on the dry (Γ = 0) weight of DNA. The points represent the mean (± 1 S.D.) of 3-5 independent experiments.

The nonproportional relationship between the yields of these base damage products and the mass of added water is likely the result of two factors that could modulate the formation of nonradical base damage products from their respective DNA radical precursors. From our earlier study of the release of unaltered bases from irradiated hydrated DNA (Swarts et al 1992), it was concluded that the radiolysis of the inner-most water molecules of the DNA hydration layer (Γ = 12-15) damaged DNA through the transfer of holes and electrons generated in these water molecules. Since the deposition of holes and electrons from these hydration waters would be in addition to those formed by direct ionization of the DNA molecule (direct effect), it has been suggested that these water molecules contribute to a quasi-direct effect (Becker et al 1994). This is unlike the DNA damage ($\Gamma > 15$) that is derived from the indirect effect where bulk radicals (e.g. OH·, e_{aq}^-, and H·) are the predominant damaging species. Consequently, it is expected that damage products which are formed predominately from the reaction of holes and electrons with the DNA should predominate in hydrated DNA ($\Gamma \leq 15$), whereas the yields of damage products which are formed primarily from the attack of bulk water radicals should increase at higher values of Γ.

The results of our base damage study support this expectation. In hydrated DNA ($\Gamma \leq$ 2.5-13.2), where the direct and quasi-direct effects have their greatest influence, the major base damage products that are found are 5,6-diHThy, 8-OH-Gua, and to a lesser extent FapyGua. This finding is significance because it correlates well with the major DNA base radicals that have been observed in ESR studies of irradiated DNA (Mroczka et al 1993, Cullis et al 1986, Gräslund et al 1971, Gregoli et al 1982, Sevilla et al 1991, Wang et al 1993, Wang et al 1994). In the irradiated frozen DNA, the thymine anion and the guanine cation are the predominant base radicals formed initially from the radiolysis of DNA and the surrounding hydration water. Protonation of the thymine anion to the oxidizing radical T(6H)· followed by a one electron reduction leads to the formation of 5,6-diHThy; a product that has been obtained in irradiated thymidine (Sharpatyi et al 1978, Shaw et al 1988). The guanine radical cation, once it undergoes an hydroxylation at the C-8 position to form a neutral guanine radical G(8OH)·, could form 8-OH-Gua via a one electron oxidation of G(8OH)·, similar to the reaction of the C-8 hydroxyl radical adduct of guanine (Steenken 1989). If this same G(8OH)· was to instead undergo a ring opening reaction followed by a one electron reduction, it would lead to the formation of FapyGua (Steenken 1989). The guanine cation can also undergo a facile deprotonation reaction to form a neutral guanine radical G(-H^+)·. This radical could conceivably lead to the formation of 8-OH-Gua through the oxidation of the G(-H^+)·, followed by hydroxylation of the oxidized species. Since the thymine anion and guanine cation are formed as a result of the localization of holes and electrons on DNA, it is not unexpected that 5,6-diHThy, 8-OH-Gua, and FapyGua are the major base damage products formed in the hydrated DNA at $\Gamma \leq 13.2$ where the direct and quasi-direct effects predominate.

The conformational change in the DNA from a pseudo-C form to the B form is also likely to be important in the formation of base damage products. This conformational change occurs as the DNA hydration is increased from Γ =2.5 to Γ = 32.8 (Saenger 1984). At $\Gamma < 15$, where a the pseudo-C conformation of the DNA predominates, holes and electrons are effectively trapped in the DNA; thereby increasing the relative amount of charged DNA base radicals. This phenomenon could explain the sharp increase in the yield of 5,6-diHThy as Γ is increased from 2.5 to 13.2, because increasing the number of electrons trapped in the DNA should result in elevated levels of the thymine radical anion and, ultimately, increase the yield of 5,6-diHThy. However, as more of the DNA assumes the B-conformation ($\Gamma > 15$), hole/electron recombinations should increase as the base stacking in the hydrated DNA begins to optimize, thus reducing the levels of charged DNA radicals and, subsequently, the yield of 5,6-diHThy. The rather sharp decrease in the yield of 5,6-diHThy between Γ = 13.2 and 22.4 suggests that changes in the base stacking occur rapidly over this relatively narrow range of hydration. Thus, the hydration dependent changes in DNA conformation appear to have a profound affect on the formation of radiation-induced base damage products.

Surprisingly, the yields of 8-OH-Gua did not show the same dependence on increasing hydration as was observed for 5,6-diHThy (Fig. 1, Table 1). If, as proposed above, the recombination of holes and electrons at Γ = 2.5 to Γ = 22.4 depends on the conformation of the DNA, then it would be expected that the yield of 8-OH-Gua should also roughly correlate with the relative amount of the charged guanine radical. One would expect the yield of 8-OH-Gua to be low at Γ = 2.5, maximizing around Γ = 10.3-13.2 and then decrease with increasing hydration. Instead the yield of 8-OH-Gua is relatively flat from Γ= 2.5 to Γ= 13.2 and then increases with increasing hydration (Fig. 1). The reason for the differences in the yields of 8-OH-

Gua and 5,6-diHThy as a function of Γ may be two-fold. In DNA hydrated to Γ = 2.5, where the availability of water is limited to the 2-3 water molecules on the deoxyribose/phosphate moiety, there may be differing efficiencies by which the thymine radical anion and the guanine radical cation are converted to 5,6-diHThy and 8-OH-Gua. Since the C-8 carbon on guanine is oriented toward these water molecules, it is likely that the guanine radical precursors involved in the formation of 8-OH-Gua are not as restricted in their ability to react with these water molecules as might be the case for the thymine radicals involved in the formation of 5,6-diHThy.

In DNA hydrated to Γ = 22.4, the increased yields of 8-OH-Gua over what are observed for 5,6-diHThy may be due to the fact that reactions of bulk water radicals (e.g. OH·) with the DNA may have already occurred, thus providing a different pathway through which 8-OH-Gua can be formed. This explanation is supported by the yield data in Table 1 for Thy glycol and the base damage products derived from Cyt Glycol (e.g. 5-OH-Ura, 5-OH-Cyt and 5,6-diOH-Ura). All of these products are believed to originate from the hydroxyl radical adducts of thymine and cytosine when DNA is irradiated under nitrogen. Note that the yields of these base damage products are low in DNA that has been hydrated to Γ = 13.2, but then increase with increasing hydration as expected, if they result from the attack of hydroxyl radicals generated by the radiolysis of the water molecules that are outside of the inner-most water molecules of the hydration layer. The degree of hydration (Γ = 13.2-22.4) in the base damage study at which reaction of bulk water radicals with DNA (indirect effect) begin to be observed is consistent with the degree of hydration (Γ = 12-15) at which these reactions appeared to begin in our base release study (Swarts et al 1992).

In conclusion, the results of this study on the formation of radiation-induced base damage products in hydrated DNA supports the idea that the inner-most water molecules of the DNA hydration layer contribute to a quasi-direct effect that produces DNA damage by processes that are more closely related to the direct effect than the indirect effect.

Acknowledgements

This work has been supported by grant #DE-FG05-86ER60404 and #DE-FG0286-ER60455 from the US DOE. Support from the NCI of the NIH (grant #CA45424) is also acknowledged.

References

Becker D, Sevilla MD (1993) The chemical consequences of radiation damage to DNA. Advances in Radiation Biology Lett JT, Sinclair WK (eds) Academic Press New York, p. 121-180
Cullis PM, Symons MCR (1986) Effects of direct radiation on deoxyribonucleic acid. Radiat Res 27:93-100
Dizdaroglu M (1993) Quantitative determination of oxidative base damage in DNA by stable isotope-dilution mass spectrometry. FEBS 315:1-6
Dizdaroglu M (1992) Measurement of radiation-induced damage to DNA at the molecular level. Int J Radiat Biol 61:175-183
Fuciarelli AF, Wegher BJ, Gajewski E, Dizdaroglu M, Blakely WF (1989) Quantitative measurement of radiation-induced base damage products in DNA using gas chromatography-mass spectrometry. Radiat Res 119:219-231
Gräslund A, Ehrenberg A, Rupprecht A, Strom G (1971) Ionic base radicals in γ-irradiated oriented DNA. Biochem Biophys Acta 254:172-186
Gregoli S, Olast M, Bertinchamps A (1982) Radiolytic pathways in γ-irradiated DNA: Influence of chemical and conformational factors. Int J Radiat Biol 61:299-313
Saenger W (1984) Water and nucleic acids. In Principles of Nucleic Acid Structure Springer-Verlag New York, p. 368-384
Sevilla MD, Becker D, Yan M, Summerfield SR (1991) Relative abundance of primary ion radicals in γ-irradiated DNA: Cytosine vs. thymine anion and guanine vs. adenine cation. J Phys Chem 95:3409-3415
Sharpatyi VA, Cadet J, Téoule R (1978) Final products obtained from the gamma radiolysis of frozen aqueous solutions of thymidine. Int J Radiat Biol 33:419-423
Shaw AA, Voituriez L, Cadet J, Gregoli S, Symons MCR (1988) Identification of the products resulting from the direct effects of gamma radiation on thymidine. J Chem Soc Perkin Trans 2, 1303-1307
Steenken S (1989) Purine base, nucleosides, and nucleotides: Aqueous solution redox chemistry and transformation reactions of their radical cations and e^- and OH adducts. Chem Rev 89:503-520
Swarts SG, Sevilla MD, Becker D, Tokar CJ, Wheeler KT (1992) Radiation-induced DNA damage as a function of hydration. I. Release of unaltered bases. Radiat Res 129:333-344
Swarts SG, Miao L, Smith GS, Wheeler KT, Effects of formic acid hydrolysis on the quantitation of radiation-induced DNA base damage products assayed by gas chromatography/mass spectrometry. Radiat Environ Biophys, in review
Wang W, Becker D, Sevilla MD (1993) The influence of hydration on the absolute yields of primary ionic free radicals in γ-irradiated DNA at 77K. I. Total radical yields. Radiat Res 135:146-154
Wang W, Yan M, Becker D, Sevilla MD (1994) The influence of hydration on the absolute yields of primary ionic free radicals in γ-irradiated DNA at 77K. II. Individual radical yields. Radiat Res 137:2-10

EFFECTS OF HYDRATION ON FREE RADICAL TRAPPING IN X-IRRADIATED FROZEN POLYNUCLEOTIDES

Nicholas Mroczka and William A. Bernhard
Department of Biophysics
University of Rochester
Rochester, NY 14642, USA

Introduction

Native DNA in aqueous solution is surrounded by a solvation layer of 20-22 water molecules per nucleotide (Saenger 1984). These waters are distinguishable from the bulk water by virtue of being more tightly associated with the DNA. The w/w ratio of water to DNA (called Γ) is ~1.0 for a fully hydrated DNA. When an aqueous DNA solution is frozen, it separates into two primary phases: ice and aggregates of hydrated DNA. Irradiation of such frozen solutions creates trapped radicals. With solid state EPR, we are able to study the distribution of free radical damage in the frozen DNA solution and thus better understand how the hydration shell influences that distribution.

Is there transfer of damage from the hydration shell to the DNA ?

One approach to answering this question has been to irradiate frozen DNA samples and measure the total yield (G) of trapped free radicals while varying the H_2O/DNA ratio. If the free radicals formed in the DNA hydration shell are trapped on the DNA polymer with the same efficiency as the radicals formed by direct ionization of DNA and none of the free radicals formed in bulk ice reach the DNA, then the G versus Γ curve should appear as shown in Figure 1. In the region $0<\Gamma<1$, G is expected to be constant due to the unchanged efficiency in trapping. For $\Gamma>1$, the yield asymptotically approaches the lower trapping yield of ice (~0.05 mmol/J at 4 K, Mroczka and Bernhard 1993). From measurements at 77 K, Gregoli *et al.* (1982) were the first to note this kind of response and propose damage transfer from water to DNA. Further support for this hypothesis was provided by Hüttermann *et al.* (1992). Wang *et al.* (1993) found that G increases in the $0<\Gamma<1$ region, from ~0.1 mmol/J to ~0.3 mmol/J, then at $\Gamma\sim1$ the curve turns downward. Mroczka and Bernhard (1993),

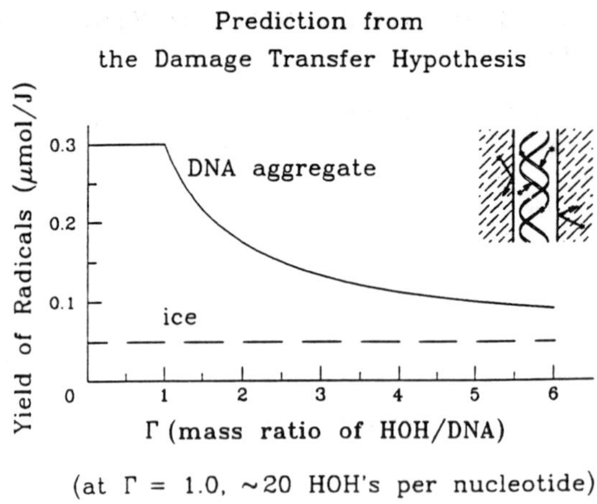

Figure 1. The inset on the right depicts the salient features of the damage transfer hypothesis: ionizations in the ice create radicals that do not escape the ice, ionizations in the DNA create radicals trapped in the DNA, and ionizations in the DNA hydration layer create radicals trapped in the DNA. If the efficiency of the latter two processes is the same, the above G vs. Γ curve is expected. G remains constant during hydration of the DNA and then asymptotically approaches the G of ice as the amount of ice increases.

studying films of DNA at 4 K, found that G decreases slightly between Γ equal to 0 and 1, from ~0.30 mmol/J to ~0.27 mmol/J, then at Γ~1.4 the slope turns more negative. In sum, these results indicate that ionization of the ice yields very little, if any, free radical trapping on the DNA and suggest damage transfer may occur from the hydration layer of DNA to the DNA polymer.

Although the measurements of G as a function of Γ are consistent with the damage transfer hypothesis, there are questions regarding the role of other important variables. Commensurate with changes in Γ, particularly for Γ<1, are changes in DNA conformation, packing, and proton transfer pathways. In spite of these confounding variables, we are rather certain that some damage transfer occurs. This is because the yield of radicals ($H^{\bullet}$, $O^{-\bullet}$, and $HO^{\bullet}$) trapped in the hydration layer itself is smaller than expected. Working at 4 K, Bernhard *et al.* (1994a) were able to put an upper limit of <0.01 mmol/J on the yield of radicals trapped in the DNA hydration layer at 4 K. For comparison, ice at 4 K has a yield of ~0.05 mmol/J (Mroczka 1993). These measurements, made at Q-band microwave frequencies, are not particularly sensitive to detection of the broad $HO^{\bullet}$ signal. Since EPR detects $H^{\bullet}$ with high sensitivity, the absence of $H^{\bullet}$ up to a Γ of 1.4 at 4 K (Mroczka 1993) is persuasive evidence on two points. 1. Electrons produced in an average hydration sphere of 28 waters per nucleotide are trapped in the DNA not in the water. 2. Electrons produced in the water beyond the average of 28 waters per nucleotide lead to the expected amount of trapped H•. Wang *et al.* (1993) arrived at a similar conclusion monitoring HOΣ trapping at 77 K using X-band microwave frequencies. More recently, Becker *et al.* (1994) found at 77 K that the yield of HO• trapped in the DNA hydration layer (~0.01 mmol/J) was substantially less than in an aqueous glass (~0.1 mmol/J). Overall, the scarcity of radicals trapped in the hydration layer is persuasive evidence for radical transfer from the hydration layer to the DNA. Still unknown is the extent of the radical transfer, particularly that of the incipient hole formed in the frozen hydration shell.

Packing has a Strong Influence on the Yield of Radical Trapping

At a fixed value of hydration, e.g. $\Gamma \sim 0.3$, the yield of trapped radicals in powder DNA (herring sperm) at 4 K can vary substantially, from 0.2 to 0.6 mmol/J (Mroczka 1993). This variation is not a consequence of systematic or random errors as can be demonstrated with DNA prepared as films where the variations are greatly reduced. This large variation in G at constant Γ is believed to be a consequence of variations in packing and/or conformation.

We have been investigating the influence of conformation and packing on free radical trapping. Although the details of this work will be reported elsewhere some of the observations are as follows. Strandedness appears to have little effect on total radical yield. The yields at 4 K for native Calf thymus DNA(Na) at $\Gamma < 0.2$ varies little between native and denatured, 0.39 and 0.37 (mmol/J), respectively (Mroczka and Bernhard, unpublished results). We are also trying to assess the influence of base stacking by studying, at 4 K, a series of polynucleotides: for example, poly(U) and poly(A):poly(U).

The results on poly(U) are provocative. On one hand, like powder DNA, powder poly(U) shows a large variation in yields. At $\Gamma < 0.5$, the yield ranges from 0.3 to 0.6 mmol/J. On the other hand, unlike DNA, the G vs. Γ curve has a distinct negative slope in the $0<\Gamma<1$ region, descending from ~0.5 to 0.3 mmol/J. If films of poly(U) are used instead of powders, the G vs. Γ curve remains constant at ~0.28

mmol/J over the same $0<\Gamma<1$ region (Mroczka and Bernhard, unpublished results). Unfortunately, we have not yet been able to measure the degree of stacking in these samples. There are other results, however, that indicate packing of the poly(U) polymers is more important to radical trapping than base stacking.

The G vs. Γ curves for poly(A):poly(U) powders and films are flat. For powders, G equals 0.27 mmol/J and, for films, G equals 0.38 mmol/J (Mroczka and Bernhard, unpublished results). Given the propensity of poly(A):poly(U) to form a double helix, this difference is more likely a consequence of changes in packing than in base stacking. The same argument holds for the results cited above on the large variability in DNA. Thus, in these two polymers, where a duplex conformation is believed to be persistent, the substantial variation in radical trapping is most likely a consequence of variations in packing.

In a broader context there is no question as to the large influence packing has on free radical trapping. The yield of free radicals trapped in single crystals of nucleic acid constituents varies by at least two orders of magnitude (Bernhard 1994b). For example, 1-methylcytosine and cytosine monohydrate are chemically very similar but form crystals with substantially different packing. The radical yields at ~ 4 K differ by a factor of 100. The values are $G<0.001$ mmol/J for 1-methylcytosine and $G\sim0.08$ mmol/J for cytosine monohydrate. Furthermore, cytosine monohydrate crystals can be dehydrated without significantly altering the packing between the cytosine molecules. Given the large change in hydration and the minor change in packing, the resulting change in yield is very small, $G\sim0.05$ mmol/J for dehydrated cytosine monohydrate.

Distribution of Damage to DNA *in vivo.*

Living organisms provide accommodations for a very lengthy complement of DNA polymer by maintaining it in a compact package. For example, the ratio of the unpacked DNA length to the longest length of the packaged DNA in interphase chromatin fibrils and chromosomes is 40-60:1; in sperm heads and bacterial nucleoids, it is ~200:1; and in T-even phage, it is 520:1 (DuPraw 1970). In the latter, the DNA concentration corresponds to that of an 80% solution.

The variability in DNA packaging is generally considered to involve compromises between minimizing storage volume and maintaining access to the required molecular functions of DNA. Eukaryotes wrap DNA about histone cores that in turn are organized further in chromatin. Organisms such as dinoflagellates algae (Bouligand 1968, Livolant 1978), and some bacteria (Gourret 1978), however, use only small amounts of basic proteins. In dinoflagellates, the DNA is thought to order spontaneously into a cholesteric phase. This cholesteric phase consists of stacked layers of DNA, each layer consisting of parallel DNA strands. There is also evidence for cholesteric packing of DNA in certain mitochondria (Brugerolle 1979).

Ultimately, the radiation chemistry of DNA must be understood in the context of these macroassemblies of DNA. The task of predicting the type and yield of DNA lesions requires an understanding of how packing influences the distribution of radiation energy immediately after deposition and the subsequent reactions. Factors which influence these processes are water of solvation, type of counter ions and packaging. At this stage, there is much to be done in sorting out the effects of these variables. What we have learned from the EPR studies of solid state system is that the average free radical density trapped by DNA can vary by at least a factor of 3 and that packaging appears to be the most significant variable.

Key Points

- Free radical damage is transferred from the hydration layer to DNA but we do not know the extent of the transfer.
- Because variations in the degree of hydration alter DNA packing and conformation, hydration effects radical trapping by more than just damage transfer.
- Packing has a pronounced effect (at least 3X) on radical trapping in solid DNA samples.
- Packing variations that occur in *vivo* have the potential of producing commensurate variations in the local density of damage.

Acknowledgments

This investigation was supported by PHS Grant 2-R37-CA32546, awarded by the National Cancer Institute, DHHS. Its contents are solely the responsibility of the authors and do not necessarily represent the official views of the National Cancer Institute.

References

Becker D, La Vere T, Sevilla MD (1994) Radiat Res (in press)

Bernhard WA, Mroczka N, and Barnes J (1994a) Combination is the dominant free radical process initiated in DNA by ionizing radiation: an overview based on solid-state EPR studies. Int J Radiat Biol (in press)

Bernhard WA, Barnes J, Mercer KR, Mroczka N (1994b) The influence of packing on free radical yields in crystalline nucleic acids: the pyrimidine bases. Radiat Res **140**, 199-214

Bouligand Y, Soyer MO, Puiseux-Dao S (1968) La structure fibrillaire et l'orientation des chromosones chez les Dinoflagellés. Chromosoma **24**, 251-287

Brugerolle G, Mignot JP (1979) Distribution et organization de l'ADN dans le complese kinetoplaste-mitochondrie chez un Bodonidé, protozoaire kinetoplastidé, protozoaire kinetoplastidé; variations au cours du cycle cellulaire. Biol Cell **35**, 111-114

DuPraw EJ (1970) DNA and Chromosomes Holt, Rinehart and Winston, New York

Gourret JP (1978) Description et interprétation des nucléoides structurés observés dans des bactéroides de Rhizobium. Biol Cell **32**, 299-306

Gregoli S, Olast M, and Bertinchamps A, (1982) Radiolytic pathways in g-irradiated DNA: influence of chemical and conformational factors. Radiat Res **89**, 238-254

Hüttermann J, Rohrig M, and Kohnlein W (1992) Free radicals from irradiated lyophilized DNA: influence of water of hydration. Int J Radiat Biol **61**, 299-313

Livolant R, Bouligand Y (1978) New observations on the twisted arrangement of Dinoflagellate chromosomes. Chromosoma **68**, 21-44

Mroczka N, and Bernhard WA(1993) Hydration effects on free radical yields in DNA X-irradiated at 4 K. Radiat Res **135**, 155-159

Saenger W, (1984) Principles of Nucleic Acid Structure Springer-Verlag, NY, 368-384

Wang W, Yan M, Becker D, and Sevilla MD (1993) The influence of hydration on the absolute yields of primary free radicals in g-irradiated DNA at 77 K. Radiat Res **135**, 146-154

γ-IRRADIATION OF THE HYDRATION LAYER OF DNA: •OH FORMATION VS. HOLE TRANSFER

David Becker, Therese La Vere and Michael D. Sevilla
Chemistry Department
Oakland University
Rochester, MI 48309

Introduction

It has often been assumed that the radiation damage in the hydration layer of DNA is rapidly transferred to the DNA in the form of electrons and holes (reactions 1,2) (Gregoli *et al.*, 1982; Hüttermann and Voit, 1990; Cullis et *al.*, 1990; Symons, 1991; Swarts *et al.*, 1992; Becker and Sevilla, 1993; Mroczka and Bernhard, 1993). However, the proton transfer reaction shown in (3) is known take place on the time scale of a molecular vibration (Klassen, 1987) and may be competitive with the hole transfer reaction (2).

$$e^{-}\bullet + DNA \rightarrow DNA^{-}\bullet \quad (1)$$

$$H_2O^{+}\bullet + DNA \rightarrow H_2O + DNA^{+}\bullet \quad (2)$$

$$H_2O^{+}\bullet + H_2O \rightarrow H_3O^{+} + \bullet OH \quad (3)$$

In order to determine the importance of (2), we have undertaken a search for •OH in the DNA hydration layer by attempting direct ESR detection of •OH at low temperature and also by H_2O_2 product analysis (Becker *et al.* 1994). It has been suggested that the hydration layer in DNA forms a glassy structure (Symons 1991). In aqueous glasses there is a large range of hydrogen bonding strengths to trapped •OH, thus, the low field component of the •OH ESR spectrum at X-band frequencies consists of a broad feature extending over ca. 250 G (Riederer *et al.* 1983). When H_2O is γ-irradiated in the absence of oxygen, the only source of H_2O_2 is reaction 4 (Buxton, 1987). Thus a way to probe for •OH in the DNA hydration layer is to analyze for its diamagnetic product, H_2O_2.

$$HO\bullet + HO\bullet \rightarrow H_2O_2 \quad (4)$$

Materials and Methods

Salmon testes DNA from Sigma Chemical Company and CsF from Aldrich were used without further purification. Anoxic hydrated DNA samples were prepared by equilibrating DNA at the appropriate relative humidity for 4 to 7 days in a nitrogen atmosphere. For H_2O_2 analysis, oxygen free samples were irradiated at 77 K, warmed and then used for experimental runs. The values of Γ reported (in water molecules per nucleotide) include 2.5 waters per nucleotide that can not be removed from the DNA by drying (Swarts *et al.* 1992). Dry DNA samples were prepared by vacuum drying DNA for 2 days and sealing under vacuum. CsF was chosen for preparation of a glassy system because only water radicals are trapped when the glass is γ–irradiated at 77 K.

All ESR spectra are digitized and stored. Spectra are recorded for annealed samples only after cooling to 90 K in order to eliminate temperature dependent changes in the ESR spectra. All H_2O_2 determinations were done on a YSI Glucose Analyzer after warming samples to room temperature. Catalase was used in each experimental run to obtain a background reading. The H_2O_2 yields reported are only the catalase sensitive portions of the readings. Vacuum dried DNA irradiated at 77 K (and then dissolved in H_2O), a variety of nucleotides, bases and sugars and acetaldehyde were tested on the glucose analyzer; all gave only background readings.

Results

Detection of •OH in the DNA Hydration Layer. Figure 1A shows the ESR spectrum at 77 K of an unannealed anoxic DNA sample (Γ = 14) irradiated at 77 K to a dose of 65 kGy. A broad (ca. 250 G) resonance is evident at low fields at high amplification. Figure 1B shows the ESR spectra of dry DNA (Γ = 2.5) under conditions equivalent to that in 1A; the broad resonance evident in Figure 1A is not present. Figure 1C shows the spectrum of •OH in a 6 M CsF glass; it shows a broad low field resonance and a resolved doublet in the g = 2.0 region. This spectrum is very similar to that of •OH in other aqueous glasses (Reiderer *et al.* 1983, Ohlmann and Hüttermann 1993) and is nearly superimposable in the low field region with that in Figure 1A. The superimposibility of the low field regions of 1C and 1A is good evidence that •OH in a glassy matrix is present in the DNA sample. Since

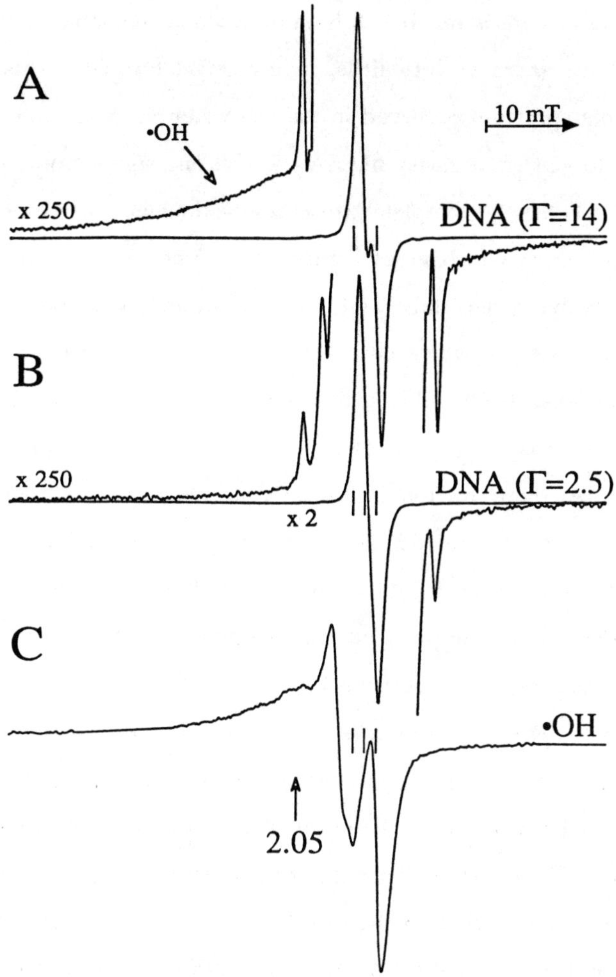

Figure 1 A. First derivative ESR spectrum of γ-irradiated hydrated DNA at 77 K. The 250 fold expansion shows the low field spectrum we assign to the •OH radical. B. First derivative ESR spectrum of γ-irradiated dry DNA at 77 K. C. The ESR spectrum of the •OH radical in γ-irradiated 6 M CsF at 77 K. The g-value marker at 2.05 is the approximate $g_{||}$ position for •OH in crystalline ice. Two lines from hydrogen atoms are subtracted from the wings in C. The total spectral extent in A, B and C is 65 mT. Reprinted with permission of Radiation Research Society (Becker *et al.*, 1994).

there is no bulk water in this sample (bulk water as ice first appears at $\Gamma = 20$) (Wang *et al.*, 1993), the •OH resonance must originate with the 14 waters (per nucleotide) of hydration. Because there is much less hydration water available in the dry DNA sample (Figure 1B), there is very little, if any, •OH formed. This is a strong indication that the spectrum observed in the DNA sample does, indeed, contain a component due to •OH in a glassy matrix. The strong signal from DNA primary radicals at $g = 2$ in Figure 1A completely obscures the doublet due to •OH.

As the sample which gives Figure 1A is annealed, the broad resonance attributed to •OH diminishes until, at 135 K, it essentially disappears. This is the behavior expected of •OH, which is well known to anneal out of frozen ices and glasses by 135 K (Kevan 1968) and is further confirmation that we are observing •OH in the DNA hydration layer.

In Figure 1A, there are two sharp on-scale resonances, situated in a near symmetrical fashion 7.1 mT on each side of the center of the DNA spectrum; these are the outer lines of the TH• (5,6-dihydrothymin-5-yl) radical spectrum (Wang and Sevilla 1994). Also, we wish to point out that the two ESR resonances from H• have been computer-subtracted from spectrum 1C.

In summary, there are three lines of evidence that •OH, originating with γ-irradiation of the DNA hydration layer, has been detected. First, is the similarity in spectra between a known hydroxyl radical in 6 M CsF and the presumed •OH spectrum in DNA. Second, is the behavior of the spectrum in DNA as the sample is annealed and third, is the lack of the hydroxyl radical spectrum in dry DNA.

Yield of H_2O_2 in Hydrated DNA. DNA samples with hydration levels $14 < \Gamma < 17$ were irradiated at 77 K and annealed to ambient conditions for H_2O_2 analysis. In the dose range 65 kGy – 195 kGy, an average yield of 0.0035 μmol H_2O_2/J was found. For these samples there is no bulk water present (as ice) at 77 K. Since reaction 4 is the only reasonable source for the formation of hydrogen peroxide in anoxic samples, the detection of H_2O_2 in the hydration layer of irradiated DNA samples is additional strong evidence for formation of •OH in the hydration layer.

G Values for •OH in Glasses and in DNA. In order to determine the low temperature yield of •OH in DNA (at $\Gamma = 13$), the spectrum of •OH in 6 M CsF

(Figure 1C) was used as a benchmark spectrum for •OH in hydrated DNA and the •OH spin concentration in DNA compared to the number of spins found in a sample of $CuSO_4$ of known concentration with the same shape and size as the DNA sample. The yield as a function of dose was then fitted to equation (5), in which C is the yield, D is the dose and $C_{max} = G/k$. (Wang *et al.* 1993).

$$C = C_{max}(1 - e^{-kD}) \quad (5)$$

For DNA with $\Gamma = 14$, the best fit to (5) was obtained with G = 0.035 mmol/J and k = 0.1 kGy^{-1}. Because of the small amount of •OH present in the DNA spectrum, these values must be regarded as estimates. For the purposes of calculating •OH yields and absorbed dose in this system, it was assumed that only the 13 molecules of water per nucleotide constitute the energy absorbing medium.

In a frozen 6 M CsF glass containing H_2O_2 as an electron scavenger (to mimic the electron scavenging of DNA) the G-value of •OH from reaction 3 was estimated to be G(•OH) = 0.13 μmol/J. Irradiation of H_2O_2 in a 6 M CsF glass results in formation of •OH due to reaction 6. In order to estimate the yields of •OH from reaction 3 only we assumed that approximately equal amounts of •OH result from reactions 3 and 6, respectively.

$$e^- + H_2O_2 \rightarrow OH^- + \bullet OH \quad (6)$$

Based on preliminary results, Figure 2 shows the approximate behavior of G(•OH) as a function of the hydration level in γ-irradiated DNA at 77 K. At low hydration levels, no •OH is found; as the hydration is increased to $\Gamma = 20$, G(•OH) reaches a maximum and appears to plateau at approximately 0.05 mmol/J.

Discussion

The primary experimental result of this work is the presentation of two lines of evidence that the hydroxyl radical (•OH) is formed in the hydration layer of γ-

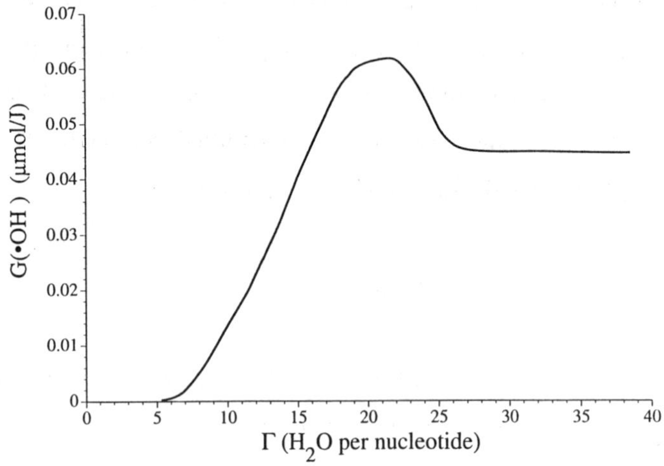

Figure 2. A schematic, based on preliminary data, of G(•OH) in the hydration layer of DNA as a function of the level of hydration. Below Γ = 5 no hydroxyl radical is found whereas after Γ = 10 significant amounts of •OH are found. Work is in progress to better quantify the relationship between G(•OH) and the amount of water of hydration.

irradiated DNA; these are the direct observation of the ESR glass-like spectrum of •OH and the detection of H_2O_2 in γ-irradiated hydrated DNA.

By using the production of •OH in 6 M CsF as a model for production of •OH in the DNA hydration layer, some inferences regarding the fate of various species can be deduced from the yields we report (Figure 3). The results with 6 M CsF suggest that the G for production of •OH in the hydration layer in the absence of a hole scavenger should be near 0.13 μmole/J. Figure 3 indicates that G(•OH) in the DNA hydration layer is ca. 0.05 μmole/J for fully hydrated DNA. This suggests that ca. 40% of the holes (H_2O^+) formed in the DNA hydration layer end up as trapped •OH and that 60% of the holes that escape recombination in spurs transfer as dry charge to the DNA. The dry charge that transfers would result in the formation of 0.08 μmol DNA$^{\bullet +}$ per joule absorbed by hydration water.

We have relied on a model aqueous glass in order to account for the fate of the H_2O^+ formed in the DNA hydration layer; the greater complexity of the DNA hydration water system may eventually demand a more sophisticated approach (Becker *et al.* 1994). For example, it has been suggested that •OH, if produced in the

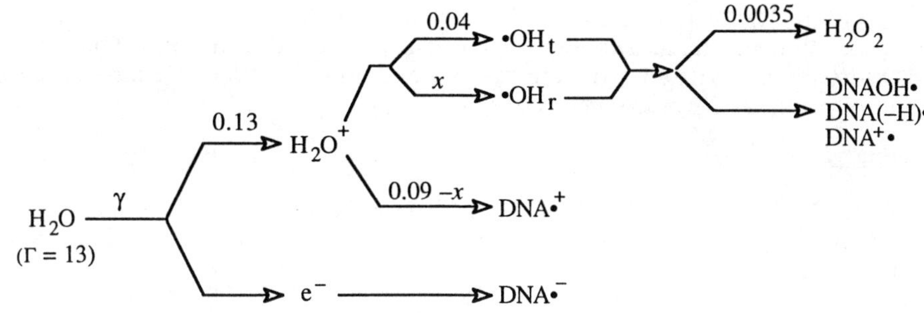

Figure 3. A schematic of early processes that originate with the DNA hydration layer after γ–irradiation of hydrated DNA at 77 K. Estimated yields in micromoles per joule are shown for each process. The values 0.05 and 0.0035 are experimental values reported in this work; *x* is presumably small. If and when they are formed, the products at the upper right would originate from reactive hydroxyl radical (•OH_r) at low temperature and trapped (•OH_t) upon warming. For clarity, we have elected not to display the many likely recombination reactions that occur. Reprinted with permission of Radiation Research Society (Becker *et al.*, 1994).

hydration layer, may oxidize DNA to form G^{+}• (Cullis *et al.* 1990) and by doing so, mimic hole transfer.

Our current results suggest that, at 77 K, over half of the oxidative damage originating in hydration water transfers to the DNA as dry charge. The first 5 waters trap no observable •OH; thus, it appears that they transfer charge completely to the DNA. At hydration levels greater than Γ = ca. 10, trapped •OH is observed; by Γ = 20, the overall yield of •OH is 0.06 μmol/J. If we assume the first 10 waters transfer charge, the yield in the next 10 is 0.12 μmol/J, which is essentially the same as that found in the 6 M CsF aqueous glass. As expected, the outer waters of hydration do not transfer dry charge as efficiently as the inner ones which are closer to the DNA double helix. Hole transfer from the inner waters of hydration may account for the unscavengable damage from the hydration layer in mammalian DNA (Ward 1991). Further work to better quantify the yield of •OH at various DNA hydration levels is in progress.

Acknowledgments

We thank the Office of Health and Environmental Research of the Department of Energy (Grant No. DEFGO2ER60455) and the National Cancer Institute of the National Institutes of Health (Grant No. RO1CA45424) for support of this work. We also thank Dorothy D. Dziedzic, John Reddan and Rose Sharma for their assistance in hydrogen peroxide analyses and Wending Wang for his assistance in hydroxyl radical G-value determinations.

References

Becker D, Sevilla MD (1993) The chemical consequences of radiation damage to DNA. Advances in Radiation Biology Lett JT, Adler H (eds) Academic Press New York pages 121-180.

Becker D, La Vere T, Sevilla MD (1994) ESR detection at 77 K of the hydroxyl radical in the hydration layer of γ-irradiated DNA. Radiat. Res. 140:123–129.

Buxton GV (1987) Radiation chemistry of the liquid state: (1) water and homogeneous aqueous solutions. Radiation Chemistry Principles and Applications Farhataziz, Rodgers MAJ (eds) VCH Publishers New York pages 321–349.

Cullis PM, Langman S, Podmore ID, Symons MCR (1990) Effects of ionizing radiation on DNA. Part VI. Effects of hydroxyl radical scavengers. J Chem Soc Faraday Trans 86:3267-3277.

Gregoli S, Olast M, Bertinchamps A (1982) Radiolytic pathways in γ-irradiated DNA: influence of chemical and conformational factors. Radiat Res 89:238-254.

Hüttermann J, Voit K (1987) Free radicals from direct action of ionizing radiation in DNA: structural assignments from EPR spectroscopy. Electronic Magnetic Resonance of the Solid State Weil JA (ed) The Canadian Society for Chemistry Ottawa Canada pages 267-279.

Kevan L (1991) Radiation chemistry of frozen aqueous solutions. Radiation Chemistry of Aqueous Systems Stein G (ed) Wiley Interscience/The Weizmann Science Press of Israel New York Jerusalem 1968 pages 21-72

Mroczka N, Bernhard WA (1993) Hydration effects on free radical yields in DNA x-irradiated at 4 K. Radiat Res 135:155-159.

Ohlmann J, Hüttermann J (1993) Reactions of water radiolysis intermediates with pyrimidine and purine bases in aqueous BeF_2 glasses: an EPR study. Int J Radiat Biol 63:427-436.

Riederer H, Hüttermann J, Boon P, Symons MCR (1983) Hydroxyl radicals in aqueous glasses: characterization and reactivity studied by ESR spectroscopy. J Magn Reson. 54:54-66.

Swarts SG, Sevilla MD, Becker D, Tokar CT, Wheeler KT (1992) Radiation-induced DNA damage as a function of hydration. I. Release of unaltered bases. Radiat Res 129:333-344.

Symons MCR (1991) The role of radiation induced charge migration with DNA: ESR studies. NATO ASI Ser Ser H 54:111-124.

Wang W, Becker D, Sevilla MD (1993) The influence of hydration on the absolute yields of primary ionic free radicals in γ-irradiated DNA at 77 K. Radiat Res 135:146-154.

Wang W, Sevilla MD (1994) Protonation of nucleobase anions in gamma-irradiated DNA and model systems. Which DNA base is the ultimate sink for the electron? Radiat Res 138:9-17.

Ward JF (1991) Mechanisms of radiation action on DNA model systems – their relevance to cellular DNA. NATO ASI Ser Ser H 54:1-16.

APPLICABILITY OF EXPERIMENTAL MODELS FOR CELLULAR RADIATION TARGETS

John T. Lett
Radiological Health Sciences
Colorado State University
Fort Collins, CO 80523 USA

Introduction

Rigorous development of a unifying theory of cellular sensitivity to ionizing radiations requires a proper understanding of the early phase of the chemical stage of radiation action where rapid chemical changes follow the deposition of radiant energy in the cellular radiation target (Lett 1990, 1992, 1994). Monte Carlo simulations of radiation transport processes and track structures can describe the physical stage of radiation action, but such analyses have not been extended as yet to the processes of charge and energy transfer associated with the early chemical events.

In terms of energy deposition, the cellular radiation target comprises the hydrated deoxyribonucleoprotein (chromatin) (Lett 1990, 1992, 1994); but for the fast chemical reactions, the roles of oxygen, local radioprotectors (e.g., SH compounds) and multivalent metallic ions have to be considered (Lett 1990, 1994). There is also the question of whether radicals (notably ˙OH), which are produced in bulk nuclear water beyond the primary hydration shells of the DNA and the constituent proteins of chromatin, produce significant levels of cellular DNA damage at radiobiologically significant doses of ionizing radiations (Lett 1990, 1992, 1994).

Over the last three to four decades, a variety of experimental models have been used to approximate, or simulate, radiation damage to nuclear chromatin *in vivo*. They fall into three main categories:

1. Hydrated DNA and/or deoxyribonucleoprotein.
2. Frozen solutions of DNA.
3. Aqueous solutions of DNA and/or deoxyribonucleoprotein containing radioprotectors believed to act primarily as radical scavengers, mainly of ˙OH.

Some of the difficulties inherent to models in categories 1 and 3 are examined here. Emphasis in the first case is upon the preparation of the model system and in the second is upon the interpretation of the actual radiation responses observed. All the information utilized is taken from the published literature.

The Cluster Hypothesis

Modern theories of radiation action at the cellular level stem from seminal experiments described in the 1960s. The developments in chemistry preceded those in physics. The original studies with "anhydrous" and hydrated DNA (Alexander and Lett 1960; Alexander et al. 1961; Lett and Alexander 1961; Lett et al. 1961) showed that:

a. Double strand breaks in DNA (dsb) were induced linearly with aerobic dose by both sparsely ionizing and densely ionizing radiations (1 - 2 MeV electrons and ^{210}Po α-particles, respectively) until the water content of the hydrated fibers exceeded that of anhydrous DNA.
b. The oxygen effect for dsb induction in the DNA fibers was small.
c. When the ratio of water: DNA in the hydrated fibers progressed beyond the value of 1:1, the relationship between the dsb induction frequency and water content changed in a manner indicative of an increasing contribution from dsb formed by random induction of single strand breaks (ssb) on opposite strands of the twin helix.

The linearity of dsb induction with aerobic dose in DNA fibers was explained with the hypothesis that *dsb are formed as independent events by clusters of ionizations*.

By 1973, the linear induction of dsb with aerobic dose in a mammalian cell had been established by Corry and Cole (1973), whose interpretation of the mechanism was analogous to that above; namely, that dsb were formed by single hits, i.e., discrete events. Although earlier papers by the same authors did indicate linearity of dsb induction with dose in metaphase chromosomes (e.g., Corry and Cole 1968), the phenomenon of speed dependence for DNA sedimentation was not understood properly at the time. Furthermore, the paper quoted also established the rejoining (repair) of radiation-induced dsb in a mammalian cell. The dsb strand-breaking efficiencies in the DNA fibers and the cells are comparable to those recorded more recently; i.e., some two to three decades later (Lett 1990, 1992, 1994).

Nowadays, the cluster hypothesis has strong support and is the basis for the Monte Carlo simulations of dsb induction in cellular chromatin. The consensus of current opinion is that dsb are formed linearly with aerobic dose in cells, at low and high LET, by clusters of ionizations, and that the clusters of damage associated with such dsb induction not only increase with LET but also become increasingly more difficult to process by the enzymatic functions which remediate dsb in cells (Lett 1990, 1992, 1994). Those concepts are integral parts of modern theories of cellular radiosensitivity, so it is crucial that the applicability of experimental models of cellular radiation target be assessed as precisely as is possible. Only the responses from aerobic exposures to sparsely ionizing radiation are examined here.

Hydrated Linear DNA

As outlined above, fibers and films of DNA with limited water contents have been useful for modelling the aerobic response of chromatin *in situ* despite the absence of proteins and the SH compounds known to contribute to the radiobiological oxygen effect by causing chemical repair of DNA damage which would otherwise be fixed by oxygen (Lett 1994). The model has also been used of late to examine early chemical changes in irradiated DNA by electron spin resonance (ESR). Careful attention must be given to the preparation and hydration of the fibers, however, if the difficulties experienced by Mroczko and Bernhard (1993) are to be eliminated.

For the original studies with DNA fibers (e.g., Lett and Alexander 1961), hydration was achieved by exposure to different relative humidities at 22 ± 40°C. There was no hysteresis: the adsorption and desorption curves were similar. Results of a broader study at 21 ± 0.5°C by Falk et al. (1962) were compared and contrasted by correspondence and discussions at the time, and it was agreed that the two sets of data were similar, except at low humidities and for a small hysteresis observed in the latter case. Over the past decade, S. A. Lindsay and his colleagues have conducted an extensive series of investigations of DNA hydration with wet-spun films (Tao et al. 1989). Hysteresis was not seen at 23 ± 2°C. The gravimetric results, together with curves for a modified form of the adsorption equation of Brunauer, Emmett, and Teller, are given in Figure 1, which is reproduced in modified form from the dissertation of Dr. T. Weidlich (see Lett 1994). From the best fit of *b*, a measure of the adsorption energy, estimates of the absolute numbers of water molecules from Raman spectra, and information obtained by other techniques, Tao et al. concluded that: the primary hydration number of DNA is *circa* 15 water molecules per nucleotide; some 3 water molecules per nucleotide are bound tightly to the DNA; the secondary hydration shell is similar to bulk water and contains the counterions. Hydration water lies mostly in grooves of the twin helix and, for our purposes, the transition from the A form to the B form, the usual biological form of DNA, can be considered to occur when the primary hydration shell is full.

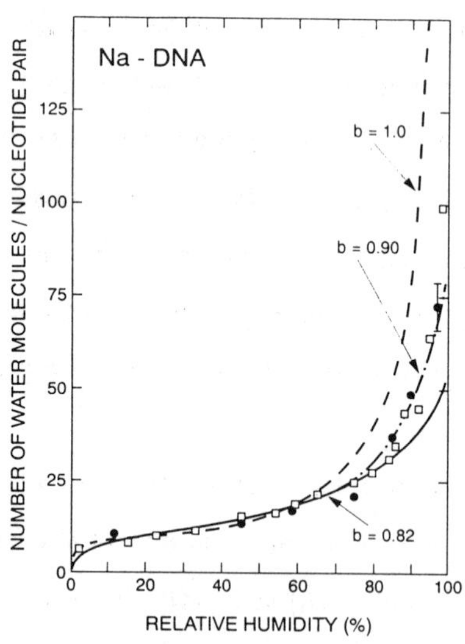

Figure 1. Hydration of Na DNA at ambient temperatures. A figure from the Ph.D. dissertation (unpublished) of Dr. T. Weidlich (Arizona State University), in which his hydration data, ❑, were compared with fits to a modified form of the adsorption equation of Brunauer, Emmett, and Teller. Data from Lett and Alexander (1961), ●, are included for comparison (see text). Reproduced and modified with permission of Dr. T. Weidlich (see Lett 1994, and Academic Press, Inc.)

The gravimetric data of Falk et al. (1962) and Hüttermann et al. (1992) lie slightly below the combined data in Figure 1, but give a curve of the same general shape, so similar responses for the hydration of fibers, films, wet-spun films, etc., were obtained in all four laboratories. For this reason, and also because collectively the properties of the DNA samples were examined by a wide spectrum of physical techniques, the composite set of data will serve as the basis for comparison here. For further information, see Lett (1994).

Water in the primary hydration shell of DNA does not crystallize into ice at low temperatures but water more distant from the macromoleculer surface does. Similar constraints presumably obtained when Hüttermann et al. (1992) and Wang et al. (1993) measured radical formation in irradiated, hydrated, DNA by ESR at 77 K or higher. Results from both laboratories were interpreted similarly: energy deposited in the (primary) hydration shell increased the formation of primary radicals in DNA by energy (charge) transfer processes; energy deposited in ice did not. The amount of hydration water in the DNA radiation target at the experimental temperature could be defined.

Evidence of "supercooling" was found when Falk et al. (1970) examined the effect of slow cooling to, and warming from, temperatures as low as about 95 K on films of partially deuterated DNA. Such "supercooling" probably explains the change in the state of some 5 water molecules and an abrupt decrease in radical yield observed by Wang et al. (1993) as DNA hydration was increased. For further details, see Lett (1994).

The above sets of results cannot be compared directly with those of Mroczko and Bernhard (1993) at 4 K because of possible differences in cooling rates, the final temperature of measurement, and inhomogeneities in DNA hydration and crystal structure (Falk et al. 1970; Tao et al. 1989). Unusual appearances of desorbed samples were reported by Mroczko and Bernhard (1993) and the hydration curves are atypical: there was a large hysteresis, and only the average values of data for the desorption curve, the reproducibility of which was poor, are comparable to the data in Figure 1. A ready cause of such variability is the use of highly swollen gels in the desorption procedure.

Apart from the variability reported by Mroczko and Bernhard (1993), the responses to sparsely ionizing radiation of hydrated DNA, frozen hydrated DNA, and frozen solutions of DNA determined by a variety of techniques (for reviews see Becker and Sevilla 1993; Lett 1990, 1992; O'Neill and Fielden 1993) are consistent with the notion that energy deposited in a primary hydration shell of some 15 water molecules per nucleotide, including the water molecules bound strongly to the phosphate groups, produces effects different from energy deposited in water outside it. A doubling of dsb induced in anhydrous DNA, and about threefold increases in DNA radicals postulated to be dsb precursors, are produced *without* significant involvement of ˙OH. Furthermore, similar levels of dsb are induced in hydrated DNA, mammalial cells, and viral chromatin *in vivo* (Lett 1990, 1992, 1994), so the hydrated DNA serves as a useful first approximation for the simulation of dsb induction in cellular radiation targets. In future experiments, all hydrated DNA samples used to model the cellular radiation target should be prepared under conditions where the effect of hydration on macromoleculer configuration already has been ascertained by a wide variety of independent biophysical and physical techniques. Refinements needed for improvement of the model are determined by the interplay among different processes in the radiation chemistry of chromatin in situ (see above and Lett 1994).

Aqueous Solutions of SV40 DNA and Chromatin: Effects of Radioprotectors

The aerobic radiation responses of mammalian chromatin *in situ* have been modeled with aqueous solutions of SV40 DNA, and its minichromosome, supplemented with radioprotectors believed to act primarily as scavengers of ˙OH generated in bulk water. The conclusion usually reached from such experiments is that DNA damage from ˙OH is the principal cause of cellular radiotoxicity (Lett 1990; von Sonntag 1987). Recently, that conclusion has been challenged (Lett 1992, 1994), and the challenge is extended here through consideration of later data on the effects of dimethyl sulfoxide (DMSO) on ssb and dsb induction in SV40 DNA and its minichromosome (Milligam et al. 1993a,b; Jones et al. 1994). Additional analysis is given elsewhere (Lett 1994).

Efforts to determine the "direct" effect of sparsely ionizing radiation from the radioprotection afforded the DNA and SV40 minichromosome by DMSO have been equivocal in the case of ssb induction (Milligan et al. 1993a,b). The more plausible of two models of nonhomogeneous reaction kinetics yielded induction frequencies for "direct" action of 2.8 x 10^{-10} ssb/Gy/Da for the minichromosome, which is also protected against ·OH attack by its constituent proteins, and 1.4 x 10^{-10} ssb/Gy/Da for SV40 DNA. Both of those results were believed to be similar and ≈2 x 10^{-10} ssb/Gy/Da, which value is not distinguishable, therefore, from those measured with hydrated DNA films *in vitro* and the SV40 minichromosome *in situ*, (see Lett 1990, 1994), and both SV40 DNA and the minichromosome (~2.5 x 10^{-10} Gy/Da) under conditions (~0. 1*M* DMSO) believed to simulate the circumstances of mammalian DNA *in situ* (Jones et al. 1993; see arrow on abscissa in Figure 2). When high concentrations of DMSO were employed, the efficiency of ssb induction dropped well below the calculated value for the "direct" effect (see upper arrow on ordinate in Figure 2).

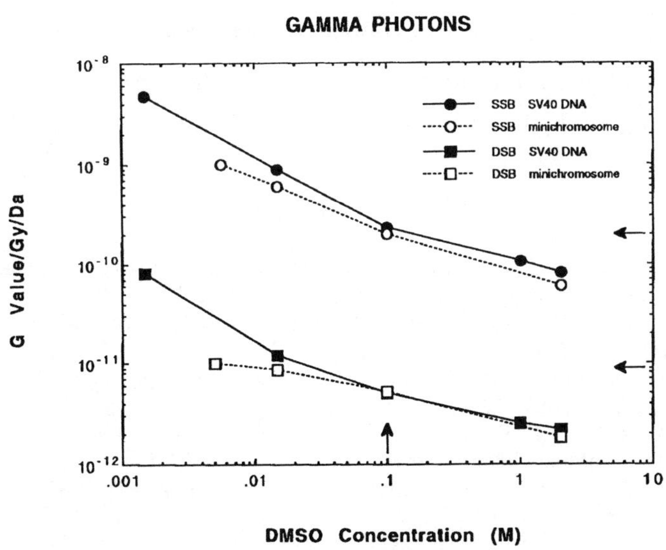

Figure 2. Effects of DMSO concentration on strand break induction in SV40 DNA and its minichromosome. Aerobic γ-irradiation of aqueous solutions at ambient temperatures. Data from the publications of J. F. Ward and colleagues (see text).

The situation for dsb induction is no better. At high concentrations of DMSO, the dsb values, 2.0 - 2.5 x 10^{-10} dsb/Gy/Da, are far below the generally accepted range of values for mammalian chromatin in situ, hydrated DNA, etc. (Lett 1994; see lower arrow on ordinate in Figure 2 for mean value).

The above evidence shows that DMSO is not acting simply as a scavenger of $^{\cdot}OH$ in the SV40 systems, so it should not be used to support definitive assertions about the usefulness of putative radical scavengers in determining the relative contributions of the "direct" and "indirect" effects to the radiation responses of mammalian cells. Previous interpretations of the effects of putative radical scavengers under anoxic irradiation conditions have also been confounded by recent observations, e.g., Milligan and Ward (1994), on the formation of long-lived DMSO radicals that react very effectively with DNA. Furthermore, intercomparisons among model systems are complicated by the findings that even the buffers used in many experimental studies can respond differently to radiation exposure at ambient and low temperatures (Cullis et al. 1993).

Acknowledgments

Preparation of this article was supported by the National Aeronautics and Space Administration with grant #NAGW-3957 and the NSCORT in Radiation Health in joint effort with the Lawrence Berkeley Laboratory.

References

Alexander P and Lett JT (1960) Role of oxygen in the crosslinking and degradiation of deoxyribonucleic acid by ionizing radiations. Nature 187: 933-934

Alexander P, Lett JT, Kopp P and Itzhaki R (1961) Degradation of dry deoxyribonucleic acid by polonium alpha-particles. Radiat Res 14: 362-373

Becker D and Sevilla MD (1993) The chemical consequences of radiation damage to DNA. Adv Radiat Biol 17:121-180

Corry PM and Cole A (1968) Radiation-induced double-strand scission of the DNA of mammalian metaphase chromosomes. Radiat Res 36:524-543

Corry PM and Cole A (1973) Double strand rejoining in mammalian DNA. Nature New Biology 245: 100-101

Cullis PM, Elsy D, Fan S and Symons MCR (1993) Marked effect of buffers on yield of single and double-strand breaks in DNA irradiated at room temperature and at 77K. Int J Radiat Biol 63:161-165

Falk M, Hartman KA Jr and Lord RC (1962) Hydration of deoxyribonucleic acid. 1. A gravimetric study. J Am Chem Soc 84:3843-3846

Falk M, Poole AG and Goymour CG (1970) Infrared study of the state of water in the hydration shell of DNA. Can J Chem 28:1536-1542

Hüttermann J, Röhrig M, Köhnlein W (1992) Free radicals from irradiated llyophilized DNA: Influence of water of hydration. Int J Radiat Biol 61:299-313

Jones GDD, Milligan JR, Ward JF, Calabro-Jones PM and Aguilera JA (1993) Yield of strand breaks as a function of scavenger concentration and LET for SV40 irradiated with ^{4}He ions. Radiat Res 136:190-196

Lett JT (1990) Damage to DNA and chromatin structure from ionizing radiations, and the radiation sensitivities of mammalian cells. Prog Nucleic Acid Res Mol Biol 39, 305-352

Lett JT (1992) Damage to cellular DNA from particulate radiations, the efficacy of its processing and the radiosensitivity of mammalian cells: Emphasis on DNA double strand breaks and chromatin breaks. Radiat Environ Biophys 31:257-277

Lett JT (1994) The renaissance in basic cellular radiobiology and its significance for radiation therapy. Adv Radiat Biol Vol 18:181-223

Lett JT, Stacey KA, and Alexander P (1961) Crosslinking of dry deoxyribonucleic acid by electrons. Radiat Res 14: 349-362

Lett JT and Alexander P (1961) Crosslinking and degradation of deoxyribonucleic acid gels with varying water contents when irradiated with electrons. Radiat Res 15: 362-373

Milligan JR and Ward JF (1994) Yield of single-strand breaks due to attack on DNA by scavenger-derived radicals. Radiat Res 137:295-299

Milligan JR, Aguilera JA and Ward JF (1993a) Variation of single-strand break yield with concentration for SV40 minichromosome irradiated in aqueous solution. Radiat Res 133:158-162

Milligan JR, Aguilera JA and Ward JF (1993b) Variation of single-strand break yield with scavenger concentration for plasmid DNA irradiated in aqueous solution. Radiat Res 133:151-157

Mroczko N and Bernhard WA (1993) Hydration effects on free radical yields in DNA x-irradiated at 4K. Radiat Res 135:155-159

O'Neill P and Fielden EM (1993) Primary free radical processes in DNA. Adv Radiat Biol 17:53-120

Tao NJ, Lindsay SM and Ruppercht A (1989) Structure of DNA hydration shells studied by Raman spectroscopy. Biopolymers 28:1019-1030

von Sonntag C (1987) *The Chemical Basis of Radiation Biology* Taylor and Francis London

Wang W, Becker D and Sevilla MD (1993) The influence of hydration on the absolute yields of primary ionic free radicals in γ-irradiated DNA at 77 K. 1. Total radical yields. Radiat Res 135:146-154

Chapter 5

Mechanisms of DNA Strand Breakage

MECHANISMS OF DNA STRAND BREAKAGE BY IONIZING RADIATION

J. R. Milligan, C. L. Limoli, C. C. L. Wu, J. Y.-Y. Ng and J. F. Ward,
Department of Radiology
University of California at San Diego,
9500 Gilman Drive
La Jolla, CA 92093-0610, USA

Introduction

1.1. Hydroxyl radical damage.

The effects of added scavengers on cellular radiosensitivity reveal that the hydroxyl radical is the major species responsible for the indirect effect (Chapman *et al.*, 1973). In consequence, several groups have investigated the reactivity of DNA with the hydroxyl radical. The reaction is mechanistically complex since many products have been identified (Fuciarelli *et al.*, 1989). One reaction product that is easily detected both *in vitro* and *in vivo* is the single strand break (SSB). The interaction between DNA and the hydroxyl radical is also kinetically complicated, as might be expected for a diffusion controlled reaction of a macromolecule (Schwartz, 1971). This non-homogeneous kinetic behavior results in a variation of the second order rate constant with hydroxyl radical lifetime. We have accumulated sufficient data to quantify this variation over a wide range of radical lifetimes.

1.2. Damage clustering.

Although characterizing the reactivity of the hydroxyl radical with DNA is important for an understanding of the radiation chemistry of DNA and ultimately of radiation biology, the microdosimetry of energy deposition by ionizing radiation suggests that attention should be paid to the possibility of multiple radical attacks from single energy deposition events. The initial distribution of chemically reactive species produced by ionizing radiation is a consequence of the spatial energy deposition pattern. When these reactive species are formed *in vivo*, they react within *ca.* 3 nm of their site of origin. This addition dispersion of the deposited energy will be reflected in the distribution of chemically modified sites in cellular macromolecules, most importantly DNA.

For low LET radiation (γ-rays, fast electrons), a significant fraction of the energy deposition events are sufficiently large to form at least 2 ion pairs (Pimblott *et al.*, 1990). This localization of reactive species (hydroxyl radicals from water radiolysis with some additional contribution of radical cations from the direct effect) would then be expected to result in the production of clustered damage, or so-called locally multiply damaged sites (LMDS). Various track structure simulations make testable predictions about the degree of clustering that might be expected with different forms of ionizing radiation. Although there is some kinetic evidence for the existence of damage clustering (Prise *et al.*, 1993), more direct experimental evidence from product analysis has proven elusive. We reasoned that linked DNA molecules may offer an alternative experimental approach to evaluate the LMDS model. Under conditions where hydroxyl radical diffusion distances are relatively large (*i.e.* low scavenger concentration), two linked plasmids should behave towards ionizing radiation as if they were unlinked. As the radical diffusion distance is decreased to the very small distance of 3 nm (or 10 base pairs) which applies within a cell (*i.e.* a high scavenger concentration), the LMDS hypothesis predicts that the products of irradiation should reflect the intimate association of the linked rings. Therefore we prepared a concatenated DNA substrate consisting of two singly linked supercoiled double stranded rings of unequal size (2817 and 1065 bp), and compared product distributions after irradiation under various conditions.

Experimental

Plasmid isolation and purification techniques, irradiation and gel electrophoresis conditions, video imaging analysis, and calculation of SSB yields have all been described previously (Milligan *et al.*, 1993). Irradiated catenane samples were assayed by 2-dimensional agarose gel electrophoresis, with 2×10^{-6} mol dm^{-3} chloroquine present during electrophoresis in the second dimension (which was perpendicular to the first dimension). These gels were otherwise loaded, run, stained, and imaged in an identical manner to the conventional 1-dimensional gels. $\gamma\delta$-Resolvase was prepared by induction with nalidixic acid of an overexpression vector (Mott *et al.*, 1985). The enzyme was purified by a modification of the original procedure (Reed, 1981), with the gel filtration step delayed until after ion exchange chromatography, and the substitution of CM-Sepharose CL-6B and Sephacryl S-100 for BioRex 70 and Bio-Gel P-60 respectively. The DNA catenane was prepared by incubation of plasmid pDH82 with γδ-resolvase (Reed, 1981). The two forms of radiation discussed here are ^{137}Cs γ-rays (662 keV) and ^{244}Cm α-particles (5.8 MeV).

Results

3.1. Hydroxyl radical damage.

Plotted in Figure 1 are the SSB yields determined at various combinations of DNA and DMSO concentrations. In the absence of any added scavenger, the SSB yield (in units of μmol J^{-1}) remains essentially constant over a wide range of DNA concentrations. In the presence of a given DMSO concentration, the SSB yield is proportional to the DNA concentration. These observations can be understood in terms of a competition between DMSO and the DNA for the hydroxyl radical. Both

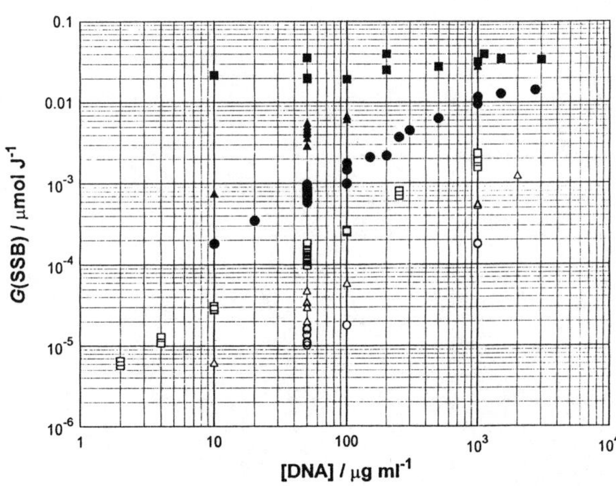

Figure 1. Effect of DNA concentration on the SSB yield in the presence of no added scavenger (■), 10^{-4} mol dm^{-3} DMSO (▲), 10^{-3} mol dm^{-3} DMSO (●), 10^{-2} mol dm^{-3} DMSO (□), 10^{-1} mol dm^{-3} DMSO (△), or 1 mol dm^{-3} DMSO (○).

DNA and DMSO react with the hydroxyl radical by second order reactions. This competition can be analyzed kinetically in the following manner:

$$\frac{1}{G(\mathrm{SSB})} = \frac{1}{G_0(\mathrm{SSB})} + \frac{k_2^{\mathrm{DMSO}}[\mathrm{DMSO}]}{G_0(\mathrm{SSB})k_2^{\mathrm{DNA}}[\mathrm{DNA}]} \qquad (1)$$

where k_2^{DNA} and k_2^{DMSO} represent, respectively, the second order rate constants for reactions for the hydroxyl radical with DNA and DMSO, and G(SSB) and G_0(SSB) represent, respectively, the SSB yields in the presence and absence of added DMSO. If this simple description is correct, then plotting $G(\mathrm{SSB})^{-1}$ against either $[\mathrm{DNA}]^{-1}$ or [DMSO] should give a straight line. In practice the former plot is linear while the latter is not (Milligan *et al.*, 1993), a result observed by others (van Rijn *et al.*, 1985; Udovicic *et al.*, 1991). In Figure 2, the variation of SSB yield with DNA concentration (assuming 325 g mol^{-1} per nucleotide) in 10^{-2} mol dm^{-3} DMSO using a double reciprocal plot according to equation (*1*). From the slope and intercept of the straight line fitted to the data, the second order rate constant for the reaction of

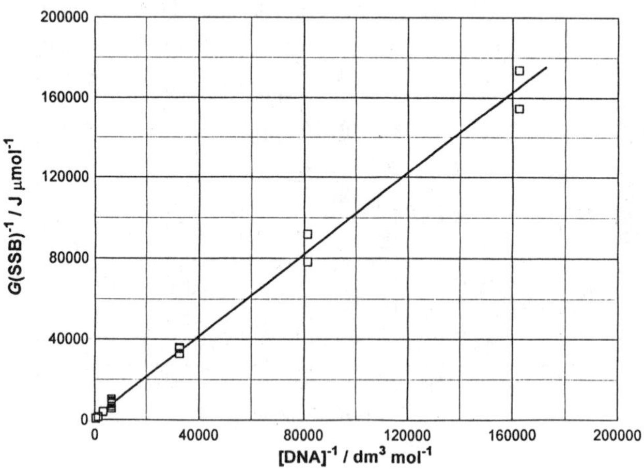

Figure 2. Effect of DNA concentration on the SSB yield. Double reciprocal plot according to competition kinetics (see equation 1) of the data at 10^{-2} mol dm^{-3} DMSO plotted in Figure 1. The data are fitted with a least mean square straight line of the form y=mx+c.

the hydroxyl radical with DNA can be calculated as 1.76×10^9 $dm^3\ mol^{-1}\ s^{-1}$. Similar plots (not shown) can be constructed from the data in Figure 1, one for each DMSO concentration.

The mean lifetimes of the hydroxyl radical at these DMSO concentrations are given by $(k_2^{DMSO}[DMSO])^{-1}$. Therefore it is possible to estimate the apparent second order rate constant as a function of the hydroxyl radical lifetime. This variation is plotted in Figure 3, where shorter lifetimes clearly result in a higher second order rate constant.

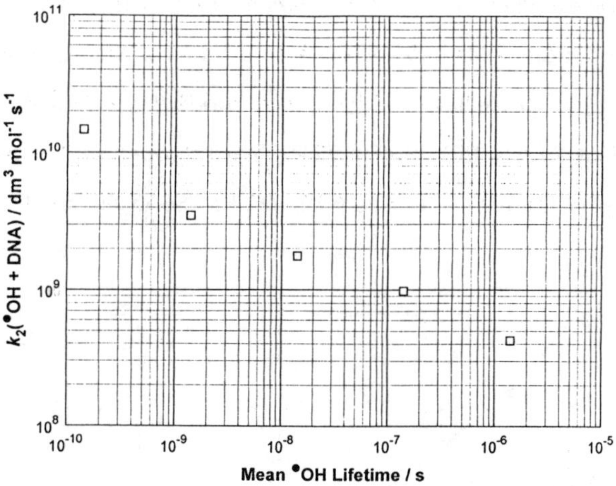

Figure 3. Effect of hydroxyl radical lifetime on the second order rate constant for reaction of the hydroxyl radical with DNA (see Results).

3.2. Catenane product distribution.

Plasmid pDH82 (3882 bp) contains two recognition sites for the enzyme γδ-resolvase, separated by a distance of 1065 bp. Incubation of pDH82 with purified γδ-resolvase forms after an intramolecular recombination event (Grindley, 1993) a catenane consisting of two double stranded DNA molecules of lengths 2817 bp and 1065 bp singly linked with one another (analogous to two links of a chain). The structure of the catenane was verified by restriction mapping (not shown). Irradiation of aqueous solutions of the catenane produces many products, resulting from the introduction of SSB and/or a DSB damage in one and/or other of the rings. While many of these can be separated by conventional gel electrophoresis, it was necessary to resort to 2-dimensional gel electrophoresis to resolve them all. The products were identified by comparison with authentic specimens prepared by incubation of the catenane with appropriate restriction endonucleases, and from the fact that only species containing at least one supercoiled ring lie off the diagonal axis of a 2-dimensional gel (not shown).

Figures 4 and 5 show the product distribution after γ-irradiation of the catenane in the presence of 1 mol dm^{-3} DMSO. Yield dose plots for the starting material and the three major products are plotted in Figure 4. The remaining fraction of the original catenane decreases exponentially with dose, as expected from experiments with regular plasmids. The major product is the supercoiled small ring linked with the open circle large ring (*i.e.* the product of a SSB in the large ring) which is formed in a *ca.* 4 fold higher yield than the open circle small ring linked with the supercoiled large ring (*i.e.* the product of a SSB in the small ring). The

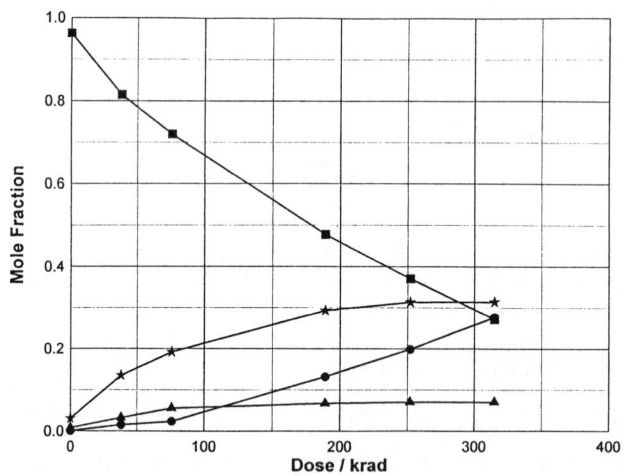

Figure 4. Yield dose plot for linked ring products of catenane irradiation. The catenane was γ-irradiated to the doses indicated in the presence of 1 mol dm^{-3} DMSO and the product yields determined after 2-dimensional gel electrophoresis. Shown here are the supercoiled small ring linked with the supercoiled large ring (ie. the original catenane, ■), open circle small ring linked with supercoiled large ring (▲), supercoiled small ring linked with open circle large ring (★), and open circle small ring linked with open circle large ring (●). Adjacent points are connected by straight lines.

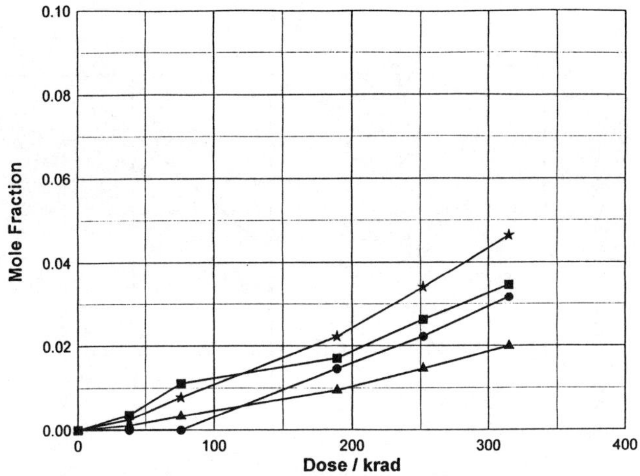

Figure 5. Yield dose plot for single ring products of catenane irradiation. Irradiation conditions were identical to those in Figure 4. Shown here are the open circle small ring (★), the linear small ring (▲), the open circle large ring (●), and the linearized large ring (■). Adjacent points are joined by straight lines.

open circle small ring linked with the open circle large ring (*i.e.* the product of a SSB in both rings) is initially formed very inefficiently, but at higher doses becomes the dominant product (not shown).

Yield dose plots for the minor products (those comprising single rings) are shown in Figure 5. The open circle small ring is formed in *ca.* 2 fold the yield of the linear small ring, while in the large ring distribution of damage between open circle and linear forms is approximately equal. Neither the supercoiled large ring nor the supercoiled small ring were detected.

Figure 6 shows the formation with dose of the linearized large ring under six different irradiation conditions. The abscissa has been normalized by dividing the radiation dose by the D_0 for the loss of the original catenane under the irradiation conditions (determined from a semi log plot, not shown). The linearized large ring is formed in *ca.* 10 fold higher yield by α-particle irradiation than by γ-irradiation. For both forms of radiation, the yield of the linearized large ring increases with DMSO concentration.

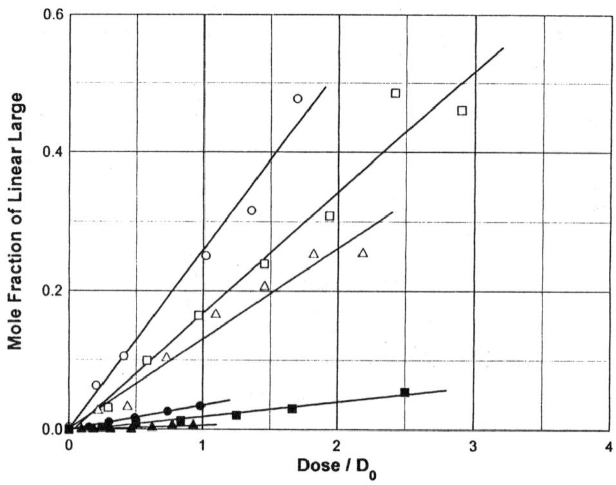

Figure 6. Yield dose plot for the formation of the linearized large ring as a function of radiation type and scavenger concentration. Irradiation conditions were in the presence of 10^{-4} mol dm^{-3} (triangle), 10^{-2} mol dm^{-3} (square), or 1 mol dm^{-3} (circle) DMSO with either α-particles (open symbols) or γ-rays (filled symbols). The data sets are fitted with least mean square straight lines of the form y=mx+c.

Discussion

4.1. Hydroxyl radical damage.

Models more sophisticated than simple homogeneous competition kinetics have been developed to describe the interaction of the hydroxyl radical with DNA (Mark *et al.*, 1989; LaVerne *et al.*, 1993). Quantitative results such as those plotted in Figure 3 should help to calibrate and further refine these models.

4.2. Catenane irradiation.

The difference in target size between the large and small rings (2817 bp *cf.* 1065 bp) is probably responsible for the difference in yield between the products of SSB damage to the large and small rings that is evident in Figure 4. The rate of formation with dose of the product bearing a SSB in both rings increases with increasing dose. This induction requirement suggests that the product is formed mainly by a sequential process involving the introduction of a SSB into an undamaged ring linked with a ring already containing a previously formed SSB. In contrast, since supercoiled single rings were not observed, the formation of the linearized large ring is almost always accompanied by a SSB or DSB in the small ring (and *vice versa*), and we assume this to be largely the result of a multiple •OH attack in a small region of the catenane (*i.e.* a LMDS).

The SSB yields under these irradiation conditions have previously been shown to be proportional to the •OH yield (Jones *et al.*, 1993; LaVerne *et al.*, 1993). Figure 6 (where the X-axis is equivalent to numbers of SSBs per catenane) reveals that higher LET radiation leads to a higher yield of the linearized large ring product. One explanation for this observation is that the greater •OH density in the region of the α-particle track (compared with that for γ-ray spurs) leads to an increased frequency of multiple radical attacks (and therefore possibly multiple breaks) relative to single radical events (*i.e.* SSBs). A similar mechanism may be invoked to account for the effect of DMSO concentration on the yield of the linearized large ring, since higher scavenger concentrations result in shorter •OH lifetimes and diffusion distances, and therefore in higher •OH densities.

Acknowledgements

We are grateful to Peter Geiduschek (Department of Biology, University of California at San Diego) and Nigel Grindley (Department of Molecular Biophysics and Biochemistry, Yale University) for samples of plasmids pDH82 and pGH286 respectively. Supported by PHS grant CA46295 and DOE grant DEFG0393ER61592.

References

Chapman JD, Reuvers AP, Borsa JP, Greenstock CL (1973) Chemical radioprotection and radiosensitization of mammalian cells growing *in vitro*. Radiat Res 56: 291-306

Fuciarelli AF, Wegher BJ, Gajewski E, Dizdaroglu M, Blakely WF (1989) Quantitative measurement of radiation induced base products in DNA using gas chromatography mass spectrometry. Radiat Res 119: 219-231

Grindley NDF (1993) Analysis of a nucleoprotein complex: the synaptosome of γδ-resolvase. Science 262: 738-740

Jones GDD, Milligan JR, Ward JF, Aguilera JA (1993) Yield of strand breaks as a function of scavenger concentration and LET for SV40 irradiated with ^4He ions. Radiat Res 136: 190-19

LaVerne JA, Pimblott SM (1993) Yields of hydroxyl radical and hydrated electron scavenging reactions in aqueous solutions of biological interest. Radiat Res 135: 16-23

Mark F, Becker U, Herak JN, Schulte-Frohlinde D (1989) Radiolysis of DNA in aqueous solution in the presence of a scavenger: a kinetic model based on a nonhomogeneous reaction of OH radicals with DNA molecules of spherical or cylindrical shape. Radiat Environ Biophys 28: 81-99

Milligan JR, Aguilera JA, Ward JF (1993) Variation of single strand break yield with scavenger concentration for plasmid DNA irradiated in aqueous solution. Radiat Res 133: 151-157

Mott JE, Grant RA, Ho Y-S, Platt T (1985) Maximizing gene expression from plasmid vectors containing the λ P_L promoter: strategies for overproducing transcription factor ρ. Proc Natl Acad Sci USA, 82: 88-92

Pimblott SM, LaVerne JA, Mozumder A, Green NJB (1990) Structure of electron tracks in water. 1. Distribution of energy deposition events. J Phys Chem 94: 488-495

Prise KM, Davies S, Michael BD (1993) Evidence for induction of DNA double strand breaks at paired radical sites. Radiat Res 134: 102-106

Reed RR (1981) Transposon mediated site-specific recombination: a defined *in vitro* system. Cell 25: 713-719

van Rijn K, Mayer T, Blok J, Verberne JB, Loman H (1985) Reaction rate of OH radicals with ϕx174 DNA: influence of salt and scavenger. Int J Radiat Biol 47: 309-317

Schwarz HA (1971) Some applications of time-dependent rate constant theory to radiation chemistry. J Chem Phys, 55: 3647-3650

Udovidib L, Mark F, Bothe E, Schulte-Frohlinde D (1991) Non-homogeneous kinetics in the competition of single stranded calf thymus DNA and low molecular weight scavengers for OH radicals: a comparison of experimental data and theoretical models. Int J Radiat Biol 59: 677-697

THE DISTRIBUTION OF SINGLE STRAND BREAKAGE AT GUANINE INITIATED BY 193nm LIGHT IS DIFFERENT FOR SINGLE AND DOUBLE STRANDED DNA

T. Melvin, M.A. Plumb, S.W. Botchway, P. O'Neill
Medical Research Council Radiobiology Unit
Chilton, Didcot, Oxon,
OX11 0RD, UK

A.W. Parker
Central Laser Facility
Rutherford Appleton Lab.
Chilton, Didcot, Oxon
OX11 0QX, UUK

Introduction

Non repairable DNA double strand breaks have been implicated in cellular inactivation by ionising radiation. (Blöcher (1988), Jenner et al. (1993), Frankenberg-Schwager (1990)). Single strand breaks occur in DNA following exposure to ionising radiation and are a result of the interaction of both water radicals and direct deposition of energy within DNA (direct effects). Although cellular double strand breakage occurs via a single track event, the majority of double strand breakage reactions occur as a result of two separate radical events. Mechanistic studies of strand breakage in DNA are therefore normally limited to the study of single strand breakage processes (O'Neill and Fielden (1993)).

Possible strand breakage pathways in DNA by ionising radiation are numerous, but can be subdivided into two types: (i) formation of the primary radical at the sugar molecule, for instance by hydrogen abstraction by a hydroxyl radical at the sugar phosphate backbone; or (ii) via first formation of a nucleic acid base radical, whereby the radical damage is transferred to the sugar molecule involving a hydrogen abstraction reaction. In both cases a number of different sugar radicals are formed, but one which finally decomposes to cleave the sugar/phosphate backbone is a prerequisite for strand breakage.

Studies of hydroxyl radical initiated strand breakage using polydeoxyribo- and polyribo-nucleotides have indicated that the yields and modes of base to sugar radical transfer processes vary significantly between the nucleic acid bases (Adinarayana et al. (1988), Bothe et al. (1982), Jones and O'Neill (1990, 1991) Melvin et al. (1994a)). This differential reactivity is a function of the ability of the different nucleic acid base radicals to abstract a hydrogen atom from a neighbouring sugar which then leads to strand breakage. Strand breakage studies of the direct ionisation of the nucleic acid bases within DNA or polynucleotides likewise indicate a differential reactivity of the products of each of the nucleic acid base radical cations toward strand breakage (Görner et al. (1992), Wolf et al. (1993)). In addition, our recent results indicate that the distribution of strand breaks from guanine radical cations in double stranded DNA is dependent on the neighbouring nucleic acid base sequence (Melvin et al. (1994b)).

It is difficult to study the species produced upon direct energy deposition in DNA by ionising radiation in an aqueous environment due to the overwhelming yield of water radicals. One approach to study the chemistry of the radical species produced upon ionisation of DNA in aqueous solution at room temperature has utilised high energy laser light (193nm-266nm) (Candeias et al. (1992, 1993, 1994)); photo-irradiation results in the mono- or bi-photonic ionisation of the DNA and its components. Several groups have shown that 193nm light induces single strand breaks in DNA by a monophotonic process with a quantum yield of ~ 5×10^{-4} in the absence and presence of oxygen and alkali labile strand breaks are formed with a yield which is less than or equal to the yield of frank breaks (Görner (1994), Kochevar (1993), Gurzardyan (1993)). The quantum yields for photoionisation of DNA and related biomolecules are significantly greater than the quantum yield for single strand breaks and are in the range 0.02-0.1 based upon the yields of hydrated electrons produced (Candeias et al. (1992), Görner (1994)). The proportion of each of the constituents which are photoionised within DNA can be estimated by using values for the extinction coefficient and the quantum yield for photoionisation of the DNA constituents; the contribution of guanine has been estimated to be ~60% for calf thymus DNA. Migration of the oxidative damage from the other nucleic acid bases will occur to a certain extent thus increasing the relative proportion of radical cations formed at the more easily oxidised site, guanine.

With 248nm light under conditions of biphotonic excitation, guanine has previously been identified (Croke et al. (1988)) as the major site for single strand breakage in DNA possibly involving singlet oxygen (Devasagayam et al. (1991), Tuite et al. (1993), Schneider et al. (1990)). Electron spin resonance studies on biphotonic ionisation of DNA in frozen aqueous solution at 77K have also identified one electron oxidised guanine (with some contribution from $A^{\cdot+}$) as the major oxidised species (Malone el al. (1994)). It was argued that hole transfer occurs from the other bases to guanine (Malone el al. (1994)). With 193nm light, the possibility of monophotonic ionisation of the phosphate group of DNA becomes possible (Candeias and Steenken (1992)) and has been postulated (Gut et al. (1993)) to account for the formation of photo-induced strand breaks with 193nm light. In contrast, the studies of Görner generally postulate the one electron oxidised bases as the precursors to single strand breakage (Görner (1994, 1993)). Our previous observations, obtained from a study of the site of strand breakage for double stranded DNA by 193nm light (Melvin et al. (1994b)), indeed indicated that (i) the strand breakage results from the product(s) of the guanine radical cation, (ii) the strand breakage is not a result of a singlet oxygen reaction with guanine (singlet oxygen could be a very minor product which may be formed by quenching of albeit short-lived triplet states (Bishop et al. (1994)), and (iii) superoxide anion, a biproduct formed from the reaction of the photo-ejected electron with oxygen, is not the cause of the non-random strand breakage of double stranded DNA. Strand breakage from the other nucleic acid base radical cations would have to be at least an order of magnitude greater than that from guanine in order to be detected.

Although a selectivity for strand breakage at the guanine sites for double stranded DNA has been found, an enhanced yield of strand breakage at specific guanines occurs with a guanine nearest neighbour and a reduced yield of strand breakage with a 3'-thymine as a nearest neighbour (Melvin (1994)). It was unrealistic to postulate a mechanism for strand breakage in light of the fact that the yield of strand breakage is so low. However, we question here whether the selectivity and distribution of strand breakage at guanine found for photoionisation of double stranded DNA with 193 nm light is also found for a single stranded DNA sample.

Materials and Methods:

5' End-labelled DNA probe was generated as described from the pSKMIP-1α plasmid, containing murine MIP-1α cytokine cDNA (Melvin et al. (1994b)).

pSKMIP-1α plasmid DNA was first linearised by digestion with EcoR1, and the 5' phosphate groups were removed by treatment with calf intestinal phosphase. The DNA (10 μg) was 5' end-labelled with T4 polynucleotide kinase in the presence of 50 μCi of [γ-^{32}P]-ATP (3000 Ci/mmol, Amersham), and then digested with Xhol. The DNA was resolved by 1.5% (w/v) low melting point agarose (LMP, GIBCO/BRL) gel electrophoresis and the 5' end-labelled cDNA plasmid insert purified and ethanol precipitated.

Insert DNA (~1 μg) was dissolved in 200 μl of an aqueous solution containing 0.05 μg/μl of yeast tRNA to give a stock solution containing ~5 ng/μl DNA and an A_{193} value of ~3 (A_{260} = 1). Prior to photolysis, 5 μl of the stock solution was diluted in 70 μl of a 0.01 μg/μl aqueous solution of yeast tRNA to give a A_{193}~1 and a DNA concentration of ~0.3 ng/μl.

The DNA sample was denatured in a boiling water bath for 5 mins, then snap frozen at -55°C. Just before irradiation, the sample was thawed and immediately transferred into the photolysis cuvette. Both the control and the denatured DNA solutions were irradiated (193nm) in a quartz cuvette (1mm pathlength, 10mm window) for 2 seconds using the ArF excimer laser at a pulse repetition rate of 10Hz, and subsequently ethanol precipitated. The nucleic acid pellet was washed with 70% ethanol, and dried under vacuum, dissolved in sample loading buffer (95% formamide, 0.01% bromphenol blue, 0.01% Xylene cyanol), and resolved by denaturing 6% (w/v) polyacrylamide/50% (w/v) urea gel electrophoresis and autoradiography (Melvin (1994b)).

Results

To determine whether 193 nm photolysis of single stranded DNA results in a similar specificity for strand breakage as for double stranded DNA, a known double stranded DNA sequence was 5' end-labelled on one strand, photolysed in both the single stranded and double stranded form under aerobic conditions and the product resolved by denaturing sequencing gel electrophoresis. Non-random cleavage of the DNA was observed for both samples, as depicted in the autoradiograph shown in Figure 1.

As previously reported and shown again here, the photoinduced cleavage of double stranded DNA shows a selectivity associated with guanine. However, neighbouring guanines generally result in bands of high intensity whereas guanine bands with a neighbouring 3'-thymine are less intense. Strand breakage at cytosine, the complimentary base to guanine, is not significant; this suggests that the guanine radical formed does not cause strand breakage on the complimentary strand or that a product of the strand breakage at guanine does not cause a second strand breakage by an interstrand process. The cleavage for the single stranded sample likewise shows non-random cleavage with almost exclusive strand breakage at guanine. However the distribution effects of neighbouring bases observed for the double stranded sample are not apparent and strand breakage occurs at almost every guanine site to a comparable degree. An exception is that no strand breakage is detected in either the single or double stranded DNA for a guanine site where both neighbours are thymine.

Figure 2 shows a magnified region of the autoradiogram, the differences in the yields of single strand breakage at the guanine sides for the control and denatured samples are as follows (and indicated in the figure):

A. The guanine sites with 3'-thymine molecules, where strand breakage is only just apparent in the double stranded sample, can be easily observed in the denatured sample.

B. Strand breakage at guanine sites is more pronounced for neighbouring guanine sites as compared to other guanine cleavage sites for the double stranded sample as compared to the denatured sample.

C. Although strand breakage occurs at most other guanine sites for double stranded DNA there are a few exceptions (those without a 3'-thymine neighbour, e.g. guanine with two adenine neighbours), but strand breakage occurs at these sites for denatured samples.

In short, strand breakage of the denatured DNA occurs at almost every guanine site to a comparable degree, whereas for the double stranded sample there is a variation in the amount of strand breakage at each of the guanine sites and this is dependent on the neighbouring base sequence.

Figure 1

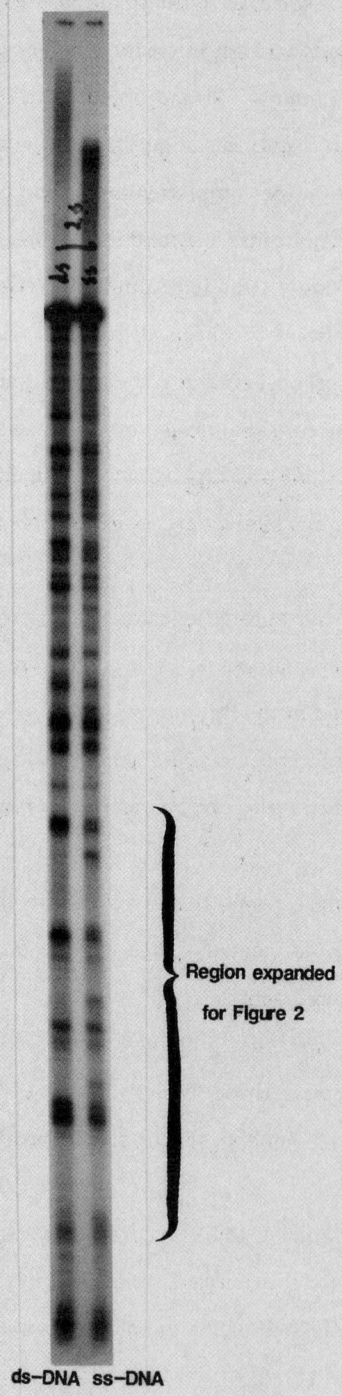

Figure 2

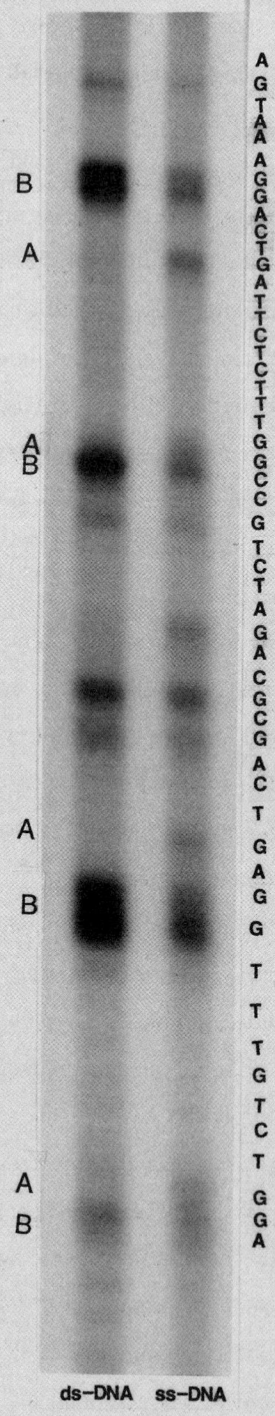

Discussion

Photolysis of DNA in aqueous solution with 193nm light has been shown to cause 'frank' strand breakage reactions in DNA with a ϕ_{ssb} of $4x10^{-4}$, strand breakage is therefore a minor product (2%) of the photoionisation 'event' (Görner (1993, 1994)). The strand breakage in both single and double stranded DNA occurs predominantly at guanine, indicating that the guanine radical cation or a subsequent product is the precursor. Using the extinction coefficients of the bases and the sugar phosphate (Candeias et al. (1993)), it is estimated for DNA that the majority of the photons (>97%) at 193nm are absorbed by the base moieties and the majority of radical cations formed by this photolysis event occur or migrate to guanine.

There are two known pathways (but there could be others) for the decay of the guanine radical cation; the major pathway for loss of the radical cation of guanine is deprotonation, but a very slow minor hydration pathway could also take place to yield a 8-hydroxyguanine radical and ultimately 8-hydroxyguanine (Kasai et al. (1992)). This product was undetected in product studies by Steenken et al. (1993). Although not detected, strand breakage is a very minor process, so the yield of precursors to strand breakage are expected to be very low. 8-Hydroxyguanine, once formed, is not known to be a precursor to strand breakage (Devasagayam et al. (1991), Schneider et al. (1990)). At present we do not know which guanine radical/diamagnetic product(s) is responsible for the single strand breakage by 193nm light.

Although strand breakage occurs at guanine, these experiments indicate that the distribution of strand breakage from the product(s) of the guanine radical cation are different in the single stranded DNA as compared to the double stranded form. This might indicate base stacking interactions in the double stranded DNA might facilitate strand breakage by the guanine radical product neighbouring a second guanine molecule; conformation overlap of a guanine radical with the neighbouring 3'-sugar may facilitate strand breakage. The overlap of the guanine radical with the neighbouring 3'-sugar when the 3'-neighbour is thymine may be such that strand breakage is not predominant.

The strandedness of the sample might also have an additional effect on the radical chemistry of the guanine radical cation. It has been postulated by Steenken (1993) that the complimentary cytosine base is a proton acceptor from the guanine radical cation in double stranded DNA. Although proton transfer is expected to cause localisation of the damage with respect to migration along the strand, the chemistry of the radical cation could also be modified with respect to that observed for the single stranded form. Our hypothesis is as follows, since the pKa values for the guanine radical cation and cytosine are similar (Steenken (1993)), the proton might be expected to exist in equilibrium between the two strands. A result of this is that the rate of the deprotonation reaction would be reduced in the double stranded sample as compared to the single stranded sample so that the radical cation in the double stranded sample is longer lived. Therefore reactions with this radical cation, such as possible hydration reactions, could be more significant for the double stranded DNA sample as compared to the single stranded sample. The fact that these experiments show that strand breakage still takes place for the single stranded sample at guanine might indicate that the equilibration of the proton between the guanine radical cation and cytosine plays an insignificant role in the fate of the products of the guanine radical cation towards strand breakage.

The guanine radical cation product causes strand breakage and is expected to cause unaltered base release as a result of formation of radicals on a common sugar molecule. The yield for unaltered base release is an order of magnitude greater for adenine, thymine and cytosine than for guanine (Görner (1994)), which is in contrast to strand breakage studies on double stranded DNA (Melvin et al. (1994b)), where with a 3'-thymine group strand breakage is reduced and with a neighbouring guanine the yield of strand breakage is at least an order of magnitude greater. If base release and strand breakage have a single unique common pathway, then the yield of guanine released would have been expected to be greater. This would indicate that several types of sugar damage occur leading to: (i) strand breakage and base release, or (ii) only base release or (iii) only strand breakage.

In summary, we report a difference in distribution of strand breakage for products of the guanine radical between single and double stranded samples of DNA after photoirradiation with 193nm light. A mechanistic interpretation of these observations cannot, at present, be made because strand breakage is a minor fate of the guanine radical cation. Time resolved light scattering measurements should help determine which guanine species is/are responsible for strand breakage and if strand breakage is frank or a result of labile diamagnetic products.

Acknowledgements

We thank the ERSERC Rutherford Appleton Laboratory (Laser Support Facility) for their assistance. This study was partly funded by the Leukaemia Research Fund and the CEC (Contract N° F13P-CT920027)

References

Adinarayana, A.M., Bothe E., Schulte-Frohlinde, D. (1988) Hydroxyl radical induced strand break formation in single-stranded polynucleotides and single stranded DNA in aqueous solution as measured by light scattering and by conductivity. *Int. J. Rad. Biol.*, **54**, 723-727.

Bishop, S.M., Malone, M., Phillips, D., Parker, A.W. and Symons, M.C.R. (1994) Singlet oxygen sensitisation by excited state DNA. *J. Chem. Soc. Chem. Commun.*, 871-872.

Bothe E., Schulte-Frohlinde, D. (1982) Release of K^+ and H^+ from poly U in aqueous solution upon γ- and electron irradiation. *Zeitschrift fur Naturforschung*, **37c**, 1191-1204.

Blöcher, D. (1988) DNA double-strand break repair determines the RBE of α-particles. *Int. J. Radiat. Biol.*, **54**, 761-771.

Candeias, L.P. and Steenken, S. (1993) Electron transfer in di(deoxy)nucleoside phosphates in aqueous solution: rapid migration of oxidative damage (via adenine) to guanine. *J. Am. Chem. Soc.* **115**, 2437-2440.

Candeias, L.P. and Steenken, S. (1992) Ionization of purine nucleosides and nucleotides and their components by 193 nm laser photolysis in aqueous solution. *J. Am. Chem. Soc.* **114**, 699-704.

Candeias, L.P., O'Neill, P., Jones, G.D.D. and Steenken, S. (1992) Ionization of polynucleotides and DNA in aqueous solution by 193 nm pulsed laser light: identification of base-derived radicals. *Int. J. Radiat. Biol.* **61**, 15-20.

Croke, D.T., Blau, W., OhUigin, C, Kelly, J.M. and McConnell, D.J. (1988) Photolysis of phosphodiester bonds in plasmid DNA by hight intensity UV laser irradiation. *Photochem. Photobiol.*, **47**, 527-536.

Devasagayam, T.P.A., Steenken, S., Obendorf, M.S.W., Schulz, W.A. and Sies, H. (1991) Formation of 8-hydroxy(deoxy)guanosine and generation of strand breaks at guanine residues in DNA by singlet oxygen. *Biochemistry*, **30**, 6283-6289.

Frankenberg-Schwager, M. (1990) Induction, repair and biological relevance of radiation-induced DNA lesions in eukaryotic cells. *Radiat. Environ. Biophys*, **29**, 273-292.

Görner, H, Wala, M., Schulte-Frohlinde, D. (1992) Strand breakage in poly (C), poly(A) and single and double stranded DNA induced by nanosecond laser excitation at 193nm. *Photochem. Photobiol.*, **55**, 173-184.

Görner, H. (1994) Photochemistry of DNA and related biomolecules: quantum yields and consequences of photoionisation. *J. Photochem. Photobiol. B: Biol*, (In Press).

Gurzadyan, G.G. and Görner, H. (1994) Photolesions in DNA upon 193 nm excitation. *Photochem. Photobiol.*, **58**, 71-80.

Gut, I.G., Farmer, R., Huang, R.C. and Kochevar, I.E. (1993) Upper excited state photochemistry of DNA. *Photochem. Photobiol.*, **58**, 313-317.

Kasai, H., Yamaizumi, Z., Berger, M., Cadet, J. (1992) Photosensitized formation of 7,8-dihydro-8-oxo-2'-deoxyguanosine (8-hydroxy-2'-deoxyguanosine) in DNA by riboflavin - a nonsinglet oxygen mediated reaction. *J. Am. Chem. Soc.*, **114**, 9692-9694.

Jenner, T.J., de Lara, C.M., O'Neill, P. and Stevens, D.L. (1993) Induction and rejoining of DNA double-strand breaks in V79-4 mammalian cells following γ- and α-irradiation. *Int. J. Radiat. Biol.*, **64**, 265-273.

Jones, G.D.D. and O'Neill, P. (1990) Kinetics of radiation induced strand breakage in polynucleotides in the presence of oxygen: a time resolved light-scattering study. *Int. J. Rad. Biol.*, **57**, 1123-1137.

Jones, G.D.D. and O'Neill, P. (1991) Kinetics of radiation induced strand breakage in pyrimidine polynucleotides in the presence and absence of oxygen: a time resolved light-scattering study. *Int. J. Rad. Biol.*, **59**, 1127-1145.

Kochevar, I.E. and Buckley, L.A. (1990) Photochemistry of DNA using 193 nm Excimer Laser Radiation. *Photochem. Photobiol.*, **51**, 527-532.

Malone, M.E., Symons, M.C.R. and Parker, A.W. (1994) DNA in glasses at 77K: high energy ionizing radiation versus UV electron ejection. *Int. J. Radiat. Biol.* (In Press).

Melvin, T., O'Neill, P., Jones, G.G.D. (1994a) The influence of nucleic acid base composition on radiation-induced strand breakage in single stranded DNA: a time resolved study. *Int. J. Radiat. Biol.* (In Press).

Melvin, T., Plumb, M.A., Botchway S.W., O'Neill, P. and Parker, A.W. (1994b) 193nm Light induces single strand breakage of DNA predominantly at guanine. *Photochem. Photobiol.*, (submitted for publication).

O'Neill, P. and Fielden, E.M. (1993) Primary free radical processes in DNA. *Adv. Radiat. Biol.*, **17**, 53-120.

Schneider, J.E., Price, S., Maidt, L., Gutteridge, J.M.C. and Floyd, R.A. (1990) Methylene blue plus light mediates 8-hydroxy-2'-deoxyguanosine formation in DNA preferentially over strand breaks. *Nucleic Acids Res.* **18**, 631-635.

Steenken, S. (1993) Electron transfer induced acidity basicity and reactivity changes of purine and pyrimidine bases, consequences of redox processes for DNA base pairs. *Free Rad. Res. Comms.* **16**, 349-379.

Tuite, E.M. and Kelly, J.M. (1993) Photochemical interactions of methylene blue and analogues with DNA and other biological substrates. *J. Photochem. Photobiol. B: Biol.* **21**, 103-124.

Wolf, P., Jones, G.D.D., Candeias, L.P. and O'Neill, P. (1993) Induction of strand breaks in polyribonucleotides and DNA by the sulphate radical anion: role of electron loss centres as precursors of strand breakage. *Int. J. Radiat. Biol.* **64**, 7-18.

A COMPARISON OF THE FAST KINETICS OF RADIATION-INDUCED DNA LESIONS FROM LOW AND HIGH LET RADIATIONS: CLUSTERED DAMAGE AND DEPENDENCE ON OER

KM Prise, M Folkard and BD Michael
Cancer Research Campaign Gray Laboratory
PO BOX 100, Mount Vernon Hospital
Northwood, Middlesex, HA6 2JR, UK

Introduction

The effectiveness of radiations of differing quality is related to the patterns of energy deposition within the ionizing particle tracks as they cross cellular DNA. For processes dependent on oxygen, involving free radicals, the localisation of energy deposition will lead to clustering of radicals close to, or on, the DNA and have important consequences for the reactivity of these radicals with modifiers and the subsequent production of molecular damage on the DNA. Several models have been suggested for the production of DNA double-strand breaks (dsb), which are thought to be important determinants of cellular response to radiation. These involve; the production of two ssb, by separate radiation tracks on opposite strands, close enough together to form a dsb (Boon et al. 1984); a radical transfer mechanism from one strand to the other to produce a dsb from a single track (Siddiqi and Bothe 1987) and dsb produced from clustered lesions or locally multiply damaged sites (LMDS), by a single track involving, in the case of a dsb, at least two radicals, one on either strand of the DNA (Ward 1981). Our previous studies, with low LET electrons, have measured the chemical repair of the oxygen free radical precursors which lead to the production of strand breaks and suggested that clustered damage may be important in the production of dsb. Using the gas-explosion system we can measure these reactions, occurring over millisecond timescales in a variety of biological systems. In isolated plasmid DNA, in the presence of glutathione (GSH), we have compared the chemical repair rates of the precursors of ssb and dsb and found differences related to the underlying multiplicity of the free radicals produced. In particular, we have found that for low LET radiation the free radical precursor of a dsb shows kinetics consistent with a paired radical (i.e. one radical on either strand of the DNA) whereas a ssb precursor consists of a single oxygen-dependent radical (Prise et al. 1993). In this study, the oxygen enhancement ratio measured for these lesions was also different with the OER for dsb being approximately twice that for ssb precursors. We have also observed similar differences in the chemical repair rates of ssb and dsb precursors in intact cells. (Prise et al. 1992). Based on these observations we have developed a multi-radical model which makes predictions regarding the effectiveness of high LET radiations and the relationship to the oxygen enhancement ratio (Michael et al. 1991; Michael et al. 1994 and Michael et al. in preparation). It suggests that the decrease in OER measured with increasing LET may be due to increased clustering of radicals on the DNA. This could be an alternative explanation to the oxygen-in-the-track and the interacting radicals hypotheses which have previously been suggested. In this paper we will present a preliminary experimental approach to measure the effect of radiation quality expressed as differences in the chemical repair kinetics of breaks precursors produced by high and low LET radiations as a means to observe increased clustering of radicals on the DNA.

Materials and Methods

Plasmid pBR322 DNA (4363 bp) was purified from *E. coli* HB101 using the alkaline lysis method and concentrated using Amicon microconcentrators into PTE buffer (10 mM sodium phosphate, 0.15 M $NaClO_4$, 1mM tris, 0.1 mM EDTA, pH 7.0). Samples were mixed with a 2X concentrated GSH solution and spread on to 13mm diameter, 0.01μm pore size polycarbonate membrane filters. Filters were placed under hypoxic conditions and irradiated with a 100 μs pulse of protons or deuterons from a 4.2 MeV Van de Graaff accelerator or 600 keV electrons from a Febetron source. At preset times immediately before or after the pulse, the sample was exposed to a 50 p.s.i. shot of oxygen using the gas explosion system (Prise et al. 1993; Michael et al. 1994). DNA was recovered and analysed using agarose gel-electrophoresis and the relative amounts of supercoiled, relaxed and linear forms of DNA determined using densitometry of the ethidium bromide stained gels.

Results and Discussion

Figure 1 shows typical timecourses for the chemical repair of damage produced by a 800 Gy 100 μs pulse of protons with an LET of 13 keV/μm leading to the production of ssb (upper panel) and dsb (lower panel) in the presence of 30 mM GSH. As the time of the oxygen shot is delayed after the pulse of protons, more of the oxygen-dependent damage can be chemically repaired by GSH. The repair kinetics have been fitted as previous described (Fahey et al. 1991) to yield first order

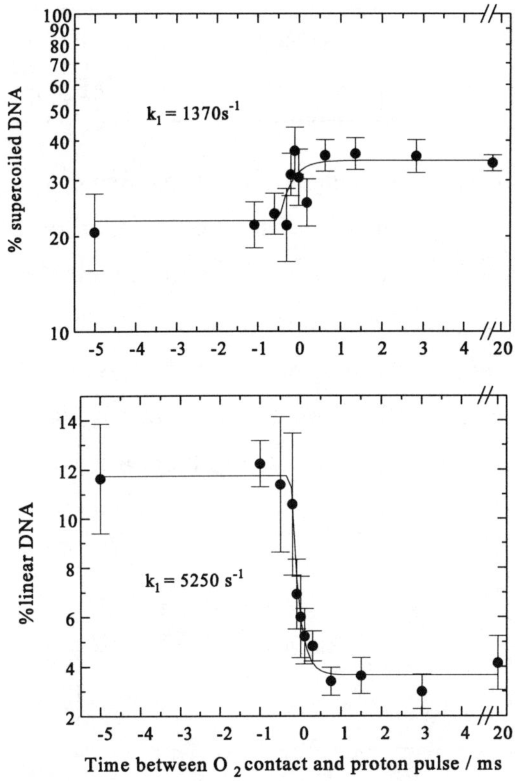

Figure 1. Induction of ssb (upper panel) and dsb (lower panel) in pBR322 DNA by 100 μs 800 Gy pulses of protons at 13 keV μm^{-1}, showing chemical repair by 30mM GSH as a function of oxygen contact time

repair rate of 1370 s^{-1} for ssb and 5250 s^{-1} for dsb. Under these conditions the OER for ssb was 1.7 and 2.8 for dsb. A range of GSH concentrations has been studied with both 13 keV/μm protons and 65 keV/μm deuterons under these conditions. From the first-order rates measured second order rates have been calculated and are shown in table 1 along with the values obtained for 600 keV electrons. The second

Table 1.

Radiation/LET	2nd order rate constant k_2 / $M^{-1}s^{-1}$	
	SSB	DSB
Electrons ~1keVμm^{-1}	$0.6 \times 10^5 \pm 0.2 \times 10^5$	$1.1 \times 10^5 \pm 0.2 \times 10^5$
Protons 13keVμm^{-1}	$1.1 \times 10^5 \pm 0.3 \times 10^5$	$2.0 \times 10^5 \pm 0.5 \times 10^5$
Deuterons 65keVμm^{-1}	$0.2 \times 10^5 \pm 0.3 \times 10^5$	$0.9 \times 10^5 \pm 0.3 \times 10^5$

order rates show an initial increase in rate with LET followed by a decrease at higher LET and at all LETs the chemical repair rates are higher for dsb than for ssb. By comparing the effectivness of the various radiations at induction of ssb and dsb when the oxygen shot is given 5 ms before irradiation or 20 ms after the radiation pulse oxygen enhancement ratios can be calculated and these are shown in figure 2. Here the expected decrease in OER with increasing LET is observed although, as seen with the rate measurements, the values are all higher for dsb in comparison to ssb.

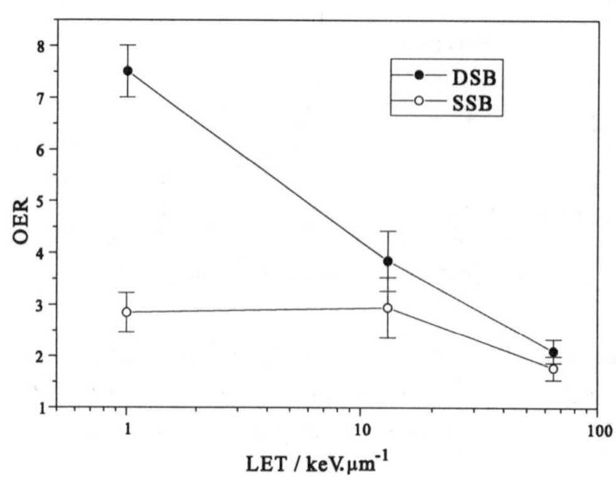

Figure 2. OER versus LET for ssb and dsb induction in pBR322 DNA irradiated in the presence of GSH.

Overall the data show two distinct differences between radiations of differing LET for both lesions, a decrease in OER with increasing LET and a change in chemical repair rates of oxygen dependent damage with increasing LET. Our modelling studies have predicted that a decrease in oxygen enhancement ratio with increasing LET could be a consequence of increased radical clustering (Michael et al. 1991; 1994). Under conditions of clustering of radicals close together over both strands the probability that even under hypoxia at least one break could be produced on either strand leading to a dsb would increase. This is dependent on the ability of a proportion of radicals to go on to produce breaks under hypoxia via a hypoxic fixation reaction. The chemical repair kinetics show an initial increase in repair rate followed by a decrease in rate. With an increase in radical multiplicity on the DNA it can be envisaged that chemical repair of a lesion consisting of several oxygen-dependent radicals will have to take place via a multistep process, leading to an apparent delay in repair and a more complex kinetic profile. Our data so far have yet to conclusively detect such a change in the kinetic profile of repair but the apparent decrease in rate at the highest LET so far studied could be consistent with repair of a spectrum of radical cluster sizes (Michael et al. 1994). The increase in rate from electrons to protons may be indicative of increased interaction of radicals as radical density increases or of the production of molecular species such as hydrogen peroxide reacting with radical sites and altering their reactivity. Overall the changes in chemical repair kinetics will probably represent a balance between these two processes.

Summary

1) Using the plasmid DNA system and fast kinetic techniques involving particle beams, the free radical processes leading to strand break production can be studied with a view to understanding the effectiveness of radiations of differing LETs and the mechanisms of dsb formation.

2) The chemical repair rates of the oxygen dependent precursors of strand breaks initially increase with LET then appear to slow, which may be indicative of a competing process at high LET as well as increased radical multiplicity.

3) With increasing LET the OER for DNA breaks decreases but the OER for dsb remains higher than those for ssb, as predicted by a multiple radical model.

4) The chemical repair rates of dsb precursors appear to remain faster than those of ssb precursors, as LET increases, supporting the concept of distinct precursor radical sites on either strand.

Acknowledgements

The authors are grateful to Boris Vojnovic and Keith Allen for the development of the pulsed proton facility and Anne Marie Whelan and James Latham for technical assistance. This work is supported by the Cancer Research Campaign and a grant from Radiation Protection Research Action of the European Community (contract F13P-CT92-0053).

References

Boon, P.J., Cullis, P.M., Symons, M.C.R., and Wren, B.W., (1984), Effects of ionizing radiation on deoxyribonucleic acid and related systems. Part 1. The role of oxygen. Journal of the Chemical Society, Perkins Transactions 2: 1393-1399.

Fahey, R.C., Prise, K.M., Stratford, M.R.L., Watfa, R.R., and Michael, B.D., (1991), Rates for repair of pBR322 DNA radicals by thiols as measured by the gas explosion technique: evidence that counter-ion condensation and co-ion depletion are significant at physiological ionic strength. International Journal of Radiation Biology, 59: 901-917.

Michael, B.D., Prise, K.M., Folkard, M., and Fahey, R.C. Radical multiplicity in radiation-induced DNA strand breaks: Implications for their chemical modification and effects. in NATO Advanced Research Workshop on Cell Biology: Early effects of radiation on DNA. 1991. Springer-Verlag, Berlin.

Michael, B.D, Prise, K.M, Folkard, M., and Vojnovic, B., (1994), An investigation of clustered radiation induced lesions in DNA. Radiation Protection Dosimetry, 52: 277-281.

Prise, K.M., Davies, S., and Michael, B.D., (1992), Measurement of fast chemical repair of radical precursors of DNA damage and lethal lesions in irradiated Chinese hamster V79 cells. International Journal of Radiation Biology, 61: 721-729.

Prise, K.M, Davies, S., and Michael, B.D, (1993), Evidence for induction of DNA double-strand breaks at paired radical sites. Radiation Research, 134: 102-106.

Siddiqi, M.A. and Bothe, E., (1987), Single- and double-strand break formation in DNA irradiated in aqueous solution: dependence on dose and OH radical scavenger concentration. Radiation Research, 112: 449-463.

Ward, J.F., (1981), Some biochemical consequences of the spatial distribution of ionizing radiation-produced free radicals. Radiation Research, 86: 185-195.

RADIATION-INDUCED STRAND-BREAKS AND DNA-PROTEIN CROSSLINKS DEPEND PREDOMINANTLY ON THE DOSE, OXYGEN CONCENTRATION AND REPAIR TIME.

K. T. Wheeler, Y. Miyagi, and H. Zhang
Department of Radiation Oncology
Bowman Gray School of Medicine
Medical Center Blvd.
Winston-Salem, NC 27157

Introduction

It is generally accepted today that, 1) some human tumors do not contain hypoxic cells, 2) some human tumors contain a few hypoxic cells, but these hypoxic cells do not limit the radiocurability of these tumors, and 3) some human tumors contain many hypoxic cells that limit their radiocurability (Gatenby *et al.* 1988, Stone *et al.* 1993). Consequently, development of methodologies for measuring the hypoxic fraction or fractional hypoxic volume of solid tumors is important for selecting those patients who would potentially benefit from therapeutic strategies aimed at eliminating radioresistant hypoxic cells.

It has been known for many years that the DNA damage produced by ionizing radiation depends upon the oxygen concentration around the DNA. For example, the number of DNA strand-breaks (SBs) formed per unit dose decreases at low oxygen concentrations, and the number of DNA-protein crosslinks formed per unit dose increases at low oxygen concentrations (Koch and Painter 1975, Fornace and Little 1977). If radiation-induced SBs and DPCs are to be useful for detecting and/or quantifying hypoxic cells in solid tumors, the formation of these lesions must depend predominantly on the oxygen concentration around the DNA. All other physical, biological, and physiological factors must either be controllable or have little influence on the assay used to measure these lesions. The following is a summary of our recent experiments to determine if the radiation-induced SBs and DPCs measured by alkaline elution may be used to estimate the hypoxic fraction or fractional hypoxic volume of solid tumors.

Materials and Methods

Subcutaneous 9L tumors: A complete description of the procedures for the *in vivo* experiments has been published (Zhang and Wheeler 1993).

Maintenance of cell lines: 9L rat brain tumor cells were grown in EMEM medium supplemented with glutamine, penicillin-streptomycin, and 10% newborn calf serum as previously described (Wheeler *et al.* 1992). Suspension cultures of CEM and CEM/VM-1 leukemia cells were grown in EMEM medium supplemented with glutamine, penicillin-streptomycin, and 10% fetal calf serum as previously described (Fernandes *et al.* 1990). Mouse mammary adenocarcinoma cells, line 66, were grown in Waymouth's medium supplemented with glutamine, penicillin-streptomycin, 6% horse serum, 6% newborn calf serum, and 3% fetal calf serum as previously described (Wallen *et al.* 1984). BG-1 human ovarian carcinoma cells were grown in McCoy's 5A medium supplemented with glutamine, nonessential amino acids, penicillin-streptomycin, and 10% fetal calf serum as previously described (Fujiwara *et al.* 1989).

9L cell experiments: A complete description of the procedures for the *in vitro* 9L experiments has been published (Zhang and Wheeler 1993). Briefly, 9L cells were grown for 2-3 days over the entire surface of 60 mm glass petri dishes in 3 ml of HEPES buffered medium (pH 7.3). After removal of 2 ml of medium, the petri dishes were placed in leak proof aluminum chambers and deoxygenated as described by Koch (1984). The cells were carefully maintained at 37°C before, during, and after irradiation. Under these conditions, O_2 consumption by both the cells and the radiation create an O_2 gradient from the center of the dish to the periphery. Thus, the O_2 concentration in the gas phase is not related to the O_2 concentration in the medium solely by its solubility at 37°C.

CEM and CEM/VM-1 experiments: CEM and CEM/VM-1 cells were harvested 2 days after seeding them in 175 cm^2 flasks. The cells were resuspended in medium containing HEPES buffer (pH 7.3) at a concentration of 1 x 10^6/ml, and then 1 ml of the cell suspension was placed in each glass petri dish that had been prewarmed to 37°C. The rest of the procedure was identical to that described for 9L cells, except that DNA repair was stopped by putting the petri dishes on ice and adding 5 ml of ice-cold Saline A.

66 P and Q experiments: Cells from stock cultures were seeded in glass petri dishes.

P cells were grown for 2-3 days in 3 ml of HEPES buffered medium; Q cells were grown for 7-10 days in 5 ml of bicarbonate buffered medium. Before P cells were placed in the chambers, all the medium was removed, and 0.8 ml was added back. Before Q cells were placed in the chambers, 0.1 ml of HEPES buffer was added, then all the medium was removed, and 0.8 ml was added back. The rest of the procedure was identical to that described for 9L cells.

BG-1 experiments: BG-1 cells were seeded in glass petri dishes and handled exactly as described above for the 66 P cells.

Irradiations: All cells were placed in aluminum chambers and irradiated with Cs-137 γ-rays at the prescribed O_2 concentrations and temperature. The dose rate determined by Fricke dosimetry was 28 Gy/min for the *in vitro* experiments and 12.7 Gy/min for the *in vivo* experiments. After irradiation, the *in vitro* cells were returned to the incubator to allow about one half of the SB damage to be repaired (9L, 5 min; CEM & CEM/VM-1, 4 min; 66 P, 3 min; 66 Q, 6.5 min; BG-1, 3.3 min). After removal from the incubator, the cells were trypsinized at 4°C, and their DNA assayed using the alkaline elution technique. For the *in vivo* experiments, subcutaneous 9L tumors were excised from rats that were killed 2 min after the irradiation began. Single cell suspensions were prepared at 4°C, and the DNA damage remaining assayed using the alkaline elution technique.

DNA assay and data analysis: A complete description of the DNA assay and the data analysis has been published (Zhang and Wheeler 1993).

Results and Discussion

The feasibility of estimating the hypoxic fraction or fractional hypoxic volume of tumors by measuring the radiation-induced SBs and DPCs remaining after approximately one half-time of SB repair using the alkaline elution technique is demonstrated in Figure 1. In these experiments, cells isolated from subcutaneous 9L tumors that had been irradiated either in air-breathing or dead rats and then allowed to repair about one half of their SB damage, were mixed in various proportions to simulate tumor samples having different hypoxic fractions. Only the percentage of the DNA retained on the filter after ≈ 24 ml of elution without proteinase K in the lysis solution, a parameter proportional to sum of the SBs and DPCs remaining after about one half-time of SB repair, was a linear function of the simulated hypoxic fraction (Zhang and Wheeler 1993). The slope of the regression equation was independent of dose from 2 Gy to 15 Gy, but the intercept depended on dose (Fig. 1). It is important to note that a 10% hypoxic fraction could be measured after a clinically relevant dose of 2 Gy (Fig. 1). Identical results were obtained when the experiments were performed with cerebellar neurons (Zhang and Wheeler 1993), so it appeared that the linear relationship between the percentage of the DNA

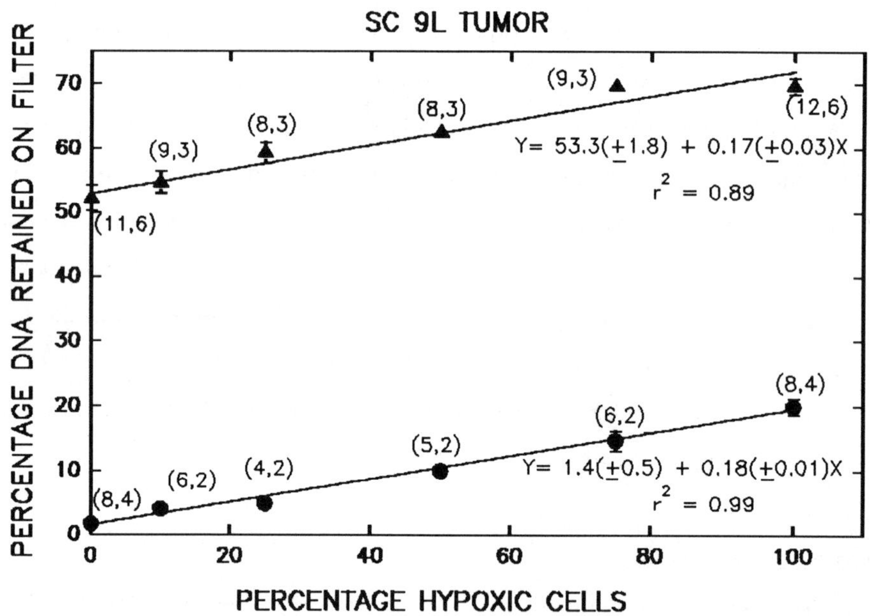

Figure 1. Percentage of DNA retained on the filter after ≈ 24 ml of elution without proteinase K in the lysis solution as a function of the percentage of irradiated hypoxic cells in the sample. Cells were isolated after about one-half time of SB repair from subcutaneous 9L tumors that were irradiated in air-breathing or dead rats, and then mixed to form the simulated hypoxic fractions. Tumors were irradiated with: ▲ 2 Gy; ● 15 Gy. The first number in the parentheses represents the number of channels averaged to get the point; the second number in the parentheses represents the number of independent experiments from which the channels came. Error bars are ± 1 SEM; where not shown, they lie within the point.

retained on the filter and the percentage of hypoxic cells in the irradiated sample might be independent of many biological and physiological factors.

The general applicability of this approach for estimating the hypoxic fraction or fractional hypoxic volume of animal and human tumors requires that the number of radiation-induced SBs and DPCs remaining after one half-time of SB repair be essentially independent of the biological properties of the tumor cells (e.g. cell type, proliferative status, protein content and composition at the nuclear matrix, *etc.*) and the physiological environment (e.g. temperature, pH, *etc.*) surrounding the tumor cells. When the repair time was adjusted to accommodate the slower SB repair at lower temperatures (e.g. Wheeler *et al.* 1992), pH, nutrient depletion, and/or growth phase did not appreciably alter the formation of SBs or DPCs in oxygenated 9L cells (Zhang and Wheeler 1994). Similar experiments have not yet been performed using cells irradiated under intermediate or fully hypoxic conditions.

To test the hypothesis that the number of radiation-induced SBs and DPCs remaining after one half-time of SB repair is relatively independent of the biological properties of the tumor cells, 9L rat brain tumor cells, proliferative (P) and quiescent (Q) mouse mammary adenocarcinoma cells (line 66), and parental CEM or mutant CEM/VM-1 human leukemia cells that differ in their topoisomerase II content, activity, and nuclear matrix binding properties were irradiated at 37°C with 15 Gy at various oxygen concentrations. After irradiation, each cell type was incubated at 37°C to complete one half-time of SB repair prior to assaying the DNA damage remaining. The oxygen dependence of SB and DPC formation was then determined using the alkaline elution assay, with and without proteinase K in the lysis solution.

In general, the sigmoidal shape of the oxygen dependence curves for SBs was identical, and the K_m values were virtually identical for all of the cells studied (Table 1). The sigmoidal shape of the oxygen dependence curves for DPCs was similar for all of the cells studied; however, it was essentially the inverse or mirror image of the oxygen dependence curves for SBs. Although the K_m values for DPC formation were identical for 9L, 66 P, and 66 Q cells (Table 1), they were about three times higher for both the CEM and the CEM/VM-1 cell lines. Thus, with the exception of the formation of DPCs in CEM and CEM/VM-1 cells, the oxygen dependence of SB and DPC formation in these irradiated cells appeared to be relatively independent of the biological factors that are most likely to complicate the use of this assay for quantifying the hypoxic fraction in solid tumors.

Table 1. Summary of the K_m values for radiation-induced DNA strand-breaks and DNA-protein crosslinks in several mammalian tumor cells[a]

Cells	DNA strand-breaks (mM of O_2)[b]	DNA-protein crosslinks (mM of O_2)[b]
9L	≈ 0.032	≈ 0.020
66 P	≈ 0.027	≈ 0.020
66 Q	≈ 0.027	≈ 0.020
CEM	≈ 0.027	≈ 0.064
CEM/VM-1	≈ 0.027	≈ 0.064

a. All K_m values were determined from data collected after one half-time of DNA strand-break repair at 37°C.

b. All K_m values are reported in terms of the oxygen concentration measured in the gas phase.

The human leukemia cell experiments are important for a number of reasons. For example, 1) these cells are of human origin rather than rodent, 2) they grow only in suspension culture, and 3) their topoisomerase II enzymes have very different characteristics; a factor that could have a profound influence on the formation of DPCs in irradiated cells. When CEM leukemia cells were treated with VM-26, a resistant line, CEM/VM-1, was isolated that contained 3.2 fold less immunoreactive topo II in their nuclear matrices (Fernandes *et al.* 1990). The topo II in CEM/VM-1 cells had 6-7 fold less decatenating activity than the topo II in CEM cells, and > 80% of this topo II could be removed from the nuclear matrices by a 1.8 M NaCl solution. Thus, the topo II in CEM and CEM/VM-1 cells differed substantially in quantity, activity, and nuclear matrix binding properties.

Although the K_m value for DPC formation was about three times greater in the CEM and CEM/VM-1 cells than the K_m value found for the other cell lines, this higher K_m value is not related to the topo II content, activity, or nuclear matrix binding properties, because the oxygen dependence curves and the K_m values are identical for both CEM and CEM/VM-1 cells. Moreover, this higher K_m value had no effect on the percentage of the DNA retained on the filter after ≈ 24 ml of elution when irradiated oxic cells were assayed (Table 2, 20-21% O_2), and it had a minimal effect when irradiated hypoxic cells were assayed (Table 2, 0.9-1.6% O_2).

Table 2. Percentage of the DNA retained on the filter after ≈ 24 ml of elution without proteinase K in the lysis solution as function of the O_2 concentration in the gas phase.[a, b]

Cells	0.9-1.6% O_2	20-21% O_2
9L	18.2 ± 4.3 (14)	1.6 ± 0.9 (13)
66 P	20.2 ± 2.9 (13)	3.0 ± 1.5 (11)
66 Q	18.1 ± 5.3 (7)	3.5 ± 1.7 (12)
BG-1	16.1 ± 3.0 (8)	1.9 ± 0.7 (6)
CEM	24.6 ± 3.3 (12)	1.5 ± 0.4 (8)
CEM/VM-1	31.0 ± 6.9 (13)	1.6 ± 0.5 (9)

a. All measurements were made after one half-time of strand-break repair.
b. Mean ± 1 SD. The values in the parentheses are the number of channels obtained from at least two independent experiments.

Preliminary experiments suggest that the slightly higher and more variable percentage of DNA retained on the filter for irradiated hypoxic CEM/VM-1 cells (Table 2) is likely caused by a slightly higher oxygen consumption rate for these human leukemia cells. A higher oxygen consumption rate would result in a lower oxygen concentration in the medium over CEM/VM-1 cells, even though the cell number in each dish and the oxygen concentration in the gas phase are identical to those in the experiments with other cell lines. Note that the BG-1 human ovarian carcinoma cells, whose oxygen consumption rate is similar to the 66 cells, have values for oxic and hypoxic cells in Table 2 that are identical to those found for the rodent cell lines. Future experiments using several animal tumor models will be required to determine if this experimental complication *in vitro* has an *in vivo* counterpart. Our unpublished data with 9L tumors and four normal tissues in rats suggest that no such complication exists *in vivo*.

In summary, our data suggest that it is feasible to estimate the hypoxic fraction or fractional hypoxic volume of solid tumors by measuring the percentage of DNA retained on polycarbonate filters after a given volume of alkaline elution without proteinase K in the lysis solution, if the measurements are made after one half-time of SB repair (Fig. 1). This parameter is proportional to the sum of the radiation-induced SBs and DPCs whose formation appears to be relatively independent of the physiological and biological factors that are often found in solid tumors (Tables 1 and 2). Although additonal work is required to completely validate this assay for estimating the hypoxic fraction or fractional hypoxic volume of solid tumors, the data at this time suggest that radiation-induced SBs and DPCs depend predominantly on the radiation dose, the repair time, and the oxygen concentration around the DNA. Within limits, the first two are controllable, so the latter should be measurable with this assay.

Acknowledgements

This work was supported by a grant from Forsyth County United Way and NIH grants RR-05404 and CA-45156. We wish to thank I. Al-Nabulsi and J. Anderson for technical support and D. Cantrell for preparation of this manuscript.

References

Fernandes DJ, Danks MK, Beck WT (1990) Decreased nuclear matrix DNA topoisomerase II in human leukemia cells resistant to VM-26 and m-AMSA. Biochemistry 29:4235-4241
Fornace AJ, Little JB (1977) DNA crosslinking induced by x-rays and chemical agents.

Biochim Biophys Acta 477:343-355
Fujiwara K, Modest EJ, Welander CE, Wallen CA (1989) Cytotoxic interactions of heat and an ether lipid analogue in human ovarian carcinoma cells. Cancer Res. 49: 6285-6289
Gatenby RA, Kessler HB, Rosenblum JS, Coia LR, Moldofsky PJ, Hartz WH, Broder GJ (1988) Oxygen distribution in squamous cell carcinoma metastases and its relationship to outcome of radiation therapy. Int J Radiat Oncol Biol Phys 14:831-838
Koch CJ (1984) A thin-film culturing technique allowing rapid gas-liquid equilibration (6 sec) with no toxicity to mammalian cells. Radiat Res 97:434-442
Koch CJ, Painter RB (1975) The effect of extreme hypoxia on the repair of DNA single-strand breaks in mammalian cells. Radiat Res 64:256-269
Stone HB, Brown JM, Phillips TL, Sutherland RM (1993) Oxygen in human tumors: Correlations between methods of measurement and response to therapy. Radiat Res 136:422-434
Wallen CA, Higashikubo R, Dethlefsen LA (1984) Murine mammary tumor cells in vitro. I. The development of a quiescent state. Cell Tissue Kinet 17:65-77
Wheeler KT, Hickman R, Nelson GB, Moore SK, Wallen CA (1992) Relationship between DNA damage, DNA repair, metabolic state and cell lethality. Radiat Environ Biophys 31:101-115
Zhang H, Wheeler KT (1993) Radiation-induced DNA damage in tumors and normal tissues I. Feasibility of estimating the hypoxic fraction. Radiat Res 136:77-88
Zhang H, Wheeler KT (1994) Radiaton-induced DNA damage in tumors and normal tissues. II. Influence of dose, residual DNA damage and physiological factors in oxygenated cells. Radiat Res 140:321-326

Chapter 6

LET Effects Influencing DNA Damage

LET-EFFECTS IN DNA

G. Kraft, G. Taucher-Scholz and J. Heilmann
Gesellschaft für Schwerionenforschung, Postfach 110552, D-64220 Darmstadt
Universität GH Kassel, D-34109 Kassel
Germany

Introduction

The use of heavy particles in radiobiological experiments provides a fundamental tool to study the influence of different ionisation densities and to prove the physical basis of models and theories. The knowledge of the interaction of high LET radiation (LET: linear energy transfer [keV/μm]) with biological matter is of great importance for the application of neutrons, protons and heavier ions in radiotherapy. It is also essential in radioprotection to estimate the risk in case of exposure to high LET radiation.

Beams of higher LET have been used for systematic experiments of cell inactivation and chromosome aberrations since the early and late sixties. First measurements on the induction of DNA damage were performed in the early seventies. Since then, the spectrum of ions used in radiobiological studies has widened substantially and ranges now from protons to uranium ions at energies up to more than 1 GeV/μ. Although the primary source of heavy ion beams, the Bevalac accelerator at Berkeley, has been shut down recently, high LET research is continued at the Unilac and SIS at the GSI in Darmstadt, the MMAC at NIRS in Chiba and to some extent at many smaller accelerators all over the world.

In this contribution, an introductory view on the physical properties of ions is given and the cellular response to high LET radiation is summarized. Then the measurements of strand break induction of DNA in solution and in intracellular DNA are reported and compared to cell survival. The possibility of changes in the quality of the lesions is discussed and finally the present status of model calculations in comparison to the experiments is given.

LET and track structure

The most important physical parameter in radiobiology is dose, i.e. the energy deposited in a mass unit. The dose deposited by particles can be calculated by the energy loss of a single particle multiplied by the number of particles traversing the volume. The numerical equivalent of the energy loss is termed linear energy transfer (LET) in radiobiology. In consequence, LET is used as the main parameter characterizing radiobiological action. According to the Bethe-Bloch equation, LET depends strongly on the atomic mass and on the energy of the particles (Fig. 1). These two quantities determine the electric charge and the interaction process of an ion passing through matter.

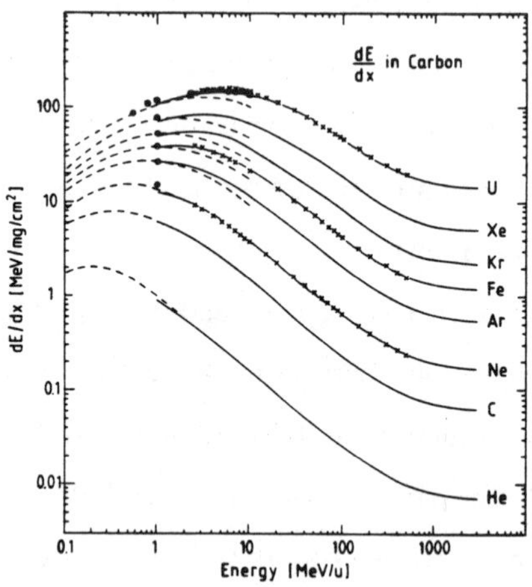

Figure 1: Energy loss dE/dx, the numerical equivalent of LET, as function of the specific energy for different accelerated particles from helium to uranium. For each Z a separate curve exists. Therefore, heavy ion action is always a function of both the atomic number and the velocity of the particles.

When heavy particles are accelerated to velocities comparable to the orbital velocities of their electrons, they strip off the outer shell electrons in their first collisions with target atoms. Due to this positive charge, the ions interact via electromagnetic forces mainly with the electrons of the target material. In these ion-electron collisions the major fraction of the energy loss of the primary particle is transferred into kinetic energy of the electron while a smaller fraction is necessary to overcome the binding energy in the ionisation process or is used for the excitation of the target atom.

According to the reaction, kinematics, the electrons are initially emitted mainly in the forward direction. The multiple scattering of these energetic electrons with other target electrons results in further ionisation processes and leads to a rapid homogeneous spatial distribution of the energy spectrum of the electrons. Multiple scattering and the energy loss of the electrons finally yield a radial dose distribution according to a $1/r^2$ law where r is the radial distance from the center of the track up to a maximal radius corresponding to the range of the most energetic electrons. According to these processes the diameter of the particle track is proportional to the velocity of the primary ion while the ionisation density, i.e. the number of ionisations per volume, is greatest at the maximum of the LET, corresponding to energies between 1 to 10 MeV/μ.

Changes in the radiobiological response are expected when the local ionisation density is increased to a level where multiple damage at the same site is likely to occur. Because the ionisation density does not depend in an unique manner from LET but from two parameters, atomic number and LET, a more complex pattern of the relative biological efficiency or action cross section from these parameters is expected.

LET effects in cells

The loss of colony forming ability as the most simple test of the DNA integrity in the complete genome of a cell has been measured in various experiments. Barendsen et al. (1963) compared the relative biological efficiency of deuterons and He ions with X-irradiation and found a maximum of RBE at a LET around 100 keV/μm. Todd (1967) predominantly used carbon beams and found a RBE maximum shifted to higher LET values between 200 and 300 keV/μm. This trend is confirmed by measurements with ions heavier than carbon as well as by more recent measurements using proton beams, where a RBE maximum was found for protons at 25 keV/μm (Belli et al. (1989)). In figure 2, RBE values from the final slope of the survival curves as well as inactivation cross sections (σ, [μm^2]) of V79 chinese hamster cells are given as a function of LET.

The complex dependence of the biological response to different particles of varying LET is clearly visible. For LET values between 5 and 100 keV/μm σ increases overproportional to LET, indicating a significant increase of RBE to values greater than one. At LET values greater than 100 keV/μm inactivation cross sections separate from the common curve and are characterized by individual σ-LET hooks indicating saturation effects of the biological response at elevated ionisation densities.

The same is true for the RBE-LET dependence. These saturation effects have been observed for the first time in experiments at the Unilac Darmstadt for various cell-types, like cultured mammalian cells, yeast and bacterial spores and named "Darmstadt hooks". However, the hook structure has been also confirmed for other biological endpoints like the incidence of chromosomal damage, the mutation-frequency in bacteria, yeast and mammalian cells or DNA strand breaks.

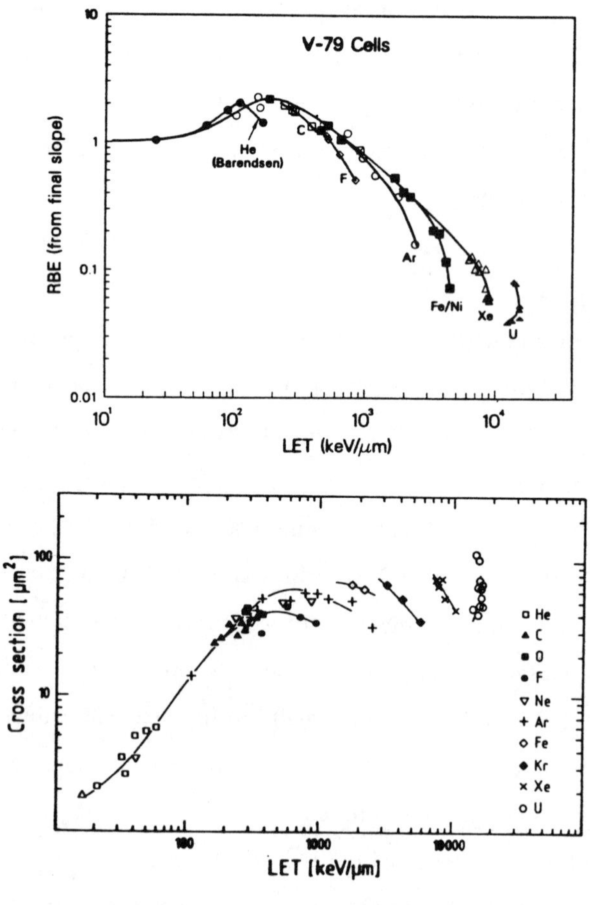

Figure 2: RBE and cross section for the inactivation of V79 cells as function of LET. For particles with Z > 2 one common curve exists up to about 100 keV/μm. For higher LET-values, functions separate for the different Z.

LET effects in DNA

Plasmid DNA

The first measurements of particle induced single (SSB) and double strand breaks (DSB) were performed by Christensen et al. (1972) using ϕX174 DNA in a highly protective buffer (Fig. 3). In this work, various ions from protons to argon at a fixed energy of 7.5 MeV/μ were used to achieve different LET-values. Similar to the inactivation measurements described in the last paragraph, an overproportional increase of the action cross section of DSB was observed with increasing LET. The cross section for SSB induction did not show such an overproportional increase, the RBE values always remained smaller than one.

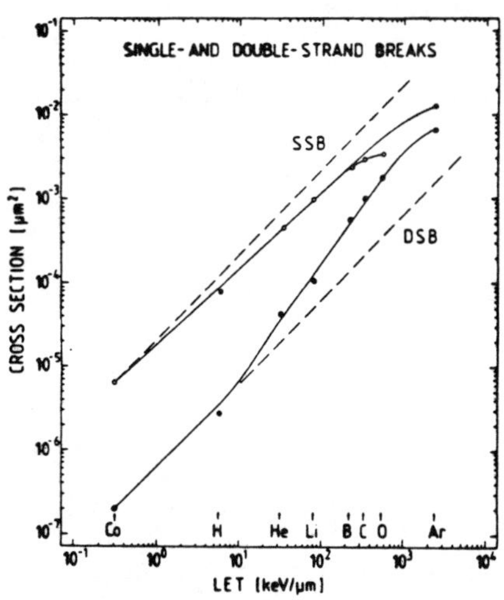

Figure 3: Cross sections for the induction of DNA single- and double strand breaks in (ϕX174 phage DNA. Irradiations were performed in highly protective solution. While the RBE for SSB-induction (open symbols) was always found smaller than unity, an overproportional increase of RBE was observed for DSB (closed symbols, data from Christensen et al. (1972)).

For SV40 DNA the induction of SSB and DSB has been measured by Roots et al. (1990) in a buffer system where the DNA is very labile (Fig. 4). Although SV40 is of comparable size to the ϕX174 plasmid, the absolute values of the cross section are two to three orders of magnitude higher than the Christensen data because of the different buffer system. In addition, no σ values were found that correspond to an RBE greater than one. Finally, in these measurements saturation effects (Darmstadt hooks) were observed for the very heavy ions from Ni to U ions at low energies.

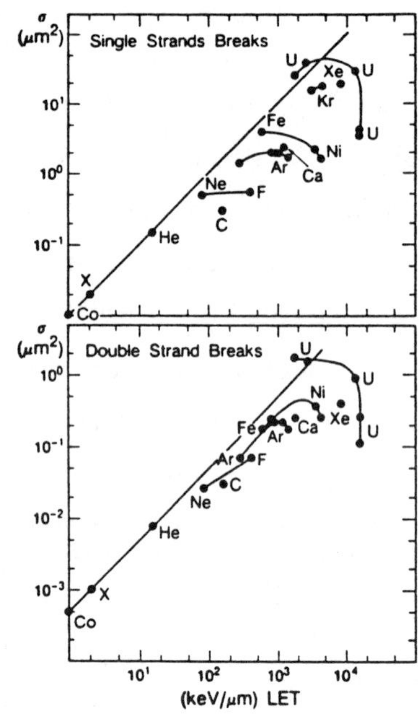

Figure 4: Data for SSB-induction (upper graph) and DSB-induction (lower graph) in SV40 viral DNA in dilute solution (Roots et al. (1990)). Individual curves (hooks) can be seen for the different particles in the high LET-range investigated here.

The most complete set of measurements on SSB and DSB induction has been published by Taucher-Scholz et al. (1992) for SV40 DNA in a Tris-EDTA buffer where DNA is stable. In this TE-buffer, cross sections of SSB and DSB follow in general the same pattern as described by Roots et al. (1990). For intermediate LET values of a few hundred keV/μm only a very slight elevation of the RBE for DSB induction to values greater than one was observed, while RBEs for SSB induction were always found smaller than one. In a recent paper by Taucher-Scholz et al. (1994), the influence of the chemical environment has been compared for various buffer systems (Fig 5). For the induction of single strand breaks, the RBE decreased in all buffer systems monotonously with LET. In a more sensitive buffer like TE the decrease with LET was steeper compared to the buffer system where the DNA is highly protected. For DSB induction, only in a highly protective buffer a RBE maximum was observed between 100 and 1000 keV/μm. In the sensitive buffer RBE decreased with LET in a similar way as for SSB induction.

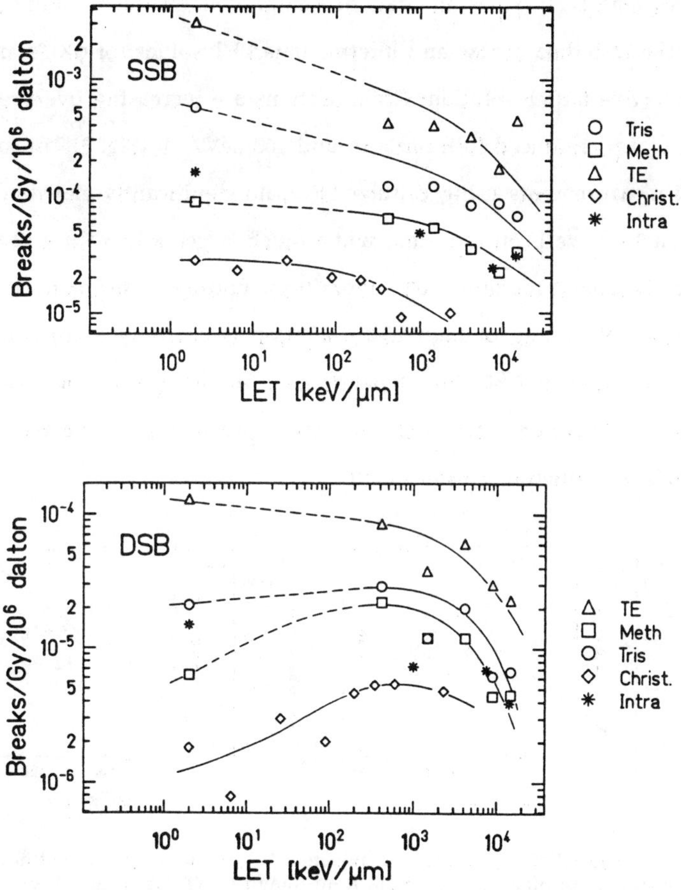

Figure 5: Biological efficiencies for the induction of SSB and DSB in SV40 viral DNA in buffer systems of different scavenging capacity (TE < Tris < Meth). Data from Christensen et al. (1972) are added for comparison as well as data measured using SV40-infected CV1-cells. With increasing scavenging capacity of the solution, efficiencies decrease. While RBE declines for SSB in any of the systems, elevated RBEs are found for DNA double strand breaks under radioprotective conditions.

DNA strand breaks in mammalian cells

For mammalian cells SSB and DSB induction has been measured in V79 cells by Kampf (1983) using the sedimentation technique. Heilmann et al. (1993) used alkaline unwinding and the filter elution techniques for strand break measurements in cultured human and bovine cells (Fig. 6). Weber and Flentje (1993) measured DSB-induction in V79 and Caski cells using neutral elution and pulse field electrophoresis. While the Heilmann data concentrate on the hook structure for high LET values, the data of Kampf et al. exhibit a similar behaviour as found for the DSB data at low and intermediate LET values for ϕX174 and SV40 in a highly protective buffer solution: Cross sections are increasing overproportional to LET forming a pronounced RBE peak around 100 keV/μm (Fig. 3). In contrast, in all other DSB measurements using cellular DNA no significantly elevated RBE values greater than 1.6 have been reported, while much larger RBE values (2-6) are found for cell inactivation at medium LET. From these findings it seems to be evident that the increased cell killing of high LET radiation is not only a consequence of the increased induction of DSB, but depends also on changes in the structure of the DNA lesions. Therefore, the post irradiation processing of the particle induced DSB has to be examined in greater detail.

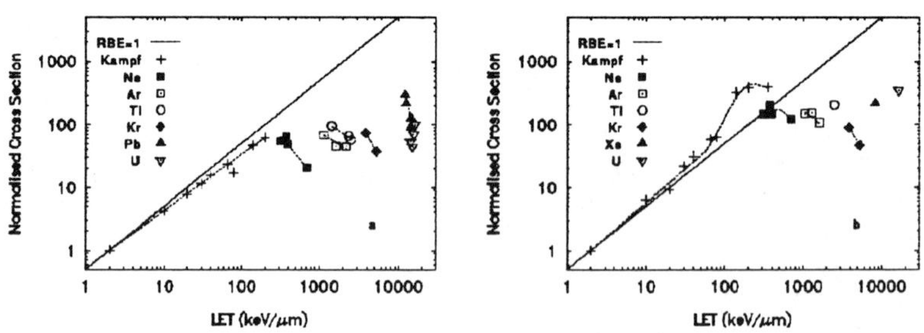

Figure 6: Normalized efficiencies for the induction of total strand breaks and double strand breaks as function of LET. Data from Kampf (1983) are included for comparison (crosses). In the high-LET range, individual curves exist for the different particles. The solid line corresponds to the efficiency of X-irradiation (RBE=l). In contrast to Kampf, efficiencies for double-strand break induction in the medium LET range are not significantly larger than unity.

Quality of the DNA lesions

From the comparison of the spatial structure of the particle track with the dirnension of the DNA as given in Fig. 7, it is evident that for high LET particles a qualitatively different structure of the DNA damage can be expected. The simultaneous emission of more than one electron at the site of DNA in nanometre-dimensions increases the probability for DSB production significantly. However, measurements showed that the induction of DSB is not largely elevated at increased LET values. This could be due to the detection methods, which are not able to

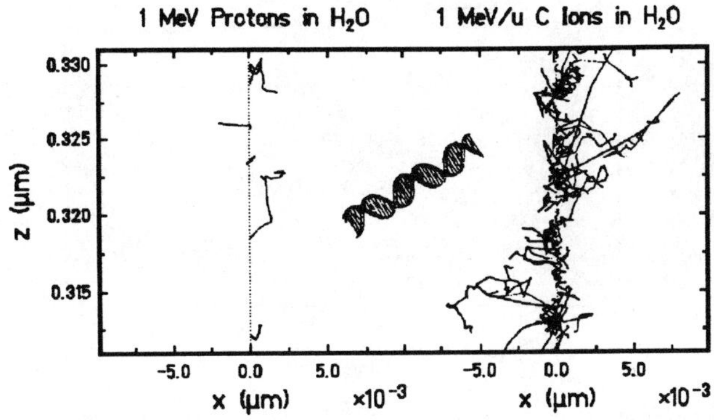

Figure 7: Calculated particle tracks of a proton and a carbon ion moving along their trajectories (dotted lines) in water. The δ-electrons produced in the primary ionisation event are scattered and generate ionisations far away from their origin. For comparison with the size of the track a DNA molecule of 25 base pairs is graphed schematically in the center of the figure.

differentiate between a simple double strand break and clustered damage restricted to a few base pairs.

Measurements of rejoining kinetics clearly indicated the different nature of particle induced lesions. Ritter et al. (1977) reported that a large fraction of particle induced double strand breaks are not rejoined. More recent measurements by Aufderheide et al. (1987), Heilmann et al. (1993), and Schneider et al. (1994) revealed, that rejoining after particle exposure depends to a large extent on the local ionisation density (Fig. 8). The same general result was found in experiments performed by Rydberg et al. (1994) and Löbrich et al. (1994) measuring rejoining and correct repair. Quantitative differences in the nature of the particle induced lesion have been also found in ESR measurements of the radical formation in dry DNA constituents. Depending on the temperature of the sample distinct differences in the radical yield and in the saturation level as function of dose were observed (J. Hüttermann, this conference). The ESR measurements and the reduced repair capacity support the assumption of a change in the quality of particle induced lesions. However, from the experiments performed to date it is not clear whether the term multiple damaged site has to be restricted to DNA damage only or whether the surrounding environment has to be included too (J. Ward, this conference).

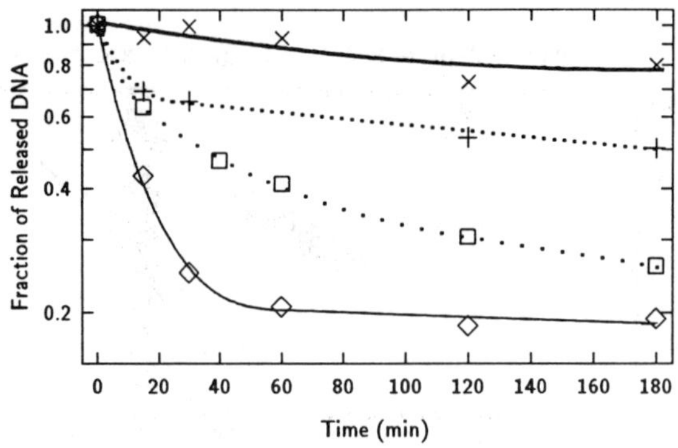

Figure 8: Rejoining of DNA double-strand breaks as measured by the constant field gel electrophoresis technique after irradiation with a) 20 Gy X-rays (◇), b) 38 Gy 390 MeV/μ Ne ions (□), c) 13 Gy 10 MeV/μ Ne ions (+), and d) 77 Gy 9.1 MeV/μ Xe ions (x). LETs correspond to 2 keV//μm, 30 keV/μm, 370 keV/μm and 7990 keV/μm, respectively.

Comparison with models

From a theoretical point of view the observed hook structure for inactivation and DNA damage has been expected on the basis of track structure calculations performed by Katz and collaborators (Katz et al. (1971)). In the Katz model the total energy deposited by a particle track is compared to the D_{37} dose as measured in x-ray experiments. For small track diameters and high LET values, doses largely in excess of the x-ray D_{37} dose are deposited and yield much more damage than necessary for cell killing. In this "overkill" regime saturation is predicated by the Katz model. However, a quantitative prediction of the individual survival curves on the basis of measurable observables is not possible. Models based on microdosimetric calculations are less adequate and do not predict saturation effects for very high LET.

Recently, a unification of the track structure model with microdosimetric ideas yielded significantly better results (Scholz and Kraft (1991)). In the unified model, the track is dissected in radial profiles of equal mean expectation values of the specific energy, i.e. of equal doses. For each dose increment, the probability of producing a lethal event is calculated according to the dose response curve as measured with x-rays. These probabilities are weighted according to their overlapping area with the critical target, i.e. the cell nucleus, and are summed up yielding the total inactivation probability. In the unified model only measurable quantities, the size of the critical target, the x-ray response curve and the radial dose distribution as measured in microdosimetric experiments are used for the calculation of the biological response to particle exposure. The results of these calculations are usually in very good agreement with measured data and reproduce the general feature of the σ-LET dependence very well vithout any additional fit parameter (Fig. 9). A similar approach has been applied by Heilmann et al. to describe DSB induction in mammalian cells as determined by electrophoretic elution methods. In their report in these proceedings they demonstrate good agreement between calculations and measurements with carbon ions.

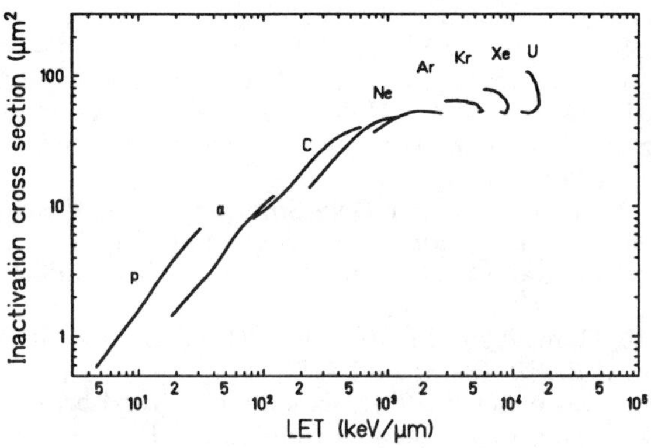

Figure 9: Calculated inactivation cross sections for V79 cells for ions as indicated. The model employed here (Scholz and Kraft (1991)) is an excellent fit to both the absolute measured cross sections and to the hook structure observed for the different ions.

Conclusions

In cell experiments distinct RBE maxima have been observed for inactivation, mutation, chromosome damage and other endpoints. The LET values of these maxima depend on the atomic number of the projectiles, and for lower atomic number the RBE maxima occur at lower LET values, indicating a strong dependence on local ionisation density and not on LET alone. However, the sharp decline in RBE from its maximum towards higher LET values due to saturation effects depends on the type of reaction studied i.e. on the various biological endpoints. In DNA measurements a similar behaviour has been observed for the induction of DSB in intracellular DNA. There, RBE values between 1 and 1.6 are found for intermediate LET values. However, drastically reduced repair and rejoining capacity of particle induced DNA lesions is found in this LET range. This indicates that the increased RBE of various cellular effects is more likely a consequence of reduced repair than of an enhanced production of DNA double strand breaks.

References

Aufderheide E, Rink H, Fheber L, Kraft G (1987) Heavy ion effects on cellular DNA: strand break induction and repair in cultured diploid lens epithelial cells *Int. J. Radiat. Biol. 51: 779-790*

Barendsen G, Walter H, Fowler J, Bewley D (1963) Effects of different ionizing radiations on human cells in tissue culture *Radiat. Res.* 18: 106-119

Belh M, Cherubim R, Finotto S, Moschini G, Sapora O, Simone G, Tabocchini M (1989) RBE-LET relationship for the survival of V79 cells irradiated with low energy protons *Int. J. Radiat. Biol.* 55: 93-104

Christensen R, Tobias C, Taylor W (1972) Heavy-ion-induced single- and double-strand breaks in ϕX-174 replicative form DNA *Int. J. Radiat. Biol.* 22: 457-477

Heilmann J, Rink H, Taucher-Scholz G, Kraft G (1993) DNA strand break induction and rejoining and cellular recovery in mammalian cells after heavy ion irradiation *Radiat. Res.* 135: 46-55

Kampf G (1983) *Die Erzeugung von DNS-Strangüchen durch ionisierende Strahlen unterschiedlicher Qualität und ihre Bedeutung für die Zellinaktivierung* Technical report Zentraiinstitut für Kernforschung, Rossendorf KfR 504 Habfil.-Schrift

Katz R, Ackerson B, Homayoonfar M, Sharma S (1971) Inactivation of cells by heavy ion bombardment *Radiat. Res.* 47: 402-425

Löbrich M, Rydberg B, Cooper P (1994) DNA double-strand breaks induced by high energy neon and iron ions in human fibroblasts. II. Probing individual NotI fragments by hybridization *Radiat. Res.* 139: 142-151

Ritter M, Cleaver J, Tobias C (1977) High-LET radiations induce a large proportion of non-rejoining DNA strand breaks *Nature* 266: 653-655

Roots R, Holley W, Chatterjee A, Irizarry M, Kraft G (1990) The formation of strand breaks in DNA after high-LET irradiation: a comparison of data from *in vitro* and cellular systems *Int. J. Radiat. Biol.* 58: 55-69

Rydberg B, Löbrich M, Cooper P (1994) DNA double-strand breaks induced by high-energy neon and iron ions in human fibroblasts. I. Pulsed-field gel electrophoresis method *Radiat. Res.* 139: 133-141

Schneider M, Taucher-Scholz G, Heilmann J, Kraft G (1994) Combination of static field gel electrophoresis and densitometric scanning for the determination of radiation induced DNA double strand breaks in CHO cells *Radiat. Environ. Biophys.* 33: 111-124

Scholz M, Kraft G (1991) A parameter-free track structure model for heavy ion cross sections. In K Chadwick, G Moschini, and M Varma (Eds.), *Biophysical Modelling of Radiation Effects* (pp. 185-192). Bristol: Adam Hilger

Taucher-Scholz G, Heilmann J, Schneider M, Kraft G (1994) Detection of heavy ion induced DNA double strand breaks using static field gel electrophoresis *Radiat. Environ. Biophys.*(In press).

Taucher-Scholz G, Stanton J, Schneider M, Kraft G (1992) Induction of DNA breaks in SV40 DNA by heavy ions *Adv. Space Res.* 12(2-3): (2)73-(2)80

Todd P (1967) Heavy-ion irradiation of cultured human cells *Radiat. Res.* Suppl. 7: 196-207

Weber K, Flentje M (1993) Lethality of heavy ion induced DNA double-strand breaks in mammalian cells *Int. J. Radiat. Biol.* 64: 169-178

ANALYSIS OF NATIVE CELLULAR DNA AFTER HEAVY ION IRRADIATION: DNA DOUBLE-STRAND BREAKS IN CHO-KI CELLS

J. Heilmann, G. Taucher-Scholz and G. Kraft
Gesellschaft für Schwerionenforschung mbH
Postfach 110552
D-64220 Darmstadt, Germany

Introduction

It is widely accepted that DNA is the crucial sensitive target of radiation at the molecular level. Especially DNA double-strand breaks (DSBs) are likely to represent the most critical type of lesions. Correct and complete repair of DSBs guarantees cellular survival, while incorrect repair and residual DNA damage may lead to cell killing, mutation, or transformation.

Several publications deal with the intracellular induction of DNA double-strand breaks after particle irradiation (Kampf, 1983; Blöcher, 1988; Belli et al., 1994; Prise et al., 1990; Weber and Flentje, 1993; Heilmann et al., 1993; Taucher-Scholz et al., 1994). They cover an LET range from 10 keV/μm to more than 10^4 keV/μm. However, there are differences both in the cell material used and the methods of DSB detection applied, making direct comparison of data difficult. In addition, there are only few data using accelerated ions in the LET region up to about 400 keV/μm. This range is of extraordinarily high interest as for LET values around 100 keV/μm the optimum efficiency of particle radiation with respect to cell killing was reported.

Therefore, a fast assay for the detection of DNA double-strand breaks was developed involving constant field gel electrophoresis (Taucher-Scholz et al., 1994) and densitometric scanning of agarose gels stained with ethidium bromide. With this technique, DSB induction was investigated after irradiation of CHO cells with carbon ions with LET values between 14 keV/μm and 400 keV/μm.

In parallel, a computer code was developed to simulate both the principle of the electrophoretic detection of DNA double-strand breaks and the action of radiations of different ionization density. The results of the experiments and the calculations are presented here and compared with each other.

Materials and Methods

Cells and Cell Culture: Mycoplasma-free CHO-KI cells were cultivated in Ham's F12 medium supplemented with 10% FCS in standard incubators at 100% relative humidity, 37°C in air with 5% CO_2. Exponential growth was ensured by regular passaging.

Irradiations: Irradiations with X-rays were performed using either a Siemens Stabiliplan or a Seifert Isovolt 320 generator. Filtering was 1mm Cu and 1 mm Al in both cases. The machines were run at 250 kV, thereby achieving dose rates of 7 Gy/min and 10 Gy/min, respectively.

For heavy ion irradiations the UNILAC and SIS facilities at GSI, Darmstadt, were employed. For particle energies up to about 16 MeV/u, cells were irradiated in petri dishes at the UNILAC, whereas the heavy ion synchrotron (SIS) served as particle source at higher energies. Here, cells were irradiated embedded in agarose (see next paragraph). Exact details of the irradiation facilities are described elsewhere (Kraft et al., 1980; Haberer et al., 1993).

Preparation of Cells for Irradiation: Cells were seeded two days before irradiation in either small petri dishes (35 mm diameter) or 75-cm^2 culture flasks in a concentration sufficient to reach plateau phase in time. The petri dishes were used directly for irradiations at low energies. For irradiations with high energy ions, cells were first harvested by trypsinization, centrifuged, resuspended in phosphate buffered saline (PBS) and embedded in 0.5% low melting point agarose at a final concentration of 2.5×10^5 cells/ml. Agarose plugs were then transferred into small microcentrifuge tubes in PBS and kept on ice until and during irradiation.

Analysis of DNA Strand Breakage: Cells irradiated at the UNILAC had to be scraped off the culture dishes on ice and embedded in agarose first. Then, the samples were treated as described below.

Agarose plugs were transferred into ice cold lysis buffer (0.5 mol/l EDTA, 2 % sodium lauryl sarcosinate) containing I mg/ml proteinase K. After equilibration for 30 min on ice, lysis was carried out for 16 h at 50°C in a water bath. The lysis buffer was then removed and replaced by TE-buffer (10 mmol/l Tris, 1 mmol/l EDTA; pH 8). The buffer was changed repeatedly to remove excess detergent.

Plugs were then cut appropriately and placed into gel slots 2 mm wide in a 0.5% agarose gel (14 cm x 14 cm). Electrophoresis was performed for 24 h at 25 V in 0.5 TBE buffer (45 mmol/l Tris, 45 mmol/l boric acid, 2 mmol/l EDTA). Gels were stained afterwards for at least 3 h with ethidium bromide (3 µg/ml) and destained overnight in distilled water. Pictures of the fluorescence signal in the gel were taken directly on a UV-transilluminator using a cooled CCD camera (Fairchild). Digitized images were transferred to a computer and fluorescence intensities in the gel slots and in the lanes were measured. Effect curves were obtained by plotting the fraction of DNA retained as function of radiation dose or particle fluence. Curves were fitted according to an equation published by Blöcher (1990):

$$F_{ret} = e^{(-x\frac{k}{n})}(1 + x\frac{k}{n}(1 - \frac{k}{n}))$$

where k is the exclusion size of the agarose gel, n is the mean size of the chromosome, and x is the number of double strand breaks per chromosome. The exclusion size is assumed to be around 6 Mbp, as was tested using *Schizosaccharomyces pombe* chromosome marker. The mean chromosome size for CHO-cells (245 Mbp) was taken from the literature (Blöcher, 1990).

Computer Code: The program is designed to create artificial fluence effect curves for different accelerated particles. The procedure is divided into several steps. In the beginning, a model chromosome is generated consisting of inactive sequences attached to the x,y-plane representing the nuclear membrane, and active sequences protruding into the lumen of the nucleus. A total of 10% of the chromosome is assumed to be active.

In the next step the chromosome is bombarded with particles. Following every particle traversal, the chromosome is walked along in nm-steps and the radial distance from the particle track is calculated for each point. The local dose is then figured out assuming a $1/r^2$ drop in dose from the center of the track to its maximum radius. Now the probability for DBS induction at this point of the chromosome is calculated. This value is used to let a random generator decide whether a strand break is set.

At the end, DNA fragments larger than the exclusion size assumed for constant field electrophoresis are added up. Repeating these steps for different particle fluences finally yields an artificial fluence-effect curve.

The only parameters needed in this code are the energy and the LET of the particles and the probability of DSB induction for X-rays. LET values in water were taken from Heinrich and his coworkers (Heinrich et al., 1991), while the probability for X-ray induced DSB breakage was (determined experimentally (see next chapter). The whole calculation follows rather strongly a model published recently to describe heavy ion action with respect to cell killing (Scholz and Kraft, 1991).

Results and Discussion

An assay using constant field gel electrophoresis followed by densitometry of ethidium bromide stained gels was applied to determine DNA double-strand break induction following X- and heavy ion irradiation. A computer simulation was tested using the experimental data.

In Fig. 1, a dose effect curve obtained with 250 kV X-rays is shown. Results from 5 independent experiments were pooled. Data were fitted using the equation cited in the materials and methods section. Furthermore, the results of a simul-

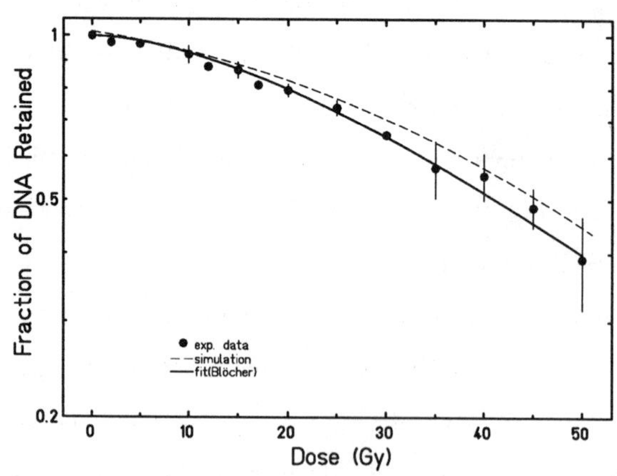

Figure 1: Dose effect curve for the induction of DNA double-strand breaks in CHO cells after X-irradiation. Mean values and errors from 5 independent experiments are given. Data were fitted with the equation given by Blöcher (1990). In addition, a computer-generated dose effect curve is plotted, showing good agreement with the experiment.

ation are given. An induction rate of 35 DSBs/Gy/cell was calculated from the experiments. The computer code generated a dose effect curve only slightly flatter than the fitted line, corresponding to a smaller induction rate for DSBs around 32 DSBs/Gy/Cell. Nevertheless, agreement between experiment and calculated data is very good. Especially the shape of the dose effect curve, which is determined by the exclusion size of the agarose gel, is reproduced well.

Heavy ion experiments were performed using accelerated carbon ions with LET-values between 14 keV/μm and 400 keV/μm. From fluence-effect curves, DSB induction rates determined and RBE values calculated for comparison with literature data. In Fig. 2 the relative biological efficiencies are plotted as a function of LET. Data published by others in this LET-regime are included (Aufderheide et al., 1987; Belli et al., 1994; Blöcher, 1988; Prise et al., 1990; Kampf, 1983; Weber and Flentje, 1993; Heilmann et al., 1993; Rydberg et al., 1994).

RBEs vary between 0.8 and 1.8 in the LET range up to 110 keV/μm. For higher ionization densities, most publications report RBE values considerably smaller than unity. One group (Heilmann et al., 1993) determined RBEs around 1 for Ne-ions with LETs between 315 keV/μm and 395 keV/μm. Only the data

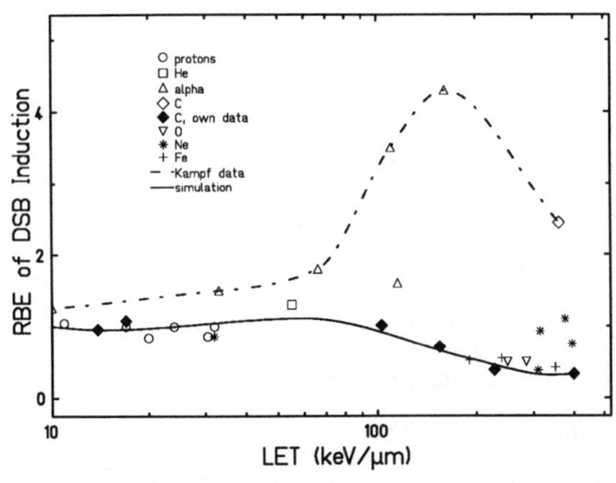

Figure 2: RBE-values of intracellular DSB induction after irradiation with various ions from protons to iron (Aufderheide et al., 1987; Belli et al., 1994; Blöcher, 1988; Prise et al., 1990; Kampf, 1983; Weber and Flentje, 1993; Heilmann et al., 1993; Rydberg et al., 1994). The RBE-LET relationship as indicated by the data published by Kampf (1983) is represented by the dashed-dotted line. The result of a simulation calculation for carbon-ions is shown, too. Besides the fact that the simulation data were scaled by a factor of two to fit the data, the general systematics of the RBE-LET relationship are described quite well with the model.

reported by Kampf (Kampf, 1983) show up with relative biological efficiencies much larger than 1 with a maximum greater than about 4 at 160 keV/μm. These discrepancies may be due to the different experimental design. As also shown in Fig 2, the computer code was applied to describe the RBE-LET relationship of DNA double-strand breakage after irradiation with carbon ions. Apart from the scaling necessary, the simulation corresponds well with the experimental results.

It is known for many years that maximum efficiency for cell killing with particle radiation is achieved in the LET region around 100 keV/μm, depending on the ions used (Barendsen et al., 1963). As can be seen from the data and the calculation shown here, a direct correlation between cell killing and DSB induction cannot be established. Considering DNA double-strand breaks as the important molecular lesions, changes in the quality of the DNA damage are thought to be responsible for the enhanced cell killing (Brenner and Ward, 1992; Goodhead, 1994). The high local ionization densities possible within particle tracks lead to multiple ionization events within nanometre regions of the target, resulting in complex damage. However this type of damage is still detected as DNA double-strand break by currently established assays and cannot be distinguished from a DSB generated e.g. by a restriction enzyme.

Clustered damage is expected to be responsible for the slow and incomplete rejoining of DNA strand breaks reported by several authors after irradiation with charged particles (Ritter et al., 1977; Aufderheide et al., 1987; Heilmann et al., 1993; Weber and Flentje, 1993). Direct measurements will be possible by detecting small fragments that are proposed as products of complex DNA lesions (Prise, 1994).

In summary, we have presented data on the intracellular induction of DNA double-strand breaks by accelerated carbon ions with LET values from 14 keV/μm to 400 keV/μm. In agreement with the majority of publications covering this LET range, relative biological efficiencies were found to be around or lower than unity, especially for LETs in excess of 100 keV/μm. A computer model describing damage induction in individual cbromosomes showed good agreement with the experimental data. Further investigations are to focus on the cellular processing of high-LET DNA-damage and the detection of complex lesions.

References

Aufderheide E, Rink H, Hieber L, Kraft G (1987) Heavy ion effects on cellular DNA: strand break induction and repair in cultured diploid lens epithelial cells. *Int. J. Radiat. Biol.* 51: 779-790

Barendsen G, Walter H, Fowler J, Bewley D (1963) Effects of different ionizing radiations on human cells in tissue culture. *Radiat. Res.* 18: 106-119

Belli M, Cera F, Cherubini R, Ianzini F, Moschini G, Sapora 0, Simone G, Tabocchini M, Tiveron P (1994) DNA double-strand breaks induced by low energy protons on V79 cells. *Int. J. Radiat. Biol.* 65: 529-536

Blöcher D (1988) DNA double-strand break repair determines the RBE of α-particles. *Int. J. Radiat. Biol.* 54: 761-777

Blöcher D (1990) In CHEF electrophoresis a linear induction of DSB corresponds to a nonlinear fraction of extracted DNA with dose. *Int. J. Radiat. Biol.* 57: 7-12

Brenner D, Ward J (1992) Constraints on energy deposition and target size of multiply damaged sites associated with DNA double-strand breaks. *Int. J. Radiat. Biol.* 61: 737-748

Goodhead D (1994) Initial events in the cellular effects of ionizing radiations: clustered damage in DNA. *Int. J. Radiat. Biol.* 65: 7 17

Haberer T, Becher W, Schardt D, Kraft G (1993) Magnetic scanning system for heavy ion therapy. *Nucl. Instr. Methods* A330: 296-305

Heilmann J, Rink H, Taucher-Scholz G, Kraft G (1993) DNA strand break Induction and rejoining and cellular recovery in mammalian cells after heavy ion irradiation. *Radiat. Res.* 135: 46-55

Heinrich W, Wiegel B, Kraft G (1991) Tables of dE/dx range and restricted energy loss of heavy ions in energy range from 1 to 3000 MeV/u. GSI-Report 91-30 GSI, Darmstadt.

Kampf G (1983) Die Erzeugung von DNS-Strangbrüchen durch ionisierende Strahlen unterschiedlicher Qualität und ihre Bedeutung für die Zellinaktivierung. Technical report Zentralinstitut für Kernforschung, Rossendorf KfR 504 Habil.-Schrift

Kraft G, Daues H, Fischer B, Kopf U, Liebold H, Quis D, Stelzer H (1980) Irradiation chamber and sample changer for biological samples. *Nucl. Instr. Meth.* 168:175-179.

Prise K (1994) Use of radiation quality as a probe for DNA lesion complexity. *Int. J. Radiat. Biol.* 65:43-48.

Prise K, Folkard M, Davies S, Michael B (1990) The irradiation of V79 mammalian cells by photons with energies below 2 MeV. Part II. Measurement of oxygen enhancement ratios and DNA damage. *Int. J. Radiat. Biol.* 58:261-277.

Ritter M, Cleaver J, Tobias C (1977) High-LET radiations induce a large proportion of non-rejoining DNA strand breaks. Nature 266:653-655.

Rydberg B, Löbrich M, Cooper P (1994) DNA double-strand breaks induces by high-energy neon and iron ions in human fibroblasts. I. Pulsed-field gel electrophoreseis method. *Radiat. Environ. Biophys.* In press.

Scholz M, Kraft G (1991) A parameter-free track structure model for heavy ion cross sections. In K Chadwick, G Moschini, and M Varma (eds.) *Biophysical Modelling of Radiation Effects* (pp. 185-192). Bristol: Adam Hilger.

Taucher-Scholz G, Heilmann J, Schneider M, Kraft G (1994) Detection of heavy ion induced DNA double strand breaks using static field gel electrophoresis. *Radiat. Environ. Biophys.* (pp. accepted for publication).

Weber K, Flentje M (1993) Lethality of heavy ion induced DNA double-strand breaks in mammalian cells. *Int. J. Radiat. Biol.* 64:169-178

LOW-LET RADIOTOXICITY FROM INCORPORATED ^{3}H OR ^{125}I DURING THE CELL CYCLE: IMPLICATIONS FOR MECHANISMS OF HIGH-LET DAMAGE

Daniel A. Geselowitz, Mary M. McManaway, Kurt G. Hofer* and Ronald Neumann
Department of Nuclear Medicine
Clinical Center, National Institutes of Health
Building 10, Room 1C401
10 Center Dr MSC 1180
Bethesda, MD 20892-1180
*Institute of Molecular Biophysics, Florida State University

Introduction

The unique cytotoxicity of DNA-incorporated Auger-emitting radionuclides offers the tantalizing possibility of DNA-targetted radiotherapies. However, much remains to be understood about the fundamental mechanism of cytoxicity of these agents. Recent reports by Hofer, Van Loon et al. (1992a), (1992b) demonstrate a puzzling aspect of the radiotoxicity of I-125 incorporated in Chinese Hamster Ovary (CHO) cells. When these cells are pulse-labelled with ^{125}IUdR in early S-phase, then frozen at various times later in the cell cycle, the radiotoxicity of ^{125}I is found to increase dramatically over the course of about 5 hours, and the shape of the log-survival curve also changes from a shouldered curve to linear during this time. The microdosimetric analysis initially reported, based on comparison to the toxicity of cells at 37° to external irradiation, suggested that the lower radiotoxicity observed at short times after the pulse could be attributable solely to the low-LET portion of the I-125 emission--in other words, there was no unique Auger toxicity at this time.

To better understand these results and better characterize the microdosimetry of incorporated nuclides, we have performed several studies on the radiotoxicity of tritium incorporated in CHO cells. The tritium beta particle actually has a similar spectrum to the low-LET portion of the I-125 electron emission, making it an excellent microdosimetric model. In the process, we have found that the cytotoxicity of a given nuclear dose from incorporated tritium is quite different from the toxicity of whole-cell irradiation. Our results answer the question about the importance of I-125's low-LET radiation, and also shed light on other issues, such as the target for induced radioresistance.

Methods

Cell-cycle dependent cytotoxicity. Chinese hamster ovary (CHO) cells ($1x10^6$), selected in mitosis and frozen, were thawed and diluted into 50 mL of complete medium (McCoy's 5a with 5% bovine calf serum) and 4.00 mL was plated into each of ten 25-cm^2 culture flasks with filter caps. After 1.0 h, aphidicolin was added to each flask to give 5 mg/mL. After an additional 7.0 h, the medium was replaced with 4.0 mL of fresh complete medium. Thirty minutes after removal of the aphidicolin, the flasks were treated in groups of three with a radiolabeled stock solution to give 1.0 mCi/mL of ^{125}IUdR, 3.0 mCi/mL of ^{3}H-TdR or 9.0 mCi/mL of ^{3}H-TdR. After 15 min of incubation the radioactive medium was removed, the culture flasks were rinsed and then replenished with 4.0 mL of complete medium containing $1x10^{-4}$ M thymidine and the flasks were returned to the incubator.

At times 30, 120 and 360 min after replacement of the radioactive medium, and at 120 min for the control flask, one flask from each treatment group was washed with PBS and treated with 0.5 mL of trypsin/versene at 37 °C for 5 min. The contents of the flask were rinsed with 3.0 mL of freezing medium (Ham's F10 with 15% bovine calf serum and 10% DMSO) and aliquoted to seven freezing vials. Three 25-mL samples were withdrawn from the first vial and plated (see below), and the vials were placed in a controlled-freezing container (Nalgene) which was placed at -70 °C. After freezing was complete, the vials were transferred to storage in a cryostat in the vapor above liquid nitrogen.

The next day, one vial from each treatment group was withdrawn, thawed, and several aliquots were plated for survival in 5 mL of McCoy's 5a supplemented with 10% fetal bovine serum, 5% calf serum, and penicillin/streptomycin in a 60-mm culture dish. Additional platings were performed over the next six weeks.

Cell survival was evaluated by the colony formation assay. The culture dishes were incubated for 7 days and were then stained with crystal violet. Viable colonies were counted visually, rejecting tiny or diffuse spots.

The concentration of total cells in the thawed freezing vial was determined using a Coulter Counter. The activity of DNA-incorporated tritium was assayed by diluting an aliquot of frozen cells into 10% trichloroacetic acid, centrifuging and removing the supernatant. For tritium, the pellet was dissolved in 1.0 mL of tissue solubilizer (Beckman BTS-450) by incubating overnight at 60 °C. The solution was added to 10 mL liquid scintillation cocktail (Beckman Ready Organic), let sit for several days, and then scintillation counted (Packard model A4530D). Efficiency correction was performed using the spectral index of an external standard and quenched primary standards. For I-125, the activity of the pellet was measured using a gamma-counter (Packard AutoGamma 5650) with efficiency determination based on the decays in the photopeak and sum peak.

Synergy of tritium decays and external irradiation. Frozen mitotic CHO cells were thawed and incubated overnight in McCoy's 5a with 10% FBS containing 0.02 μCi/mL of ^{3}H-TdR. The cells were then harvested and frozen as above at about 50000 cells/vial. Unlabelled cells were also frozen as controls. The vials were randomized and groups were treated with external X-ray doses of 0, 0.5, 1.0 1.5, 2.0, 5.0 or 10.0 Gy, and placed in cryogenic storage.

At various times, one vial from each group was thawed and plated for colony formation. Total cells were counted, and survival of the cells was calculated relative to the control treatment receiving no tritium and no irradiation.

Results and Discussion

The timetable for the cell cycle experiment is shown below. The

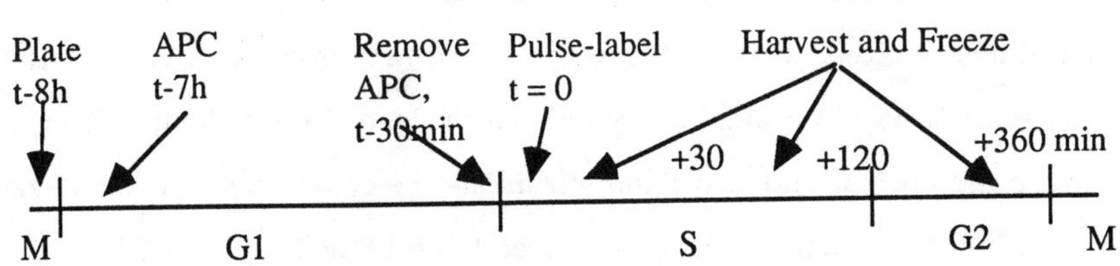

cells harvested at 30, 120 and 360 min post label are in early S-phase, mid-S and G2, respectively. The log-survival plots as a function of accumulated decays for cells labelled with ^{125}IUdR and for those labelled with ^{3}H-TdR are shown in figures 1 A and B. The results of figure 1A confirm the results reported by Hofer and coworkers,

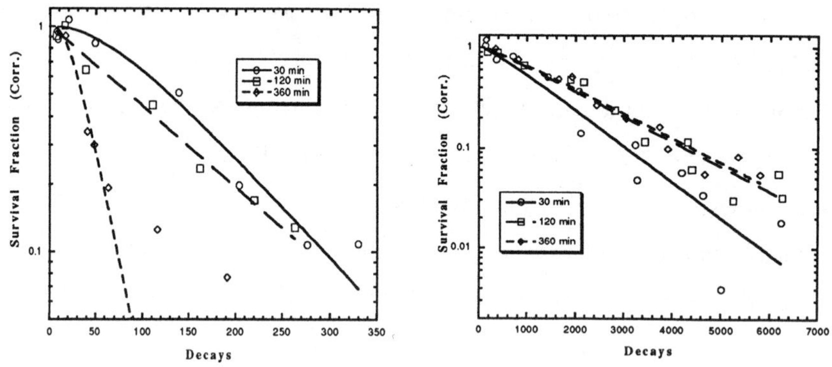

Figure 1. Log-survival plots of synchronized CHO cells pulse-labeled with ^{125}IUdR (left) or ^{3}H-TdR (right), then frozen at indicated times after labelling for decay accumulation.

in which the full radiotoxicity of I-125 appears gradually during the five hours after labelling. In figure 1B, the radiotoxicity of tritium can be seen to behave quite differently, with the cells somewhat less sensitive to tritium decays in the later time points.

The tritium data can now be used to understand the microdosimetry of I-125. The high-LET damage caused by the numerous electrons of the ^{125}I decay is probably limited to a range of about 3 nm, and is caused mainly by electrons with energies of less than 0.5 keV. About 10% of the electron energy of the I-125 decay, (0.31 fJ out of 3.1 fJ), is from particles in this energy range. Approximately 45% (1.3 fJ) of the total energy is carried by electrons with energies of 3-5 keV and range 0.4-0.7 μm, which is deposited almost entirely in the cell nucleus. The remaining 45% is in electrons with energies of 22-35 keV (range 11-23 nm) which deposit only about 15% of their energy in the cell nucleus and also slightly irradiate the cytoplasm. The total energy depostion within the nucleus has been calculated to be 1.8 or 1.9 fJ for a 261 or 637 fL nucleus, respectively(Hofer, Van Loon et al. (1992)). These volumes are the measured volumes for frozen CHO cell nuclei in early S and G2 phases.

Tritium beta particles have a maximum energy of 18 keV (range 5 μm), but 85% of the energy is delivered by electrons of less than 10 keV, and roughly half by electrons of energy less than 5 keV. Thus the tritium beta, which delivers an average of 0.91 fJ per decay, mimics fairly well the middle region of the iodine spectrum which is responsible for almost all of the low-LET dose delivered by I-125 to the nucleus. For a nuclear volume of 261-637 fL, the total nuclear dose from an I-125 decay is 2.0-2.1 times that of a tritium decay, and iodine and tritium decays are easily compared.

The average nuclear dose resulting in 37% cell survival can now be compared for iodine and tritium for the three freezing times (Table I). For tritium, which emits a low-LET beta, the D_{37} value falls from 5.2 to 3.0 Gy between the 30-min time (early S-phase) and 360 min (G2). By comparison, the dose of external irradiation to give 37% survival is about 15 Gy. Plainly, the assumption that the radiotoxicity of external irradiation will equal that of the same average nuclear dose of incorporated radionuclides is simply not valid. These data now reveal that the radiotoxicity of I-125 decays is plainly not attributable to the low-LET portion of the spectrum, and must be attributed to the Auger effect. While this does not explain the curious delayed onset of sensitvity to I-125 decays, it eliminates the hypothesis that the Auger effect is totally mitigated at the earlier time.

Table I. Cytotoxicity of incorporated tritium or ^{125}I label

Treatment	D_{37}	energy to nucleus	V_{nuc}	D_{37} (avg. nuclear dose)
^{3}H-TdR Labelled				
30 min	1500 decays	1.4 pJ	261 fL	5.2 Gy
120	2100	1.9		
360	2000	1.8	637	3.0
^{125}I-labelled				
30 min	165	0.30	261	1.1
120	116	0.21		
360	40	0.076	637	0.12

As for the fact that a given average nuclear dose from tritium decays is so much more toxic than the same external dose, there are two likely explanations: 1. The energy deposited by the tritium decay is not uniformly deposited throughout the cell nucleus, but is concentrated in critical target regions. 2. The dose delivered by tritium fails to activate an induced radioresistance mechanism which occurs in cells treated with external irradiation.

Explanation 1 seems fairly reasonable. Since the range of a median tritium beta particle is about 0.7 μm, about the diameter of a chromatid, and DNA is not distributed homogeneously throughout the nucleus, the radiation from incorporated tritium may be concentrated in the DNA.

To test the second explanation, we looked for radioresistance to tritium decays induced by external x-rays. CHO cells accumulating a given number of tritium decays while frozen were also given various doses of external irradiation. The results are shown in Figure 2. Note how, over a range of accumulated tritium decays, a distinct dip in survival is seen at 0.5 Gy, followed by recovery at 1.5-2.0 Gy, with survival then decreasing at higher doses. The implication is that doses of 1.0 to 2.0 Gy, which are less than 15% of the D_{37} value, turn on a radioresistance mechanism which has not been triggered by even 8000 tritium decays, about four times the D_{37} value. It does not appear that this effect is large enough to account for the all of discrepancy between average nuclear dose and external dose. However, it does shed light on the target for induction of the radioresistance. Since the incorporated tritium decays irradiate the cell nucleus but do not penetrate well into the cytoplasm, the target must be external to the nucleus. Possible target locations include the cytoplasm and cell membrane, and extracellular effects cannot be ruled out.

It is interesing to compare these results to those reported for radiation-induced division delay, which may be associated with a mechanism for repair of radiation damage. Schneiderman and Hofer (1980), studying division delay induced by incorporated I-125 in CHO cells, found evidence that the target for the delay was not the DNA. Since the dose to the cytoplasm by I-125 decays is small, they hypothesized that the nuclear envelope is the target.

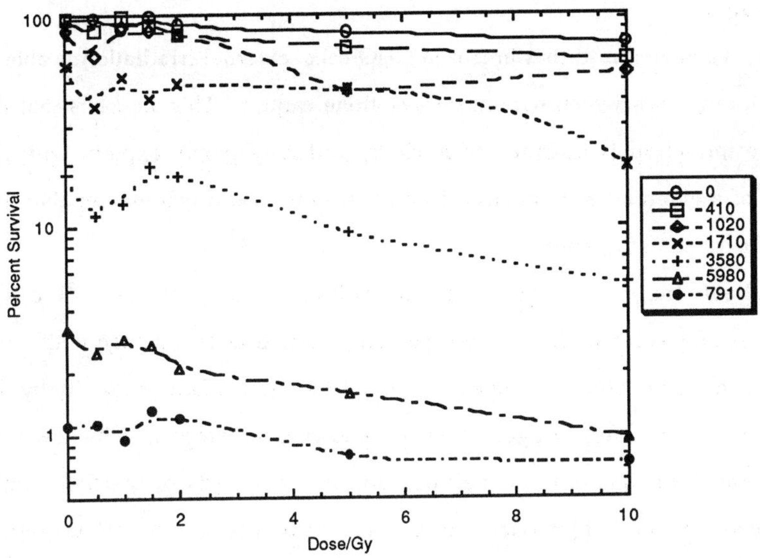

Figure 2. Survival of CHO cells receiving doses of external irradiation and accumulating tritium decays while frozen. Log survival is plotted versus the external irradiation dose for seven treatments ranging from 0 to 7910 decays/cell.

Summary

In order to better understand the relative importance of low- and high-LET radiation in the radiotoxicity of incorporated I-125, we have determined the toxicity of decays of ^{3}H-TdR incorporated into CHO cells. Studies with synchronized cells pulse-labelled in early S-phase and frozen at 30, 120 or 360 min afterward have D_{37} values of 1500, 2000 and 2100 decays, corresponding to average nuclear doses of 5.2 and 3.0 Gy at the 30- and 360-min times. The D_{37} for asynchronous cells is similar (2000 decays). As the D_{37} value for external irradiation under these conditions is approximately 15 Gy, these results indicate the pitfalls of using external irradiation dosimetry to predict the biological effects of incorporated low-LET decays.

We have confirmed the reported delayed onset of high radiotoxicity of cells pulse-labeled in early S-phase with ^{125}IUdR and frozen afterwards. Even in the 30-min samples, the average nuclear dose D_{37} value of 1.1 Gy indicates that the high-

LET Auger electrons account for most of the cytotoxicity. A previous analysis had suggested that the radiotoxicity at this time was no greater than that expected from low-LET radiation.

Finally, we have found that in frozen CHO cells, external irradiation is able to induce a radioresistance which tritium decays alone cannot. This suggests that the target for the induction is outside the nucleus, and may partly explain why the radiotoxicity of a given average nuclear dose of tritium is so much greater than the same dose of external irradiation.

We believe that the Auger effect in radionuclides such as ^{125}I offers fascinating possibilities for radiotherapy based on targetted DNA damage within the cell. However, there are still many questions about the mechanism of toxicity by the Auger electrons. Moreover, Auger emission is always accompanied by low-LET radiation. Better understanding of the micordosimetric aspects of toxicity should allow a more rational approach to design and discovery of targetted radiotherapeutics.

References

Hofer KG, van Loon N, Schneiderman MH, Charlton DE (1992a). High and low-LET cytocidal effects of DNA-associated iodine-125. Biophysical Aspects of Auger Processes, AAPM Symposium Series No. 8. Eds. R. W. Howell, V. R. Narra, K. S. R. Sastry and D. V. Rao. Woodbury, NY, American Institute of Physics. 227-248.

Hofer KG, van Loon N, Schneiderman MH, Charlton DE (1992b). The paradoxical nature of DNA damage and cell death induced by ^{125}I decay. Radiat Res **130**(1): 121-4.

Schneiderman MH, Hofer KG (1980). The target for radiation-induced division delay. Radiat Res **84**(3): 462-76.

Chapter 7

Structural Perturbations in DNA Introduced by Monochromatic Light Sources

USE OF SYNCHROTRON MONOCHROMATIC LIGHT: A PROBE FOR MECHANISTIC STUDIES IN RADIOBIOLOGICAL RESEARCH

Katsumi Kobayashi
Photon Factory
National Laboratory for High Energy Physics
Oho, Tsukuba 305
Japan

Synchrotron radiation

Synchrotron radiation (SR) is a radiation emitted from high-energy, high-speed electrons which are accelerated, for example, by a magnetic field. This situation inevitably happens with circular-shape particle accelerators used for high-energy physics. Electrons or positrons emit SR very efficiently during the acceleration in the synchrotron. Due to the characteristics shown below, demands for using SR from researchers in the various scientific fields had increased and finally resulted in the construction of electron/positron storage rings dedicated for the use of SR. Characteristics of SR are summarized below.

a) SR has a very wide spectrum ranging from infrared to hard X-rays. This spectrum contains vacuum ultraviolet (VUV), ultrasoft X-rays and soft X-rays regions, where there had previously been no tunable light source. There are no complicated structures in its spectrum, which makes spectroscopic studies easier.

b) Intensity is strong enough for irradiation experiments, even after monochromatization. For instance, SR monochromatic X-rays dispersed with a crystal monochromator ($E/\delta E \geq 1000$) have an intensity of about 10 C/kg/min (40 kR/min) at 10 keV from the 2.5 GeV electron/positron storage ring at the Photon Factory, Tsukuba, Japan. Available intensities at several experimental stations for irradiation are summarized in Table 1.

It should be noted that the change in X-ray quality resulting from changes in the applied high voltage or by insertion of filters is not considered monochromatization.

c) SR beam is emitted tangentially to the particle orbit, and nearly parallel. Accordingly, the intensity distribution in the beam is uniform in the horizontal direction, and Gaussian in the vertical direction. The beam size is determined by the aperture or slits of the beamline and typically a few mm high and several cm wide when a focusing optics is not adopted. In order to irradiate wide samples with such small beams, we developed scanning irradiation stages both in air (Kobayashi et al, 1987) and in vacuum.

Table I. Typical intensity of monochromatized SR beam obtained at several stations for irradiation at the Photon Factory

Station	Energy range (keV)	Bandwidth	Typical intensity (/min)	Irradiation
BL-4A	15 ~ 4	0.1 %	~ 40 kR	in air
BL-1B	3.3 ~ 1.7	0.1 %	~ 10 kR	in air or vac.
BL-11B	3.3 ~ 1.7	0.1 %	~ 300 kR	in air or vac.
BL-12C	0.7 ~ 0.05	~ 2 %	~ 10^{13} photon*	in vacuum
BL-27A**	6 ~ 1.5	0.1 %	~ 60 kR	in air
			~ 400 kR	in vacuum
BL-27B**	15 ~ 4	0.1 %	~ 10 kR	in air

* Total number of the photons on the sample
**Radioisotopes usable as tracers

d) Pulse structure and polarization of the beam are also characteristics of SR.

Besides the Photon Factory, there is anther SR light source in Tokyo, to which we connected a Wadsworth type monochromator and a sample chamber in order to irradiate biological specimens with monochromatic vacuum ultraviolet (VUV) light of the energy range up to about 25 eV (Ito et al, 1984, Hieda et al, 1986a). Presently, we can irradiate with monochromatic photons of desired energy in the region from UV (300 nm or 4 eV) to X-ray regions (20 keV), although some gaps remain to be filled (Hieda, 1994).

Initial processes to be investigated using synchrotron radiation

Today, radiobiology is attracting much attention, not only due to basic scientific interest, but also due to the practical points of view, such as the radiotherapy of cancer using heavy particles, or the protection from environmental radiation of low dose rates. Heavy-particle beams deposit their energy densely along tracks. To quantitatively represent this energy deposition density, LET is defined as the amount of energy deposited per unit length of a track. Heavy particles exhibit high LET, while Compton electrons generated by high-energy photons show a low LET. The irradiation of very low-dose radiation over the course of many years from environmental radionuclides is an extreme case from the viewpoint of energy deposition in time scale. In both cases, the difference is in the distribution of energy deposition in space and time. Differences in the following de-excitation processes are considered to cause biological effects which are characteristic of both types of irradiation. In order to predict these effects logically, we need to completely understand the processes leading to the effects.

Radiobiological effects are induced with energy deposited by radiation. This deposition events occur randomly and discretely. The average amount of energy deposited per one event is about 30 eV, which produces two ion pairs on average. The fundamental question that we have to answer is the consequences of each energy deposition event. Through the de-activation processes, molecular changes occur to produce the biological damage. The second is how these consequences change depending upon the density of the energy deposition events in space and time through the interaction of the reactive species produced by them.

In order to answer the above questions experimentally, we need to be able to deposit any amount of energy that we desire into biological samples. One way would be to irradiate with low-energy monochromatic photons. Since vacuum ultraviolet photons of about 40 nm wavelength would give an energy quantum of about 30 eV around the absorption site, we could study the processes and their efficiencies in producing molecular changes induced with this energy. In more energetic regions, monochromatic soft X-rays are absorbed photoelectrically, mainly by K-shell electrons of the oxygen in water molecules contained in living cells, and producing a shower of almost monoenergetic electrons in the cells. By changing the photon energy, we can change the energy of electrons, since the energy of the photoelectrons is equal to the energy of the photon minus the binding energy of the K-shell electrons of oxygen. Electrons exhibit different LET depending upon their energy. This means that we can study the LET dependence of the radiobiological efficiency. Extremely high density of the deposition events could be achieved by the Auger effect which can occur at inner shell-ionized atoms in the sample. Inner-shell ionization or excitation can efficiently be induced by using the photons having the energy corresponding to the inner shell absorption edges. From these points of view, we require a tunable light source which is sufficiently strong even after monochromatized. Synchrotron radiation is the only strong light source having a continuous spectrum from UV to X-rays.

<u>Some of the important results obtained using synchrotron radiation</u>

Germicidal lamps are commonly used for the light source of monochromatic 254 nm (4.9 eV) photons. Energy of these photons is not enough to ionize molecules of biological interest: the photons can excite the molecules. Consequences of the absorption of the photon have been well investigated. Lethal effects of photoabsorption of more energetic photons (energy deposition) have been investigated and compared with that of 254 nm photon in yeast cells (Ito et al, 1983). Using plasmid DNA, it was revealed by Hieda et al (1986b) that the base damage recognized by an UV endonuclease (with specificity for pyrimidine dimers) was not produced below 180 nm, and that single strand breaks were produced more effectively with shorter wavelength (higher energy) photons. It was noted that the action spectrum of ssb production was similar to that of the absorption of the sugar-phosphate moiety of DNA, indicating that ssb might be produced along with the photoabsorption of the backbone(sugar) of DNA. Further studies using dinucleotides have revealed that ssb is produced associated with the destruction of the sugar moiety, releasing fragments having a 5' phosphate end (Ito and Saitou 1991). It was also shown that the yield of strand breaks reached a plateau above 10 eV region (Saitou 1993), suggesting the amount of energy necessary to produce a strand break is less than 10 eV, although a lower efficiency of breakage was observed with less energy (longer wavelength photons). Recently, Hieda et al (1994b) studied the energy dependence for the production of ssb and of dsb in pBR322 DNA in the energy region between 60 and 550 eV. It was demonstrated for the first time that the yield of DSB increased with the increase in photon energy between 60 eV and 550 eV, when compared with the yield for SSB. This suggests the photon energy needed for induction of DSB is around 300 eV.

In more energetic regions, the interactions between energy deposition events could be simulated since the photon energy is deposited very locally around the inner-shell photoabsorption site through the low-energy Auger electrons. The Auger effect is known to occur at inner-shell ionized atoms, which produces low-energy Auger electrons leaving the atoms multiply ionized. This type of study is directly related to the high RBE observed with high LET radiation. Most work has focused on photoabsorption at the K-shell of phosphorus in DNA due to the strategic position in the biologically important molecule, using the advantage of the complete tunability of SR. This is considered to cause severe damage in DNA. Also advantageous for research is that the DNA molecule was found to have a sharp resonance absorption peak at the K-shell absorption edge of phosphorus, due to the formation of the phosphate moiety (Kobayashi et al, 1991). This increases the probability of photoabsorption by the phosphorus atom, and the biological effects of photoabsorption by phosphorus become more observable. The enhancement in the biological effects observed along with the irradiation of the resonance absorption peak can be considered to be initiated by the photoabsorption by phosphorus. The enhancement of biological effects with the phosphorus photoabsorption is observed in several organisms including lethality and induction of genetic changes in yeast (Kobayashi et al, 1991), lethality in bacteria (Maezawa et al, 1988), chromosome aberration in mammalian cells (Saigusa et al, 1992), and mutation and killing in mammalian cells (Watanabe et al, 1992a).

In order to know the molecular damage produced by the phosphorus photoabsorption, the production yields of ssb and dsb were investigated with a plasmid DNA irradiated in solid. It was shown that both yields do not seem to change (Hieda et al, 1994a) This suggests that the damage responsible for the enhancement of observed biological effects would be a minor fraction of the damage produced by the phosphorus photoabsorption.

As a close simulation of aqueous biological systems, SSB and DSB were measured with a plasmid DNA irradiated in aqueous solutions. Strand break yields were shown to be enhanced by the photoabsorption of phosphorus in the DNA molecule (Takakura et al, 1994). Similar enhancements were observed for the release of adenine from ATP in a concentrated aqueous solution (Watanabe et al,

1992b). These data suggest that water molecules may play an important role in the production mechanisms of the damage in aqueous systems.

Oxidation of ferrous ions in a Fricke solution was measured in the soft X-ray region down to 1.8 keV, which showed a decrease of the yield along with the decrease of the photon energy (Watanabe et al, 1994).

Concluding remarks

All the results shown above which were obtained using SR are very important in elucidating the mechanism of biological effects of radiation. These data demonstrate that SR is a powerful probe in the mechanistic studies in radiobiology. Experimental results from countries other than Japan are being accumulated recently and radiobiology using SR will soon become more popular.

References

Hieda, K., Maezawa, H., Ito, A., Kobayashi, K., Furusawa, Y. and Ito, T. (1986a) Choice of Coatings for the Optical Elements in the Irradiation System of Vacuum-Ultraviolet Radiation above 50 nm. Photochem. Photobiol., 44, 417-419

Hieda, K., Hayakawa, Y., Ito, A., Kobayashi, K. and Ito, T. (1986b) Wavelength Dependence of the Formation of Single-Strand Breaks and Base Changes in DNA by the Ultraviolet Radiation above 150 nm. Photochem. Photobiol., 44, 379-383.

Hieda, K. (1994) DNA damage induced by vacuum and soft X-ray photons from synchrotron radiation. Int. J. Radiat. Biol. (in press)

Hieda, K. Hirono, T., Azami, A., Suzuki, M., Furusawa, Y., Maezawa, H., Usami, N., Yokoya, A. and Kobayashi, K. (1994a) Single- and double-strand breaks in pBR322 plasmid DNA by monochromatic X-rays on and off the K-absorption peak of phosphorus. Int. J. Radiat. Biol., submitted

Hieda, K., Suzuki, K., Yokoya, A. and Kobayashi, K. (1994b) Single- and double-strand breaks induced by monochromatic ultra-soft X-ryas from 60 to 550 eV. J. Radiat. Res. in press

Ito, T., Ito, A., Hieda, K. and Kobayashi, K. (1983) Wavelength Dependence of Inactivation and Membrane Damage to *Saccharomyces cerevisiae* Cells by Monochromatic Vacuum-UV Radiation (145-190 nm). Radiat. Res., 96, 532-548

Ito, T., Kada, T., Okada, S., Hieda, K., Kobayashi, K., Maezawa, H. and Ito, A. (1984) Synchrotron System for Monochromatic UV Radiation (> 140 nm) of Biological Material. Radiat. Res., 98, 65-73.

Ito, T. and Saitou, M. (1991) Effects of vacuum ultraviolet radiation on deoxyoligonucleotides in solids in the wavelength region around and above ionization potential with special reference to the chain scission. Radiat. Phys. Chem. 37, 681-690

Kobayashi, K., Hieda, K., Maezawa, H., Ando, M. and Ito, T. (1987) Monochromatic X-ray Irradiation System(0.08nm-0.4nm) for Radiation Biology Studies Using Synchrotron Radiation at the Photon Factory. J. Radiat. Res., 28, 243-253.

Kobayashi, K., Hieda, K., Maezawa, H., Furusawa, Y., Suzuki, M. and Ito, T. (1991) Effects of K-shell X-ray Absorption of Intracellular Phosphorus on Yeast Cells. Int. J. Radiat. Biol., 59, 643-650

Maezawa, H., Furusawa, Y., Hieda, K., Kobayashi, K., Mori, T., Suzuki, K. and Ito , T. (1988) Killing on *E. coli* and *Deinococcus radiodurans* by monochromatic X-rays at resonance energy of phosphorus K-absorption edge. Photon Factory Activity Report, 6, 90

Saigusa, S., Ejima, Y., Kobayashi, K. and Sasaki, M.S. (1992) Induction of chromosome aberrations by monochromatic X-rays with resonance energy of phosphorus K-shell absorption edge. Int. J. Radiat. Biol., 61, 785-790

Saitou, M. (1993) Wavelength dependence of the photodecomposition of 2'-deoxythymidylyl-(3'-5')-2'-deoxythymidine, dTpdT, irradiated by monochromatic vacuum-UV photons above 50 nm. Photochem. Photobiol., 57, 950-957

Takakura, K., Maezawa, H., Kobayashi, K. and Hieda, K. (1994) Strand breaks in DNA in buffered solution induced by monochromatic X-rays around the K-shell absorption edge of phosphorus. Synchrotron Radiation in Biosciences.(eds. T. Sasaki et al), Oxford University Press, Oxford, pp756-764

Watanabe, M., Suzuki, M., Watanabe, K., Suzuki, K., Usami, N., Yokoya, A. and Kobayashi, K. (1992a) Mutagenic and Tranforming Effects of Soft-X-rays with Resonance Energy of Phosphorus K-absorption Edge. Int. J. Radiat. Biol., 61, 161-168

Watanabe, R., Ishikawa, M., Kobayashi, K. and Takakura, K. (1992b) Damage to Adenosine-triphosphate Induced by Monochromatic X rays Around the K Shell Absorption Edge of Phosphorus. Biophysical Aspects of Auger Processes, AAPM Symposium Series No. 8, (eds. R.W. Howell, V.R. Narra, K.S. Sastry and D.V. Rao), American Institute of Physics, New York, 24 - 36

Watanabe, R., Usami, N. and Kobayashi, K., (1994) Oxidation yield of the ferrous ion in a Fricke solution irradiated with monochromatic synchrotron soft X-rays in the 1.8-10 keV region. Int. J. Radiat. Biol., submitted

THE YIELDS OF FREE RADICALS INDUCED BY MONOCHROMATIC SOFT X-RAYS WITH ENERGY OF THE K-ABSORPTION EDGE OF BROMINE IN BrdU/dThd COMPLEXES

M. Kuwabara,[$] S. Sawamura,[#] O. Inanami,[@] and K. Kobayashi[*]
[$]Faculty of Veterinary Medicine and [#]Faculty of Engineering, Hokkaido University, Sapporo 060, [@]Faculty of Agriculture, Iwate University, Morioka 020, *Photon Factory, National Laboratory for High Energy Physics, Tsukuba 305, Japan

Introduction

Biological Auger effects have been found not only in Br-substituted plasmid DNA (Menke *et al.* 1991) but also in cells with Br-DNA when they are exposed to soft X-rays with energies above and below the K-absorption edge of Br (Halpern and Mütze 1978; Shinohara *et al.* 1985; Maesawa *et al.* 1988a, b; Usami *et al.* 1991). These biological Auger effects were sometimes correlated to enhanced DNA damage such as single- or double-strand breaks in these studies. Free radicals induced in DNA are regarded as precursors of base damage and strand breaks (Kuwabara *et al.* 1988, 1989, 1993). Therefore, it is of interest to examine whether the Auger effects are also reflected in free-radical formation in Br-substituted DNA when they are exposed to soft X-rays with energy corresponding to the K-absorption edge of Br. In the present study BrdU·dThd complexes were employed as Br-substituted DNA models, and the yields of free radicals were measured by ESR after irradiating them in the solid state with soft X-rays having energies above and below the K-absorption edge of Br.

Materials and Methods

Preparation of solid complexes between deoxythymidine (dThd) and 5-bromo-2'-deoxyuridine (BrdU). BrdU and dThd were dissolved in aqueous solutions with weight ratios of BrdU to BrdU plus dThd of 0, 0.1, 0.2, 0.4, 0.6, 0.8 and 1.0. The solutions were dried by centrifugal evaporation. The powders were then transferred to polyethylene tubes (i. d. 0.18 cm). About 50 mg of powder was suitable for the beam size of X-rays from the synchrotron. The tubes were evacuated with a mercury pump to 1×10^{-6} torr in Pyrex glass tubes followed by admission of Ar gas and then sealed.

Irradiation of the complexes by monochromatic soft X-rays on both sides of the K-absorption edge of Br. Synchrotron radiation at the Photon Factory of the

National Laboratory for High Energy Physics (Tsukuba, Japan) was used as a source of soft X-rays. Monochromatic X-rays at 13.45 and 13.52 KeVs corresponding to below and above the Br K-edge, respectively, were obtained using a channel-cut silicon (111) crystal. Details concerning the systems were previously described (Kobayashi *et al.* 1987). The tube containing powder was exposed to X-rays by being placed in front of an X-ray beam. The exposure dose was measured with a free-air ionizing chamber.

ESR measurements. After irradiation, the powder was transferred to the opposite side of the polyethylene tube, which was not exposed to X-rays. The tube was inserted into a cylindrical quartz ESR tube and placed in the cavity of an X-band ESR spectrometer (JES RE1X, JEOL, Japan). ESR measurements were made as described in the legend of Figure 1. The ESR signal intensities were obtained by doubly integrating the ESR spectra via a computer technique. Estimation of the radical concentrations was carried out by comparing them with the ESR signal intensity of DPPH and corrected by the weight of powder inside the tube.

Photon mass attenuation and energy absorption coefficients of elements. Photon mass attenuation (μ/ρ) and energy absorption (μ_{en}/ρ) coefficients of H, C, N and O for soft X-rays at 13.45 and 13.52 KeV were determined by a log-log interpolation from coefficients of photons at 10 KeV to those at 15 KeV by employing the data of Hubbell (1982). The coefficients of Br at 13.45 and at 13.52 KeV were determined by extrapolating the coefficients of photons with energies lower than 13.49 KeV (the K-absorption edge of Br) to 13.45 KeV and with energies higher than 13.49 KeV to 13.52 KeV, respectively.

Photon mass attenuation and energy absorption coefficients of BrdU, dThd and air. Mass attenuation and energy absorption coefficients of dThd, BrdU and air were obtained by adding the coefficients of H, C, N, O and Br in each compound with suitable weighting (Johns and Cunningham 1980).

Photon and energy fluences per roentgen. Photon fluences per roentgen (photons/cm^2R) and energy fluences per roentgen (erg/cm^2R) were obtained by the equations $86.9/(\mu/\rho)_{air}/1.6\times10^{-6}/h\upsilon(MeV)$ and $86.9/(\mu_{en}/\rho)_{air}$, respectively.

Radical yields per photon fluences. The mean free path ($1/\mu$) of photons with energy of 13.52 KeV was found to become shorter than the dimension of the powder (0.18 cm) when the BrdU content in the BrdU·dThd complexes was 0.4~1.0. Therefore, free radicals induced in the powder within the mean free path were assumed to contribute to actual ESR signal intensities. Then ESR signal intensities were converted to the number of free radicals induced per gram, roentgen and photon fluence.

Radical yields per energy absorption. Photon energy absorbed in powder was obtained by assuming that photon fluence exponentially decreased with increasing powder depth and, therefore, total energy absorbed was calculated by integrating the photon fluence over the whole powder and multiplying it by the photon energy and energy absorption coefficient. Finally, radical yield per energy absorption was obtained by the method described above.

Results and Discussion

Photon mass attenuation coefficients (μ/ρ) (cm^2/g) and energy absorption coefficients (μ_{en}/ρ) (cm^2/g) of dThd, BrdU and air for photons with energies below and above the Br K-absorption edge and ^{60}Co γ-ray energy of 1250 KeV are listed in Tables 1a and b, respectively. Table 1 shows that high values for both coefficients were obtained in BrdU at photon energies above the K-edge of Br (13.52 KeV). This suggested that strong absorption of photon energy occurred in BrdU·dThd complexes with high BrdU content.

The ESR spectra obtained are shown in Figure 1. Samples for ESR spectra (a), (b) and (c) contained 100% BrdU and samples for ESR spectra (a'), (b') and (c') contained 100% dThd. If one takes the exposure doses and ESR gains into consideration, one is forced to accept that the intense ESR spectra in BrdU and very weak ESR spectra in dThd were obtained when they were exposed to X-rays at 13.45 and 13.52 KeV (see legend of Figure 1). Exposure of BrdU to doses of more than 100 KR at E = 13.52 KeV showed saturation of radical formation. A high exposure dose of γ irradiation (2 MR) was needed to produce the intense ESR spectra in both samples. The three ESR spectra of BrdU were similar regardless of the difference of

Table 1: (a) Photon mass attenuation coefficients μ/ϱ (cm^2/g) and (b) energy absorption coefficients μ_{en}/ϱ (cm^2/g) of deoxythymidine (dThd), 5-bromo-2'-deoxyuridine (BrdU) and air obtained by adding coefficients for elements in each composite material with suitable weighting.

(a) μ/ϱ (cm^2/g)

Photon Energy (MeV) \ Compds	dThd	BrdU	Air
0.01	3.132	15.121	5.016
0.01345	1.547	6.717	2.16
0.01352	1.525	38.885	2.125
0.015	1.145	29.749	1.581
1.25	0.061	0.0583	0.058

(b) μ_{en}/ϱ (cm^2/g)

Photon Energy (MeV) \ Compds	dThd	BrdU	Air
0.01	3.192	14.355	4.46
0.01345	1.241	5.763	1.83
0.01352	1.221	21.212	1.80
0.015	0.877	17.402	1.30
1.25	0.028	0.0267	0.0265

photon energies. Though exact analysis of radical species induced in the powder state was impossible, this meant that the Auger processes produced no specific radicals even though these processes contributed to radical formation.

From the ESR signal intensities, the numbers of radicals per photon fluence at 13.45, 13.52 and 1250 KeV were calculated in each BrdU·dThd complex as described in Materials and Methods and plotted against BrdU content (Figure 2). It was observed that the higher the content of BrdU in the complexes the more the radicals

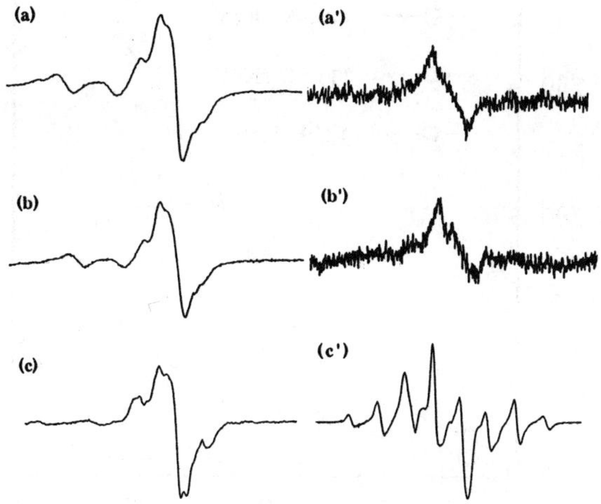

Figure 1. ESR spectra of solid BrdU exposed to a dose of about 100 KR of soft X-rays with (a) 13.52 KeV, (b) 13.45 KeV and (c) to a dose of 2 MR of ^{60}Co γ-rays. ESR spectra of solid dThd exposed to a dose of about 100 KR of soft X-rays with (a') 13.52 KeV, (b') 13.45 KeV and (c') to a dose of 2 MR of ^{60} Co γ-rays. ESR conditions were 20 mW microwave power, 100 kHz field modulation of 0.2 mT, a 0.1 sec time constant and gains of 100 for spectra (a) and (b), 79 for (c), 1000 for (a') and (b'), and 125 for (c'). All spectra were obtained as first derivatives at room temperature.

were produced at 13.45 and 13.52 KeVs. The increase of radical concentrations with increasing content was quite low in the case of γ-irradiation. The maximum values were obtained from the complexes with 80% BrdU and were about 25~50 times those of 100% dThd at photon energies above and below the K-edge of Br. γ-Irradiation of 100% BrdU induced about 2~3 times more radicals than γ-irradiation of 100% dThd did. The value obtained at 13.52 KeV in the 80% BrdU complex was about 3 times that obtained at 13.45 KeV in the 80% BrdU complex.

The next step was to obtain the yields of free radicals per absorbed energy. Usually, the conversion of the exposure dose (R) to the absorbed dose (rad or Gy) is performed by employing the factor, $f = 0.869\times(\mu_{en}/\rho)_{med}/(\mu_{en}/\rho)_{air}$. This factor is applicable only in targets in which electronic equilibrium holds and their photon absorption is quite low. Considering the coefficients at E = 13.52 KeV (data not shown) and the thickness of the powder (0.18 cm), it was difficult to apply this factor

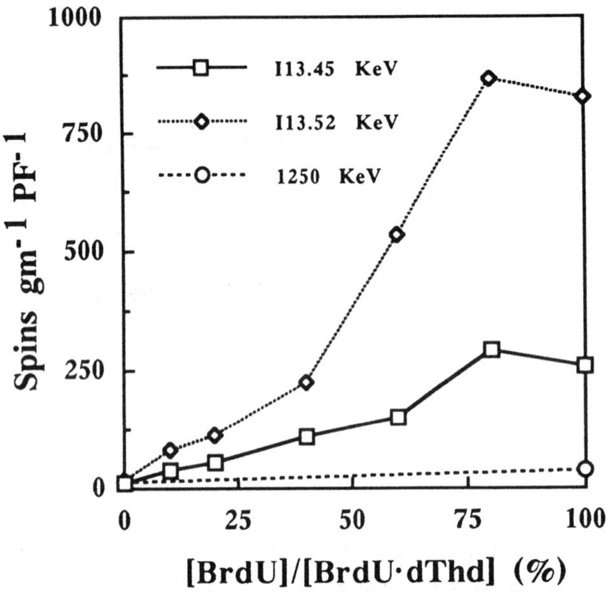

Figure 2. Yields of free radicals per photon fluence (cm^{-1}) in BrdU·dThd complexes at photon energies below and above the K-absorption edges of Br as well as at the photon energy of ^{60}Co γ-rays.

in powder with high concentrations of BrdU. In fact, the mean free paths of photons with energy at E = 13.52 KeV in the complexes consisting of 40~100% of BrdU were shorter than the thickness of the sample. The yields of products are generally expressed as the yield per photon (quantum yields) and the G-value per 100 eV or Gy for low energy photons like UV-light and high energy photons like ^{60}Co γ-rays, respectively. In the case of soft X-rays, in which inner-shell photoelectric absorption is contained, it seems difficult to express the yield not only in terms of quantum yield but also in terms of the G-value. In the present study we assumed that the majority of interactions of photons with targets were photoelectric processes and that, therefore, photon fluence exponentially decreased with the penetration depth. Thus, energy absorbed by the thin layer of powder (dx) at the depth of x (cm) could be obtained by the equation $(\mu_{en}/\rho)\times\Phi\times\exp(-\mu x)Sdx$, where, Φ is the energy fluence, μ the linear attenuation coefficient, and S the exposed area of powder. Since S is a common factor for all samples, it can be omitted from the equation. Considering the shape of the powder, we integrated the equation from 0 to the maximum powder size. A replot of Figure 2 in terms of the absorbed energy (erg/g) is shown in Figure 3. Higher yields of radicals at 13.52 KeV than at 13.45 KeV were observed in each complex as before, but the yield of the 80% BrdU complex at 13.52 KeV was reduced to about twice that at 13.45 KeV. The presence of Br in the complex resulted in a marked increase of radical yields compared with the yields in the absence of Br at 13.52 KeV but a more gentle increase was observed at 13.45 KeV. In the case of γ-irradiation a slight increase of radical yield due to the presence of Br was observed. The results obtained from irradiation at 13.45 KeV suggested that Br enhanced the efficiency of radical formation, probably due to the interference of the recombination processes between released electrons and cationic holes as well as the radical formation accompanied by the release of Br anions after electron capture. In the case of irradiation at 13.52 KeV, Auger processes seemed to participate in the radical formation in addition to the usual enhancement processes. That is, Auger processes liberated several electrons from the BrdU molecule and concomitantly left several cationic holes in it. The liberated electrons have the ability to further ionize the other molecules, whereas the holes migrate to distribute in the molecule and may extract electrons from neighboring molecules. These processes may further induce free radicals. Arguments similar to those for irradiation at 13.45 KeV seem to be applicable to γ-irradiation. However, enhancement of the radical formation by Br was quite low. This might be explained by the fact that the majority of interactions of photons at 13.45 KeV with electrons of the matter were photoelectric, whereas the majority of interactions at 1250 KeV consisted of Compton effects. The mechanism of energy deposition was very different between soft X-rays and high energy γ-rays. It might not be very informative to compare the yields obtained by soft X-irradiation with those obtained by γ-irradiation.

It was noted that the difference in the yields between 13.45 and 13.53 KeVs was small in the complexes with less than 40% BrdU. Using plasmid DNA in which 55% dThd was substituted for by BrdU, Menke *et al.* (1991) measured yields of the strand breaks by irradiation with monoenergetic X-rays on both sides of the K-absorption edge. The difference between the two energies was less clear. Maesawa *et al.* (1988b)

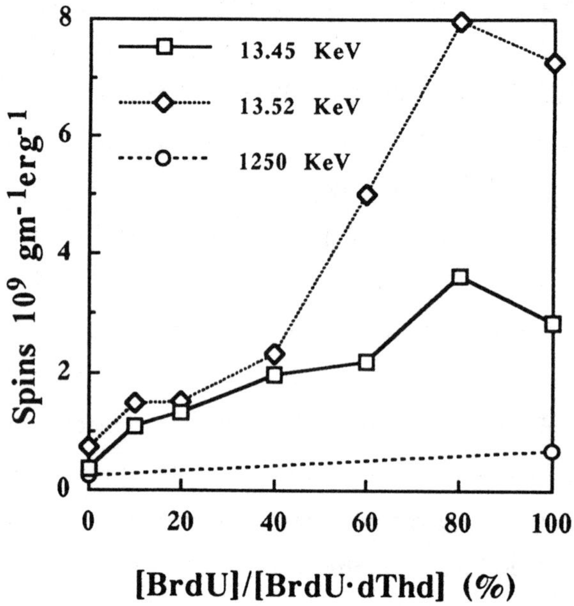

Figure 3. Yields of free radicals per energy (erg) which BrdU·dThd complexes absorbed at photon energies below and above the K-absorption edges of Br as well as at the photon energy of ^{60}Co γ-rays.

also observed no Br-substituted effects on plasmid DNA. Substitution of BrdU for 55% dThd corresponded to 15% of DNA (w/w). Figure 3 shows that the difference in the radical yields of the complexes with less than 20% BrdU was small between the two irradiations. This may be one reason why the situation is less clear in the case of brominated DNA.

Acknowledgments

The authors would like to express their gratitude to Dr. T. Ito for helpful discussion and also to the staff of the Photon Factory of the National Laboratory for High Energy Physics for technical assistance.

References

Halpern A, Mütze B (1978) Irradiation of micro-organisms with mono-energetic X-rays; biological consequences of the Auger effect. Int J Radiat Biol 34: 67-72

Hubbell JH (1982) Photon mass attenuation and energy-absorption coefficients from 1 keV to 20 MeV. Int J Appl Radiat Isot 33: 1269-1290

Johns HE, Cunningham JR (1980) The Physics of Radiology Chapter VI, Charles C Thomas Publisher Springfield

Kuwabara M, Inanami O, Endoh D, Sato F (1987) Spin trapping of precursors of thymine damage in X-irradiated DNA. Biochemistry 26: 2458-2465.

Kuwabara M, Hiraoka W, Sato F (1989) Evidence for the formation of strand-break precursors in hydroxy-attacked thymidine 5'-monophosphate by the spin trapping method. Biochemistry 28: 9625-9632.

Kuwabara M, Ohshima H, Sato F, Ono A, Matsuda A (1993) Spin-trapping detection of precursors of hydroxyl-radical-induced DNA damage: Identification of precursor radicals of DNA strand breaks in oligo$(dC)_{10}$ and oligo$(dT)_{10}$. Biochemistry 32: 10599-10606

Maezawa H, Hieda K, Kobayashi K, Furusawa Y, Mori T, Suzuki K, Ito T (1988a) Effects of monoenergetic X-rays with resonance energy of bromine K-absorption edge on bromouracil-labeled E. coli cells. Int J Radiat Biol 53: 301-308

Maezawa H, Hieda K, Kobayashi K, Ito T (1988b) Effects of Auger cascades of bromine induced by K-shell photoionization on plasmid DNA, bacteriophages, E. coli and yeast cells. <u>DNA Damage by Auger Emitters</u> Baverstock KF Charlton DE (eds) Taylor and Francis London pp. 135-146

Nelson WR, Hirayama H Rogers DWO (1985) The EGS4 Code System. SLAC-Report-265 Stanford University Sanfordpp. 92-93

Shinohara K, Ohara H, Kobayashi K, Maezawa H, Hieda K, Okada S, Ito T (1985) Enhanced killing of HeLa cells pre-labeled with 5-bromodeoxyuridine by monochromatic synchrotron radiation at 0.9 Å: an evidence for Auger enhancement in mammalian cells. J Radiat Res 26: 334-338

Usami N, Kobayashi K, Maezawa H, Hieda K, Ishizaka S (1991) Biological effects of Auger processes of bromine on yeast cells induced by monochromatic synchrotron X-rays. Int J Radiat Biol 60: 757-768

CRITICAL ENERGIES FOR SSB AND DSB INDUCTION IN PLASMID DNA: STUDIES USING SYNCHROTRON RADIATION

B.D. Michael,[1] K.M. Prise,[1] M. Folkard,[1] B. Vojnovic,[1] B. Brocklehurst,[2] I.H. Munro[3], and A. Hopkirk[3]

[1]CRC Gray Laboratory
Mount Vernon Hospital
Northwood, Middlesex
HA6 2JR, UK

[2]Department of Chemistry
University of Sheffield
Sheffield, S3 7HF, UK

[3]DRAL Daresury Lab,
Warrington, Cheshire
WA4 4AD, UK

Introduction

The most frequent energy depositions along the charged-particle tracks associated with high-energy ionizing radiations are on the order of 10's of eV. This has been shown in experimental studies (Rauth and Simpson, 1964; Johnson and Rymer, 1967) and in Monte-Carlo simulations of the interactions of charged particle tracks with DNA (Goodhead and Nikjoo, 1989; Charlton *et al.*, 1989). While simulations of this kind are being increasingly refined, there continues to be a need for experimental data that relate the magnitudes of energy-deposition events to the molecular damages that result from them. The energy range between a few and several hundred eV is of particular importance. In addition, the determination of the yields of various types of damage induced in DNA as functions of the energy deposited in it ("action spectra") may provide useful insights into the initial pathways of lesion induction. With much interest currently focussed on the role of lesion complexity in relation to biological effectiveness, it is important to know to what extent single energy deposition events can independently initiate damages involving more than one molecular site, for example a DNA dsb.

Early work in this field used the discrete excitation energies available either from metastably excited gases (Lücke-Huhle and Jung, 1973a, 1973b) or from the vacuum ultraviolet light (VUV) emitted by gaseous discharges (Wirths and Jung, 1972; Sontag and Dertinger, 1975) to study damage and inactivation in dry (ϕX174 DNA. More recently, studies by Bothe *et al.* (1990) and by Gurzadyan and Görner (1993) have used pulsed laser light at ~5 or ~6 eV to study ssb and dsb induction in aqueous solutions of calf thymus or plasmid DNA. The broad spectral range available from synchrotron VUV sources has recently been utilised by Ito (1992),

Hieda (1994) and by ourselves (Michael *et al.*, 1994) to measure ssb and dsb induction in dry pBR322 plasmid DNA. Also, we have used low-energy electrons to measure the energy dependences for ssb and dsb induction in dry pBR322 plasmid DNA (Folkard *et al.*, 1993). This paper reports some of our experimental methods and initial data, also with dry plasmid DNA, obtained using VUV from a synchrotron radiation source at selected energies in the range 7 to 25 eV, determining action spectra for ssb and dsb induction.

Methods

Vacuum UV irradiation: Irradiations were carried out at station 3.1 (dipole magnet) of the 2 GeV Daresbury electron synchrotron (DRAL Daresbury Laboratory, Daresbury, Cheshire, U.K.) operating in the multi-bunch mode with a stored current of ~200 mA. Photon energies in the range 7 to 25 eV were selected using a Seya-Namioka mount grating monochromator. Higher-order diffracted light was filtered out using windows of LiF (<11.5 eV) (1.75 mm thick, VUV grade, Harshaw Corp., Solon, OH 44139, USA), In:Ti (12 - 16.5 eV) (VF-111-b, Luxel Corp., Friday Harbor, WA 98250, USA) and AI:SI (>17 eV) (VF-11 1-a, Luxel Corp.). The arrangement used to irradiate samples is shown in Figure 1. The ultrahigh vacuum of the synchrotron

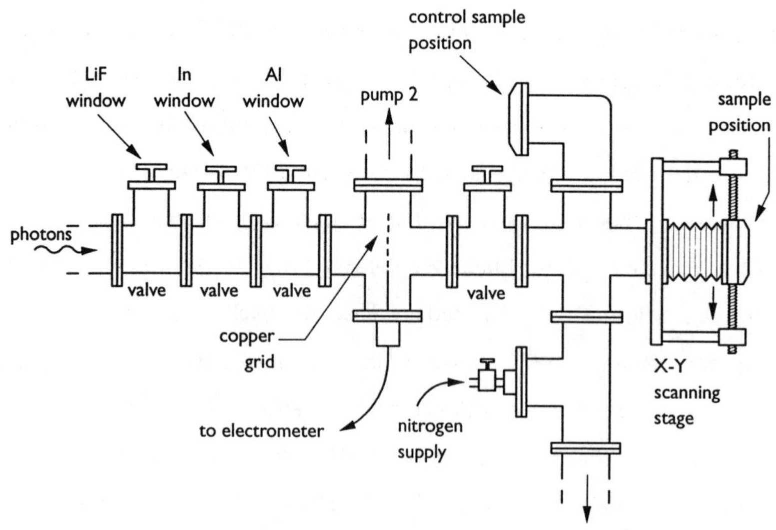

Figure 1. Apparatus used to irradiate plasmid DNA with synchrotron photons

and monochromator (~10^{-9} mbar) was isolated from the poorer vacuum of the DNA sample chamber (10^{-7} - 10^{-5} mbar) by whichever window was in place. The windows were selected and inserted without breaking the vacuum using an in-line assembly of three window valves in which they were mounted. After the window assembly, the beam passed through an electrically insulated copper grid. The photocurrent from the grid was integrated on an electrometer (Model 616, Keithley Instruments) and the resulting charge used to monitor the photon flux to which each sample was exposed. The grid was biassed to ~18 V to saturate its photocurrent. For dosimetry, a 0.9 cm^2 bright gold photodiode was placed at the sample position, with similar bias and current integration as were used for the monitor grid. Values for the quantum efficiency for photoelectric emission from gold were taken from Sampson (1967). At energies below 9 eV the quantum efficiency for gold becomes low and uncertain therefore in this region fluorescence from sodium salicylate was used as a measure of the UV flux. The fluorescence was detected by a photomultiplier, the anode current of which was integrated. This arrangement was calibrated against the gold photodiode at a higher energy, with reference to literature values for the energy dependence of salicylate fluorescence (Sampson, 1967). In addition, we have in the most recent work used a silicon photodiode, of area 1 cm^2, designed for use in the range 5 - 250 eV (type AXUV-100, International Radiation Detectors, Torrance, CA 90505, USA) from a batch with quantum efficiencies calibrated throughout this energy range by NIST. The quantum efficiency of the diode used by us is expected to be within 15% of the NIST calibration (personal communication, Dr. R. Korde, IRD). The diode photocurrent was on the order of ~1 nA and was measured directly at the input of a Keithley 616 electrometer operated in the "fast" mode. At energies where they could be intercompared, good agreement was found between the the gold and silicon diode methods of dosimetry.

The exposure rate at the sample position was on the order of 10^{10} photons cm^{-2} s^{-1}. To ensure uniform exposure across the samples, independent of the transverse distribution of beam intensity, the sample holder was scanned continuously during irradiation in a raster pattern in a plane at right angles to the beam axis. A vacuum bellows provided the necessary mechanical flexibility for the scan and the raster parameters were adjustable under computer control. There was also provision to include unirradiated control samples in a section which was not exposed to the beam but was subjected to exactly the same evacuation cycle as the irradiated samples. All parts of the system were evacuated using turbomolecular pumps. Irradiation was started when the sample chamber pressure was below 10^{-5} mbar. The irradiation times ranged from ~1 minute to several hours, depending on the exposures required and the VUV light levels available at the various photon energies selected. After irradiation, the sample chamber was refilled with nitrogen before removing the control and irradiated samples to atmosphere. Further details of the methods will be given in a future publication (Folkard *et al.*, in preparation).

Sample preparation and assay: Our initial studies used pBR322 plasmids (4363 bp), as described previously (Folkard *et al.*, 1993; Michael *et al.*, 1994). In general, the plasmids as prepared were >90% in the supercoiled form. The aim was to produce a dry monolayer of plasmids for exposure to the VUV beam. A 10 µl drop of solution containing 0.25 µg of DNA and 0.025 µg of salt (in the form of EDTA) was placed on a 13 mm diameter polycarbonate filter (10 nm pore size, Poretics Corp., Livermore, CA 94550, USA) and spread continuously over an area about 9 mm in diameter while drying took place to atmosphere. The filters were supported on glass cover slips. This was similar to the procedure used in our studies with electrons (Folkard *et al.*, 1993) and, as before, the shapes of the dose-effect curves were indicative of a partially shielded response due to failure to achieve the desired uniform monolayer of DNA and EDTA. The methods used to rehydrate and recover the samples after exposure and to assay for damage using gel electrophoresis have been described previously (Folkard *et al.*, 1993).

In our most recent work, we have improved the sample preparation methods to increase the sensitivity of the system, which was especially necessary at the lower energies studied. The improvements include freeze-drying the plasmid DNA directly on to glass coverslips to produce a more even monolayer and to reduce the background damage due to the evacuation process. Also, we decided to use a larger plasmid, pMSG-CAT (8404 bp), to increase the responsiveness of the system by virtue of the increased target size.

Results

Dose-effect relationships for ssb and dsb induction at 8 eV in pMSG-CAT are shown in Figure 2(a) and (b) respectively. As described below, the ssb:dsb ratio is >10:1 and therefore loss of the supercoiled form is predominantly due to ssb induction. Control samples that were exposed to the same evacuation cycle as the irradiated samples, but not to VUV, showed an almost constant level of damage independent of the evacuation (and exposure) time, indicating that prolongation of evacuation does not increase the background level of damage detectably. However, the proportion of the supercoiled form in the control samples was 10 - 20% lower than that of the stock solution, consistent with a constant small amount of ssb induction due to the handling and evacuation procedures. Figure 2(b) shows the production of the linear form due to dsb induction. As with ssb, control samples showed no significant increase in dsb with evacuation time.

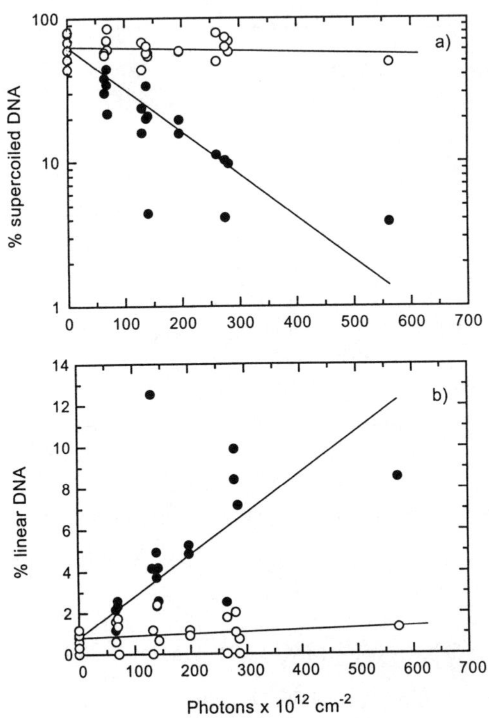

Figure 2. Induction of ssb and dsb in dry pMSG-CAT plasmids by 8 eV synchrotron photons: (●) control, (O) irradiated samples.

According to theory (Hempel and Mildenberger, 1987), loss of the supercoiled plasmid should be exponential with dose (i.e. a straight line in Figure 1(a)) and conversion to linear of the form $\alpha D.exp(-\alpha D)$ (i.e. curved slightly downwards in Figure 1(b)). As shown, both sets of data have been fitted with straight lines. However, with these and other data, there is a strain to the fit suggesting that the underlying response is curved as if the sensitivity decreases with increasing exposure. Similar behaviour was found previously in our studies with low-energy electrons (Folkard *et al.*, 1993) and photons (Michael *et al.*, 1994) and in earlier work with low energy photons (Wirths and Jung, 1972; Sontag and Dertinger, 1975) and has been attributed to partial shielding effects. Where appropriate, allowance has been made, as in our electron work, and a value derived for the responsiveness in the low-dose region. Figure 3 shows the yields of ssb and dsb calculated from the dose-response curves obtained between 7 and 25 eV using pMSG-CAT plasmids on glass. The yields are expressed as breaks per plasmid per incident photon.

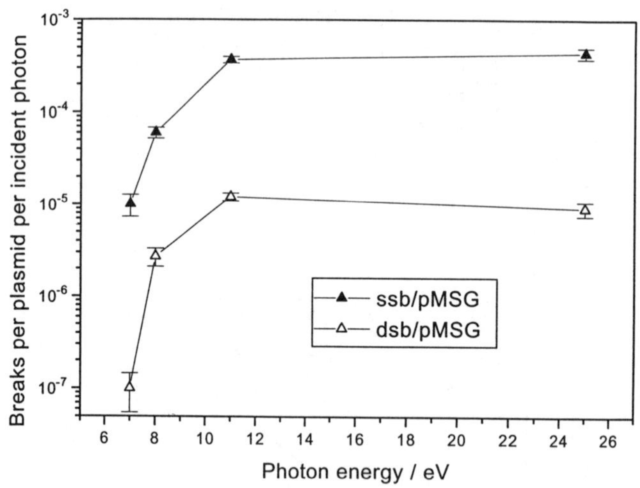

Figure 3. Induction of ssb and dsb in pMSG-CAT plasmids versus photon energy

Discussion

The data presented here for pMSG-CAT plasmids are in general agreement with those we reported earlier for pBR322 (Michael *et al.*, 1994). Use of a larger plasmid together with improved methods of sample preparation have allowed us to extend measurements down to 7 eV and demonstrate that the yields of ssb and dsb both decline quite steeply below 8 eV. Use of a VUV-sensitive silicon photodiode of known sensitivity has improved our confidence in the dosimetry, especially at the lower energies. The data presented here reinforce our earlier observation that the shapes of the ssb and dsb action spectra appear to be the same. In the earlier study with pBR322 it appeared that the ssb:dsb yield ratio was between 12:1 and 20:1, but the present data and other data not shown (Prise *et al.*, in preparation), both with pMSG-CAT, indicate a ratio of ~30:1. The higher ratio indicated by these more recent data may reflect the improved methods of sample preparation. As postulated with our earlier photon data, the similarity of the shapes of the action spectra for ssb and dsb suggest that both lesions predominantly have a common precursor in this dry DNA system. Other studies have proposed the existence of a single-event pathway to dsb induction (Boon *et al.* 1984; Siddiqi and Bothe, 1987; Krisch *et al.*, 1991), but it has also been shown that multiple-event induction may occur (Ward, 1981; Krisch *et al.*, 1991; Prise *et al.*, 1993). Bothe *et al.* (1990) have shown ssb and dsb induction in aqueous solutions of calf thymus DNA by laser pulses at 248 nm (~5 eV). Gurzadyan and Görner (1993) have reported these and other forms of damage in solutions of plasmid and calf thymus DNA by laser pulses at 193 nm (~6 eV).

In our earlier work with pBR322 and low-energy electrons (Folkard *et al.*, 1993), we found that while the threshold energy for ssb induction was below 25 eV, that for dsb induction lay between 25 and 50 eV. For dsb the value is much higher than we find with photons, indicating that in this energy range damage induction depends quite strongly on the nature of the energy transfer. Modelling studies by Charlton *et al.* (1989) indicate a threshold energy for induction of ssb at 17.5 eV, which is higher than indicated by these and the earlier VUV data. Their model indicates that dsb are not induced by energies below about 40 eV, which is in closer accordance with our earlier observations for electrons than with the present data for photons.

Acknowledgments

The authors wish to acknowledge the support of the Cancer Research Campaign, the Engineering and Physical Sciences Research Council and a grant from Radiation Protection Research Action of the European Community.

References

Boon, PJ, Cullis, PM, and Symons, MCR, and Wren, BW (1984) Effects of ionising radiation on deoxyribonucleic acid and related systems Part I: The role of oxygen *J Chem Soc: Perkin Trans II.* 1393-1399

Bothe, E, Görner, H, Opitz, J, Schulte-Frohlinde, D, Siddiqi, A, and Wala, M (1990) Single- and double-strand break formation in double-stranded DNA upon nanosecond laser-induced photoionization *Photochem Photobiol* **52:** 949-959

Charlton, DE, Nikjoo, H, and Humm, JL (1989) Calculation of initial yields of single and double strand breaks in cell nuclei from electrons, protons and alpha-particles *Int J Radiat Biol* **56:** 1-19

Folkard, M, Prise, KM, Vojnovic, B, Davies, S, Roper, MJ, and Michael, BD (1993) The measurement of DNA damage by electrons with energies between 25 eV and 4000 eV *Int J Radiat Biol* **64:** 651-658

Goodhead, DT, and Nikjoo, H (1989) Track structure analysis of ultrasoft X-rays compared to high and low LET radiations *Int J Radiat Biol* **55:** 513-529

Gurzadyan, GG, and Görner, H (1993) Photolesions in DNA upon 193 nrn excitation *Photochem Photobiol* **58:** 71-80

Hempel, K, and Mildenberger, E (1987) Determination of G-values for single- and double-strand break induction in plasmid DNA using agarose gel electrophoresis and a curve-fitting procedure *Int J Radiat Biol* **52:** 125-138

Hieda, K (1994) DNA damage induced by vacuum and soft X-ray photons from synchrotron radiation *Int J Radiat Biol* (in press)

Ito, T (1992) The effects of vacuum-UV radiation (50 - 190 nm) on microorganisms and DNA *Adv Space Res* **12:** 249-253

Johnson, CD, and Rymer, TB (1967) Existence of collective-excitation energy losses from an electron beam passing through biological materials *Nature* **213:** 10451047

Krisch, RE, Flick, MB, and Trumbore, CN (1991) Radiation chemical mechanisms of single- and double-strand break formation in irradiated SV40 DNA *Radiat Res* **126:** 251-259

Lücke-Huhle, C and Jung, H (1973a) Energy requirements for damaging DNA molecules I Energy transfer from metastable states of excited gases *Int J Radiat Biol* **24:** 221-228

Lücke-Huhle, C, and Jung, H (1973b) Energy requirements for damaging DNA molecules II Energy transfer from metastable states of excited gases to double-stranded (ϕX174 RF DNA *Int J Radiat Biol* **24:** 339-349

Michael, BD, Prise, KM, Folkard, M, Brocklehurst, B, Munro, IH, and Hopkirk, A (1994) Action spectra for single- and double-strand break induction in plasmid DNA: Studies using synchrotron radiation *Int J Radiat Biol* (in press)

Prise, KM, Davies, S, and Michael, BD (1993) Evidence for induction of DNA double-strand breaks at paired radical sites *Radiat Res* **134:** 102-106

Rauth, AM, and Simpson, JA (1964) The energy loss of electrons in solids *Radiat Res* **22:** 643-661

Sampson, JAR, (1967), *Techniques of Vacuum Ultraviolet Spectroscopy* (New York: John Wiley & Sons)

Siddiqi, MA, and Bothe, E (1987) Single- and double-strand break formation in DNA irradiated in aqueous solution: Dependence of dose and OH radical scavenger concentration *Radiat Res* **112:** 449-463

Sontag, W, and Dertinger, H (1975) Energy requirements for damaging DNA molecules III The mechanisms of inactivation of bacteriophage (ϕX 174 DNA by vacuum ultra-violet radiation *Int J Radiat Biol* **27:** 543-52

Ward, JF (1981) Some biochemical consequences of the spatial distribution of ionizing radiation-produced free radicals *Radiat Res* **86:** 185-195

Wirths, A, and Jung, H (1972) Single-strand breaks induced in DNA by vacuum ultraviolet radiation *Photoahem Photobiol* **15:** 325-330

COMMENTS ON THE RADIATION ENERGY PARAMETERS OF DNA STRAND BREAKS AND SSB/DSB RATIOS

Takashi Ito
Hagiyama-cho 4-14-21
Higashimurayama-shi
Tokyo 189 Japan

There are various energy parameters in use for DNA strand breaks: the energy required per strand break, the energy selected from the correlation of strand-break yields and deposited energy frequencies in a defined volume, the energy chosen from the oscillator strength of DNA, or the threshold energy to be determined from the experiments by low-energy monochromatic radiations (Table 1). The primary objective of this note is to examine the relevance of these parameters to the DNA strand breaks from the mechanistic viewpoints. For example, a doubt may be immediately cast on the quantity of 1750 eV per DSB (Table 1), i.e. how this large energy is expended to result in the scission of the two strands.

In the current standard method the yield of strand breaks, on which the first parameter in Table 1 is based, is calculated based on the respective DNA fractions corresponding to SSB and DSB incidences over the total DNA irradiated, and expressed in the number of strand breaks (SSB or DSB)/Gy Da, providing the estimated absorbed dose in the DNA. The ratio of SSB to DSB is then calculated either as the ratio of the respective yields, or of the required energies per SSB and DSB, or of the respective G values (Table 2), all being essentially the same in the sense that all use the estimated absorbed dose, of which the calculations depend solely on the element composition and the density of DNA.

Then, it follows that, in the current SSB/DSB in solids, the strand breaks are assumed to occur in a homogeneous medium, composed of certain elements constituting DNA, as a random event with no reflection of molecular parameters of DNA itself. A plausible presumption is that such events are identical in terms of energy expended, that is, the precursors produced by the primary energy absorption are assumed to be the same. Otherwise, under the situation where many different electronic states are expected to be produced in different parts of DNA constituents, the SSB/DSB value so calculated would be too arbitrary. Indeed, Neary et al (1970) held such a view that the SSB/DSB ratio (28 in their case, see Table 1) meant a simple probability that led to the respective breaks. This view, however, may be seriously questioned as to the obscurity of the molecular basis.

Table 1. Radiation energy parameters for DNA strand breaks

Description of parameter	SSB (eV)	DSB (eV)	Key reference
Energy required per strand break in solids based on the conventional absorbed dose	64	1750	Neary et al 1970
Energy assigned from track-structure simulation and strand-break yield in vivo based on conventional dose	≥30[a]	≥100	Goodhead and Nikjoo 1989 [a]Estimated from Fig. 2 assuming SSB/DSB=15
Energy selected from the oscillator strength of DNA	29.9	60[b]	Holley et al 1990 [b]Assumed two SSB events for DSB
Threshold energy determined from monochromatic radiation experiments in solids	~7		Hieda et al 1986 Excluded exceedingly low events
	7-8		Ito et al 1987 Scission of dApdA
	<25	25-50	Folkard et al 1993 (electrons)

Additional information: a 2-5 ionization cluster inferred for DSB (Brenner and Ward 1992); 8.3 eV (SSB/DSB=301) and 20.7 eV (SSB/DSB=133) (Suzuki et al 1993); about 8 eV for both SSB and DSB (ratio 20) (Michael 1994, personal communications). Last two from photon experiments in solids.

All of those inexplicable figures mentioned above are derived from the fact that the current treatment for the quantification of strand breaks are based on the radiation dose expressed as a homogeneous energy distribution over the medium. For the approaches which avoid the use of this conventional absorbed dose three categories can be considered based on: (a) incident photon fluence, (b) microdosimetric energy deposition events, and (c) oscillator strength information of DNA for the major absorption process due to the passage of a charged particle. This latter approach seems most promising in the sense that, unlike the other approaches, molecular properties of DNA are explicitly introduced.

Table 2. Compiled SSB/DSB ratios using the yield or the induction constant (based on the conventional absorbed dose) of strand breaks with high energy radiations

SSB/DSB Ratios					
In vitro			In vivo	Remarks	Reference
Solution	Humid	Dry			
	28				(1)
26				Near direct conditions	(2)
			10	SV40 in CV-1 cells	(3)
103				No scavenger	(4)
16				DSB from radical transfer	(5)
			50	V79 cells under hypoxia	(6)
	10				(7)
20-25				Low scavengers	(8)
46			20	No scavenger in solution	(9)
21	11	18		No scavenger in solution	(10)
			9.8[a)]	Rat thymocytes (N_2)	(11)
		12[a)]			(12)

[a)]These electron experiments are added for comparison, the energy is 10 MeV (11) and 4 keV (12), respectively. (1) Neary et al 1970; (2) Christensen et al 1972; (3) Roots et al 1985; (4) Hempel and Mildenberger 1987; (5) Siddiqi and Bothe 1987; (6) Kampf 1988; (7) Baverstock and Will 1989; (8) Roots et al 1990; (9) Krisch et al 1991; (10) Ito et al 1993; (11) Lennartz et al 1973; (12) Folkard et al 1993.

In case (a), the quantity obtained as the cross-section is difficult to interpret at present because numerous ionized states and corresponding relaxation pathways would be involved in the production of strand breaks. This, in turn, makes the SSB/DSB ratios derived from the yields hopelessly complicated. Although the track-structure simulation analyses categorized in case (b) were successful in finding a specific energy parameter for DSB through correlations with the frequency of the strand-break yield (Goodhead and Nikjoo 1989), the approach was after all not directed to the elucidation of chemical processes for the strand breaks. It, however, produced a 100 eV-energy deposition, in a volume mimicked the DNA segment size (2 nanometers), as being responsible for DSB (Table 1). It is quite interesting here to note a great reduction from 1750 eV per DSB in early homogeneous dose treatment mentioned above. Other event-based approaches do not seem to provide appropriate dose parameters (however, e.g. see Watt 1989). In one approach, perhaps categorized in a modification of case (c), an absorption event of 29.9 eV was simply adopted from the absorption spectrum of DNA as the basic parameter (see Holley et al 1990 in Table 1). This parameter is obviously tentative in the absence of sufficient knowledge. In carrying out the case (c) approach in a more rigorous way, we must undertake measurements of cross-sections of SSB and DSB along with the various, highly excited electronic states over the energy range up to 100 eV or so where the effective absorption would occur. It would lead, if fortunate, to the threshold energies as well, to the better defined SSB/DSB ratio using the efficiencies of the respective relaxation pathways as a function of photon energies. The direct measurement of the threshold energies for SSB and DSB (last part of Table 1) may be regarded as part of case (c). It has only recently began with the monochromatic low-energy photons from the synchrotron radiation. One of the immediate concerns is whether or not the threshold energy is common to both SSB and DSB. Available data seem not enough to judge it with confidence. Simple electron bombardment experiments in the extremely low-energy region (Folkard et al 1993) involve many troublesome problems for the determination of the threshold energy.

As has been well-known the SSB/DSB ratios, calculated conventionally as stated above, vary depending greatly on the conditions, especially with solution systems. Even under no scavenger conditions the values range from 20 to over 100. In solid and cellular systems they seem to tend to converge to a range from 10 to 20 (Table 2). This may be explicable on the grounds that the energy absorption processes in the latter systems would be less variable as to the condition, even if the responsible precursor states are not identifiable at present. With high-LET radiations, of which one needs a separate discussion, a tendency of lower SSB/DSB ratios is predictable based on the energy deposition structure of increased density. Nonetheless, irrespective of LET, the actual SSB/DSB "values" have never been given a reasonable molecular explanation except in one case (Siddiqi and Bothe 1987). Thus the SSB/DSB ratio remains at best a relative parameter only useful for the purpose of comparisons under a certain set of conditions. Generally, as has been discussed, it involves too many unresolvable efficiencies associated with yet unknown relaxation pathways that lead to the breaks.

In summary, the current energy parameters of DNA strand breaks, including the SSB/DSB ratios that are derived directly from them, have been examined from the mechanistic viewpoints. It indicates the inadequacy of using the homogeneous dose treatments widely employed at present. It also indicates the need for understanding of the behaviors of radiation-induced precursor states that lead to the strand breaks via dissociations.

Acknowledgements

I am greatly benefitted from discussions with Professor N. Oda in many respects relating to this paper.

References

Baverstock KF, Will S (1989) Evidence for the dominance of direct excitation of DNA in the formation of strand breaks in cells following irradiation. Int J Radiat Biol 55:563-568

Brenner DJ, Ward JF (1992) Constraints on energy deposition and and target size of multiply damaged sites associated with DNA double-strand breaks. Int J Radiat Biol 61:737-748

Christensen RC, Tobias CA, Taylor WD (1972) Heavy-ion-induced single- and double-strand breaks in ϕX-174 replicative form DNA. Int J Radiat Biol 22:457-477

Folkard M, Prise KM, Vojnovic B, Davies S, Roper MJ, Michael BD (1993) Measurement of DNA damage by electrons with energies between 25 and 4000 eV. Int J Radiat Biol 64:651-658

Goodhead DT, Nikjoo H (1989) Track structure analysis of ultrasoft X-rays compared to high- and low-LET radiations. Int J Radiat Biol 55:513-529

Hempel K, Mildenberger E (1987) Determination of G-values for single and double strand break induction in plasmid DNA using agarose gel electrophoresis and a curve fitting procedure. Int J Radiat Biol 52:125-138

Hieda K, Hayakawa Y, Ito A, Kobayashi K, Ito T (1986) Wavelength dependence of the formation of single-strand breaks and base changes in DNA by the ultraviolet radiation above 150 nm. Photochem Photobiol 44:379-383

Holley WR, Chatterjee A, Magee JL (1990) Production of DNA strand breaks by direct effects of heavy charged particles. Radiat Res 121:161-168

Ito T, Saito M, Taniguchi T (1987) A survey of photoproducts of an irradiated oligonucleotide by monochromatic photons with the energy ranged from 6.5 to 22.5 eV. Photochem Photobiol 46:979-984

Ito T, Baker SC, Strickley CD, Peak JG, Peak MJ (1993) Dependence of the yield of strand breaks induced by gamma-rays in DNA on the physical conditions of exposure: water and temperature. Int J Radiat Biol 63:289-296

Kampf G (1988) Induction of DNA double-strand breaks by ionizing radiations of different quality and their relevance for cell inactivation. Radiobiol Radiothera 29:631-658

Krisch RE, Flick MB, Trumbore CN (1991) Radiation chemical mechanisms of single- and double-strand break formation in irradiated SV40 DNA. Radiat Res 126:251-259

Lennartz M, Coquerelle T, Hagen H (1973) Effects of oxygen on DNA strand breaks in irradiated thymocytes. Int J Radiat Biol 24:621-625

Michael BD (1994) Private communication

Neary GJ, Simpson-Gildmeister VFW, Peacocke AR (1970) The influence of radiation quality and oxygen on strand breakage in dry DNA. Int J Radiat Biol 18:25-40

Roots R, Kraft G, Gosschalk E (1985) The formation of radiation-induced DNA breaks: the ratio of double-strand breaks and single-strand breaks. Int J Radiat Biol 11:259-265

Roots R, Holley W, Chatterjee A, Irizarry M, Kraft G (1990) The formation of strand breaks in DNA after high-LET irradiation: a comparison of data from in vitro and cellular systems. Int J Radiat Biol 58: 55-69

Siddiqi MA, Bothe E (1987) Single- and double-strand break formation in DNA irradiated in aqueous solution:dependence on dose and OH radical scavenger concentration. Radiat Res 112:449-463

Suzuki K, Kobayashi K. Hieda K (1993) Photon energy dependence of the induction of SSB and DSB in plasmid DNA in UV and soft X-ray regions. Proc 36th Annu Meeting Japan Radiat Res Soc 3-B-14, in Japanese

Watt DE (1989) An approach towards a unified theory of damage to mammalian cells by ionising radiation for absolute dosimetry. Radiat Protec Dosimetry 27:73-84

Chapter 8

Mass Spectrometry and Nuclear Magnetic Resonance Spectroscopy for the Determination of Structural Perturbations in Irradiated DNA

THE STUDY OF RADIATION INDUCED DNA-PROTEIN CROSSLINKS BY ELECTROSPRAY IONIZATION MASS SPECTROMETRY

Mary S. Lipton, Alfred F Fuciarelli, David L. Springer and Charles G. Edmonds.
Material and Chemical Sciences, and Biology and Chemistry Departments.
P.O. Box 999 M.S. P8-19
Pacific Northwest Laboratory
Richland, WA 99352 USA

Abstract

We have used peptide-thymine and histone-thymine solutions to model protein-DNA cross-linking chemistry induced in intact chromatin by low dosage of g-irradiation. Induced thymine crosslinking to model peptide systems has been evaluated by on-line liquid chromatography-electrospray ionization mass spectrometry (ESI-MS) and tandem mass spectrometry (MS-MS) with sensitivity comparable or superior to conventional GC-MS determinations. Radiation damage at doses as low as 0.1 Gy can be detected by this method. Additionally, thymine modified H2B can also be examined by ESI-MS and tandem-MS of the intact protein and proteinase digests. Limited information on the sites of thymine crosslinking can be obtained by tandem mass spectrometry on the intact multiply charged molecular species. More detailed information on the sites of thymine-protein crosslinking is obtained by on-line LC-ESI-MS of selective proteolysis products of the modified histones. Further MS-MS experiments on the selective proteolysis products will reveal specific modified amino acids and their sequence location. These methods reveal the nature, extent and site of radiation induced modification of the oligopeptides. Studies are being extended to the examination of the radiation induced covalent interactions between histones and oligonucleotides in higher states of organization. The eventual object is to study DNA-protein crosslinking interactions in model and native genomic nucleosome systems.

Introduction

Extensive research has been devoted to the understanding of the effect of ionizing radiation on DNA and proteins in living cells. The interaction of ionizing radiation with cellular water results in the formation of OH radicals and other free radicals in living systems (1). These free radicals can be formed in spurs (2), which contain a number of radical pairs generated by the reactions of OH radicals with DNA and the proteins components. In aqueous solutions, these OH radicals are known to cause a variety of modifications to DNA which include sugar and base damage as well as covalent DNA protein crosslinks (3, 4). Little is known about the structure and chemical nature of DNA-protein crosslinks in mammalian cells and about the role of these lesions in cellular responses to ionizing radiation.

Dizdaroglu and coworkers developed and characterized a model system for radiation induced DNA-protein crosslinks by irradiating the nucleobase thymine with the amino acid tyrosine using gas chromatography-mass spectrometry (5-7). Additionally, the formation of crosslinks with other amino acids and nucleobases such as lysine and cytosine have been studied in intact chromatin or in nucleohistones. (8-12), and covalent DNA-protein crosslinks in intact chromatin has been used to study the interaction of the DNA with histones. (13-15). This and other work has focused on the use of gas chromatography-mass spectrometry (GC-MS) to characterize the crosslinking moiety. Because GC-MS necessitates the digestion of intact biopolymers to their monomeric constituents followed by preparation of stable derivatives, sequence information is lost. It would be advantageous therefore to develop a method to detect radiation damage in substantially intact biopolymers i.e. nucleosomes and/or nucleohistones. ESI-MS permits the direct analysis of the intact modified proteins and oligonucleotides providing the potential for unambiguous evaluation of the nature and extent of natural and induced modifications. In this work we have described the characterization of nucleobase-amino acid crosslinks in a model system of the peptide angiotensin and the nucleobase thymine.

Materials and Methods

Materials: Angiotensin I and thymine base were purchased from Sigma.

Histone Isolation: 1g chick erythrocyte cells were lysed with 10% triton solution and the nuclei was recovered by centrifugation. After nuclear lysis with 70 mM NaCl, the histones were separated from the DNA by addition to hydroxyapatite resin. The proteins were eluted with 2.5 M NaCl. Individual histones were separated with a semi-prep BioRad 318 reversed phase C-18 column on a Waters 310 HPLC and elution was detected with a Waters 990 detector. Histone eluted on a gradient of 40% to 50% acetonitrile with 0.2% TFA.

Irradiation Procedures: The peptide solutions containing 20 μM angiotensin and 200 μM thymine base in a total volume of 1 mL were irradiated in a Co-60 gamma source to total doses ranging from 1 to 10 Gy. Solutions were saturated with nitrous oxide for 20 min before and during irradiation. The dose rate of (1 Gy/min) was determined by Fricke dosimetry (16). Irradiated and untreated samples lyophilized and dissolved (1 μg/μL) in 5% aqueous acetic acid for analysis.

Electrospray Ionization Mass Spectrometry: Protein solutions were analyzed by electrospray ionization mass spectrometry using a TAGA 6000E tandem quadrupole instrument (Sciex, Thronhill, Ontario, Canada) equipped with an ion source and atmosphere/vacuum interface previously described (17). Peptide solutions were analyzed with a Finnigan TSQ 7000 tandem quadrupole mass spectrometer. For the peptide and protein solutions, 20 spectra were acquired at intervals of 0.1 m/z or 0.2 m/z, respectively over the mass range of m/z 50 to 1400. The spectra were averaged, the centroid of each ion manually selected and the relative molecular masses calculated according to standard algorithms.

Results

Upon exposure of both peptide and histone solutions to ionizing radiation, crosslinked species between the nucleobase thymine and the amino acid tyrosine were obtained. The partial electrospray ionization mass spectrum of the angiotensin mixture irradiated to 75 Gy (figure 1) contains three distinct peaks at mass/charge 649, 711, and 720. These values in the 2+ charge state (i.e. $(M+2H)^{2+}$), correspond to the unadducted angiotensin and two different crosslinked species respectively. The crosslinked species which have mass increments of 124 Da (thymine) and 141 Da (thymine plus a hydroxyl group), have been previously studied by GC-MS and NMR. This type of damage has been detected in this peptide system at doses as low as 0.1 Gy (18).

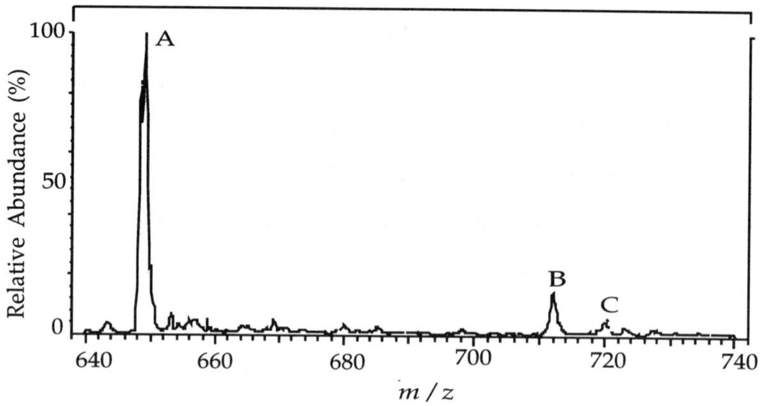

Figure 1: Partial ESI-MS spectra of angiotensin irradiated to 75 Gy showing the 2+ charge state of the unchanged angiotensin (A) and thymine (B) and hydroxythymine (C) adducted peptides.

When the histone H2B thymine mixture was exposed to increasing doses of ionizing radiation, signals in the resulting mass spectra (figure 2) become increasingly broad and complicated indicating that modifications to the protein have occurred. Panel A in figure 2 corresponds to the unirradiated histone while panels B and C correspond to the histone mixture irradiated to 5 and 10 Gy respectively. Higher resolution spectra of the 20+ charge state of the unirradiated and irradiated histone (figure 3) reveal a number of peaks for molecular species

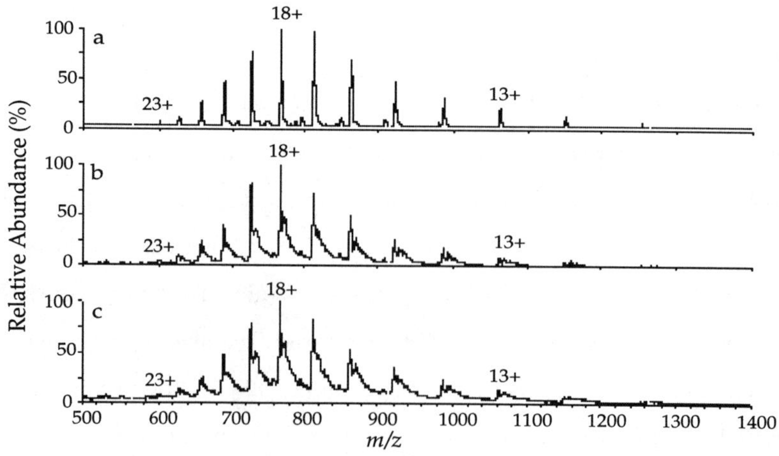

Figure 2: Electrospray ionization mass spectra of histone H2B irradiated in the presence of thymine to 0 Gy (a), 5 Gy (b) and 10 Gy (c).

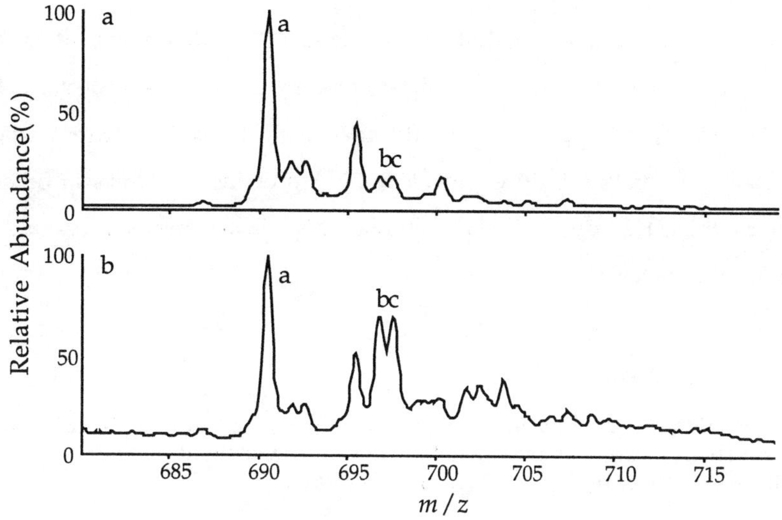

Figure 3: Partial electrospray ionization mass spectrum showing the 20+ charge state of histone H2B irradiated in the presence of thymine to 0 Gy (a) and 5 Gy (b).

corresponding to post translational and artifactual modifications (e.g. band c in unirradiated samples). Peaks b and c, while present in the unirradiated sample, are substantially larger in the irradiated sample and correspond to mass increments of 124 Da and 141 Da which are assigned to thymine and hydroxythymine adducts. As in the peptide system, these peaks are probably due to the covalent crosslinking to thymine to the protein. Further studies on this system involving proteolytic digestion of the protein and tandem mass spectrometry are needed to confirm both the location and the type of crosslinking modification.

Conclusions

In conclusion, ESI-MS is a valuable tool to characterize radiation damage in intact biopolymers. Some of the advantages over previous techniques are the possibility of analyzing radiation effects at low doses without acid hydrolysis and chemical derivitization techniques. Since the biopolymer remains intact during ESI-MS analysis, this technique will enable us to identify the location of and characterize the type of damage. This will be accomplished by the use of selective proteolytic digestion of the protein and on line LC/ESI-MS to identify the peptide fragment to which the crosslink has occurred. The determination of the exact location of the modification can be determined by tandem mass spectrometry of the peptide fragments. We are currently attempting to extend these studies to the detection of radiation induced covalent DNA-proteins crosslinks in other histone proteins, H2A/H2B dimer, H3/H4 tetramer, histone octamer, and eventually model nucleosome systems.

Acknowledgements

Work supported by the National Institutes of Health (GM47423). The Pacific Northwest Laboratory is operated for the U.S. Department of Energy (Contract DE-AC06-76RLO 1830) by Battelle Memorial Institute.

References

1 Halliwell, B., and Gutteridge, J.M.C. (1989) Free Radicals in Biology and Medicine, 2nd ed., Clarendon, Oxford. C. von Sonntag, "The Chemical Basis of Radiation Biology." Taylor & Francis, London, 1987
2 Mozumder, A. and Magee, J.L. (1966) Radiat. Res. 28 203-214.
3 C. von Sonntag, "The Chemical Basis of Radiation Biology." Taylor & Francis, London, 1987.
4 Oleinick, N.L., Chiu, S., Ramakrishnan, N., and Xue, L. (1987) Br. J. Cancer 55 (suppl. VIII), 135-140.
5 Dizdaroglu, M. (1984) J. Chrom. 295 103-121.
6 Simic M.G. and Dizdaroglu, M. (1985) Biochemistry 24 233-236.
7 Margolis, S.A., Coxon, B., Gajewksi, E., Dizdaroglu, M. (1988) Biochemistry 27 6353-6359.
8 Dizdaroglu, M., Gajewski, E., Reddy, P. and Margolis, S.A. (1989) Biochemsitry 28 3625-3628.
9 Gajewski, E., Fuciarelli, A.F. and Dizdaroglu, M. (1988) Radiat. Res. 54 445-459.
10 Dizdaroglu, M. and Gajewski, E. (1989) Cancer Res. 49 3463-3467.
11 Gajewski, E. and Dizdaroglu, M. (1990) Biochemistry 29 977-980.
12 Olinski, R., Nackerdien, Z. and Dizdaroglu, M. (1992) Arch. Biochem. Biophys. 297 193-143.
13 Solomon, M.J. and Varsharvsky, A. (1985) Proc. Natl. Acad. Sci. USA 82 6470-6474.
14 Shich, V.V., Belyavsky, A.V. and Mirzabekov, A.D. (1985) J. Mol. Biol. 185 329-339.
15 Angelov, D., Stefanovsky, V.Y., Dimitrov, V.R., Keskinova, E. and Pashev, I.G., (1988) Nucl. Acid Res. 16 4525-4538.
16 Fricke, H. and Hart, E. J. (1966) in Radiation Dosimetry (Attix, F.H. and Roesh, W. C., eds) Vol II p 167-239. Academic Press, New York.
17 Smith, R.D., Loo, J.A., Barinaga, C.J., Edmonds, C.G. and Udseth, H.R. (1990) J. Am. Soc. Mass Spectrom. 1 53-65.
18 Lipton, M.S.W., Fuciarelli, A.L., Springer, D.L. Hofstadler, S.A., Smith, R.D. and Edmonds, C.G., in preparation.

STRUCTURE DETERMINATION OF OLIGONUCLEOTIDES CONTAINING MODIFIED BASES

G. Victor Fazakerley
CEA, Service de Biochimie et de
Génétique Moléculaire, Bat. 142, DSV
Centre d'Etudes de Saclay
91191 Gif sur Yvette Cedex, France

Lawrence C. Sowers
Division of Pediatrics
City of Hope National Medical Center
1500 E. Duarte Rd.
Duarte, CA 91010 USA

Introduction

Radiation damage to DNA creates a complicated array of products, making it difficult to assess the structural consequences of any individual adduct. In order to determine the consequences of specific DNA damage products, oligonucleotides have been constructed containing chemically characterized modifications at defined sites. Using NMR spectroscopy, the structural and dynamic changes induced in the DNA duplex by the defined damage may be determined. Ideally, biological insights would determine both the types of damage products and the sequence contexts chosen for study. Structural information obtained would then be correlated with the biochemical properties of each damaged nucleoside including potential mispair formation during DNA replication. Below, we describe studies on several selected adducts which have been studied in DNA by NMR spectrosopy.

Materials and Methods

Structural studies on DNA duplexes containing damaged bases requires the synthesis of significant quantities of the appropriate oligonucleotide strands. In general, oligonucleotides containing damaged bases have reduced thermal stability therefore duplex T_m's must be considered. Thus, it is generally advisable to study duplexes containing only one modified residue which should be inserted near or at the center of the helix. Depending upon the extent of destabilization duplexes of 8 to 12 base pairs will be required. Ideally the duplex concentration should be ca. 4 mM which means mg quantities of each strand. Further these should be at least 95% pure, as minor conformational species may be naturally present (see below).

The methods for the determination of 3D structures in solution by NMR are now well established. They rely on exploiting nuclear Overhauser effects (NOEs), the dipolar interaction between pairs of protons. These measurements must be made in one of two ways. In conditions where only direct interactions are observed, very short mixing times in NOESY spectra, the magnitude of the interaction can be related to the proton-proton distance, although the signal to noise ratio is always limiting. Spectra recorded with longer mixing times give better signal to noise but the observed NOEs are no longer pure. During the mixing time, usually several hundred ms, the NOE can be transferred from one proton to the next and so on such that NOEs can be observed between protons which are far apart in space by the mechanism of spin diffusion. Under good experimental conditions, direct NOEs can be observed up to ca. 4.5 Å. This is a limiting factor for NMR studies as the local structure can be well defined, but long-range effects, such as DNA bending, are not. Further important structural information can be obtained from analysis of scalar coupling constants, either ^{1}H-^{1}H or ^{1}H-^{31}P. Those give valuable information for the determination of dihedral angles.

These studies require access to high field instruments generally operating at 500 or 600 MHz. As oligonucleotides give rise to rather large proton resonances, spectral resolution, even with two-dimensional spectroscopy, this limits the size of the duplexes which can be studied in a quantitative way to about 20 base pairs. Three dimensional spectra with a uniformly labelled (^{13}C or ^{15}N) could help to solve specific problems but the material is not commercially available.

Results

The abasic site: We have studied a model abasic site incorporated into a 9 mer duplex opposite each of the four natural bases when the sequences are aligned. The model site chosen was one in which the base was replaced by a proton. While the natural abasic site is a hemiacetal, this exists in a mixture of α and β configurations. The model chosen provides us with another proton which helps in determining the structure and is chemically stable during the ca. 6 months necessary to carry out the study.

We first studied the case of an A opposite the abasic site (1) as this is the base most frequently incorporated. The sequence was 5'd(CGTGdr5GTGC).d(GCACA14CACG) in which dr is the abasic site.

As the abasic site has no hydrogen bonding potential with the A opposite it there is the possibility that either or both of the dr and A residues could be extrahelical or if intrahelical that this part of the helix would be unstable with respect to the surrounding Watson Crick base pairs. The total resonance assignment was achieved by standard 2D NMR experiments and these showed clearly that both the dr and A residues are intrahelical. Subsequently we followed the temperature dependence of the chemical shifts of the methyl groups of the two T residues, which are not adjacent to the mismatch site, and that of the A H8 resonance (Fig. 1).

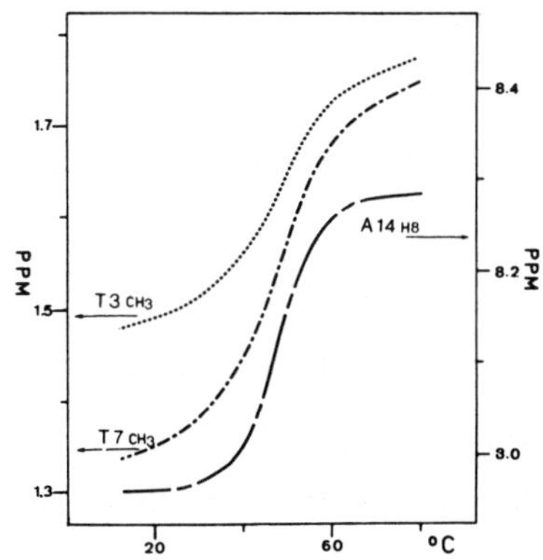

Figure 1. Chemical shifts as a function of the temperature for T7(CH3), T3 (CH3) and A14 H8.

We found, surprisingly, that the T_m of all three are the same within experimental error, 50°C. The non-hydrogen bonded A residue does not melt out of the helix before the hydrogen bonded T residues. The stacking of the A residue in the helix is normal and the position of the dr residue close to that which would be adopted by the sugar of a T residue at this position. Fig. 2 shows one of a family of observed structures.

With a G in front of the abasic site in the same sequence we observed, at low temperature, a structure similar to that for the A residue (2). The G is stacked into the helix and is in an *anti* conformation. The global stability of the duplex is almost the same as for the dr.A system, the T_m measured for the T methyl groups is 49°C. However, the G is less strongly stacked into the helix as we observed a T_m for the

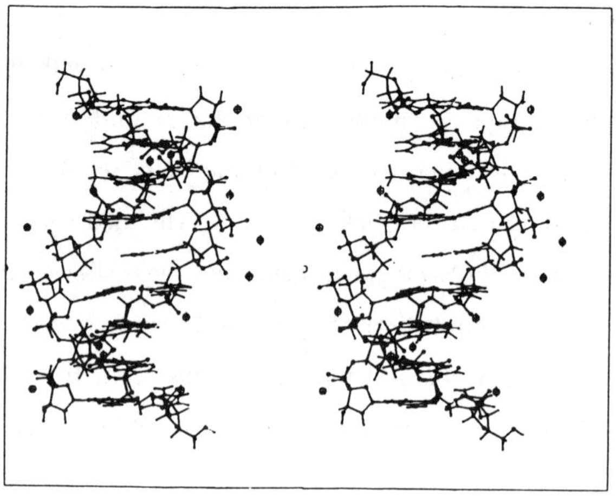

Figure 2. Stereo view of a typical structure viewed from the minor groove.

base of 36°C. Thus for the dr.G mismatch when incorporated into a longer DNA sequence with a higher global T_m the presence of an extrahelical G, and also dr, must be considered.

This intra-extrahelical equilibrium was most clearly observed for the dr.T system. We observed that the exchange for the resonances of the T residue between intra- and extrahelical forms was slow on a proton NMR time scale. Separate resonances were observed for the two forms as shown in Fig. 3.

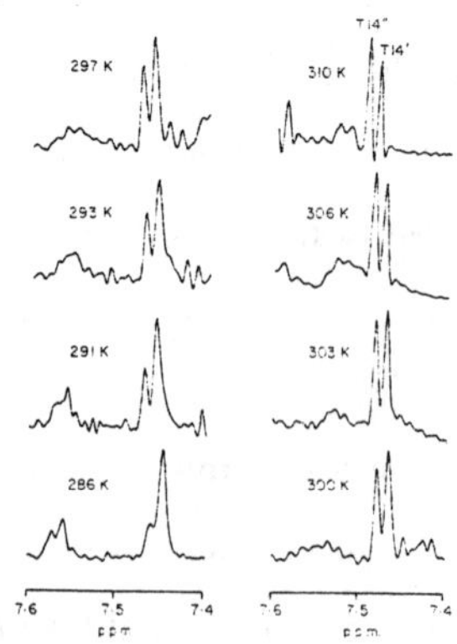

Figure 3. Part of the ID resolution enhanced spectrum of the dr.T duplex.

At low temperature a minor form corresponding to the extrahelical conformation is observed whose intensity increases with temperature. At 30°C, ca. 50% is observed extrahelical.

The dr.C system is the extreme case. Even at low temperature we observed only a conformation in which both the dr and C residues were extrahelical and the adjacent G.C base pairs were stacked over each other.

Qualitatively the same conclusions have been reached in other sequences or with other analogues (3, 4) and in studies with the hemiacetal form (5).

These four systems can be described in terms of an equilibrium between the two structures shown in scheme 1. For the dr.A system the equilibrium lies predominantly towards structure I whereas for the dr.C system we observe only structure II. The position of the equilibrium must arise from the differences in the stacking energies. We have calculated these (2) which show that the total energy is favorable for the intrahelical structure when there is a purine opposite the abasic site, whereas the collapsed form, structure II, is preferred when there is a pyrimidine opposite the abasic site.

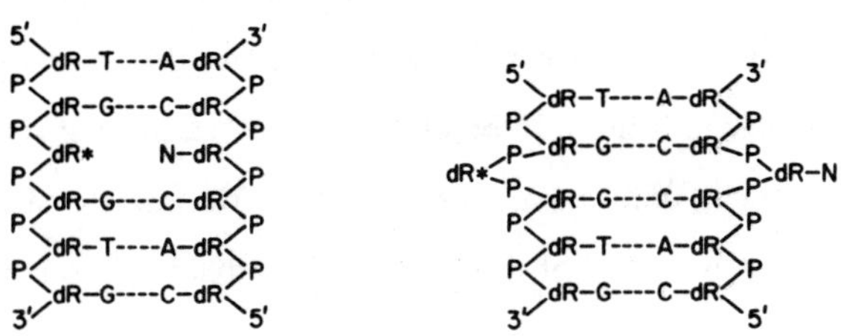

I Scheme 1 II

Based upon our structural studies we have proposed a model to explain the behavior of DNA polymerase when it encounters an abasic site on the template. In the single stranded form stacking interactions should force the abasic site out, as shown in scheme 2, structure 1. This could provide a steric block to the progression of the polymerase. This structure would be non-productive for base incorporation.

In order for replication to proceed across the abasic site a dNTP must bind to the template primer complex in a productive configuration. Interstrand stacking interactions with a purine dNTP could wedge the template bases apart, bringing the

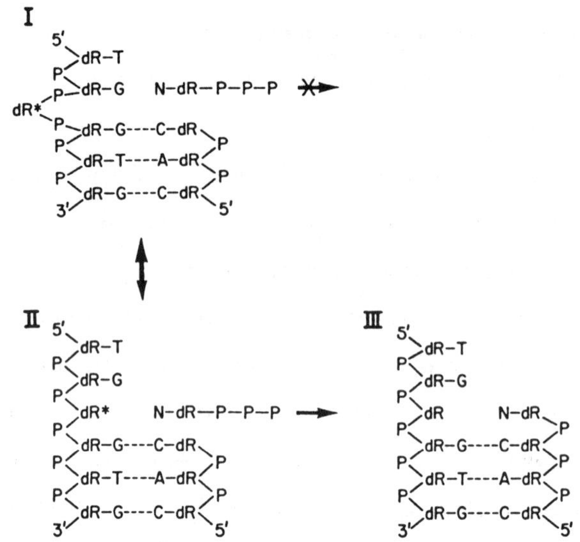

Scheme 2

abasic sugar back into the helix and forcing the structure into a productive B form (Scheme 2, structure II to III). Pyrimidine dNTPs on the other hand would likely be excluded from the helix and thus from the catalytic pocket.

If the polymerase were to bypass the abasic site and continue base incorporation this would give rise to an abasic frameshift. For the reasons given above, the abasic site would initially be extrahelical. We have studied this case in a different sequence (6, 7) and have observed that the abasic site is only found extrahelical. This does not disrupt the helix as the adjacent bases are stacked upon each other normally (Fig. 4).

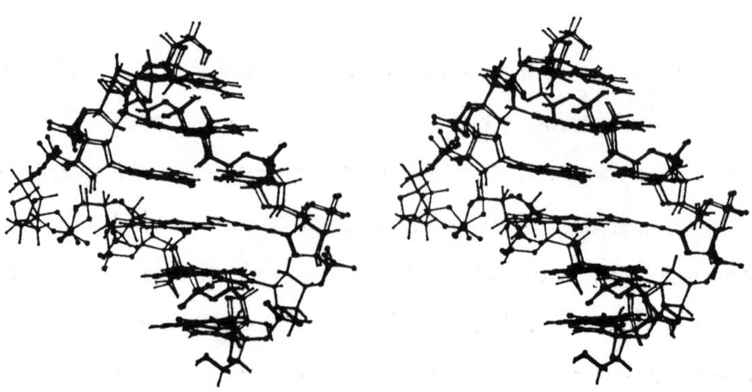

Figure 4. Stereoscopic view of two superposed structures of the abasic frameshift.

5-(hydroxymethyl)-2'-deoxyuridine: Oxidation of the methyl group of thymine residues in DNA is known to result in the formation of 5-(hydroxymethyl)-2'-deoxyuridine (HMdU). We have incorporated this base into the following sequence 5'd(CAGHGGC).d(GCCACTG) to study the HMdU.A base pair (8). We observed both bases intrahelical, base paired in Watson Crick geometry and the duplex adopts a normal B form. The hydroxymethyl group lies in the major groove, but unlike a T methyl group it is not freely rotating. We have investigated possible hydrogen bonding of the hydroxymethyl group by measuring proton-proton distances and by molecular modelling. The hydroxymethyl group lies on the 5' side of the HMdU residue with the oxygen atom pointing away from the H6 proton. Hydrogen bonding, which could explain why the group adopts a unique conformation, could occur to its own O4 carbonyl or to the N7 of the G residue in the 5' direction. Our data suggests that it is probable that the intermolecular hydrogen bonds forms. This does not require any helix distortion. Our proposed structure is shown in Fig. 5. An analogous structure has been proposed for another thymine oxidation product, thymidine glycol (9).

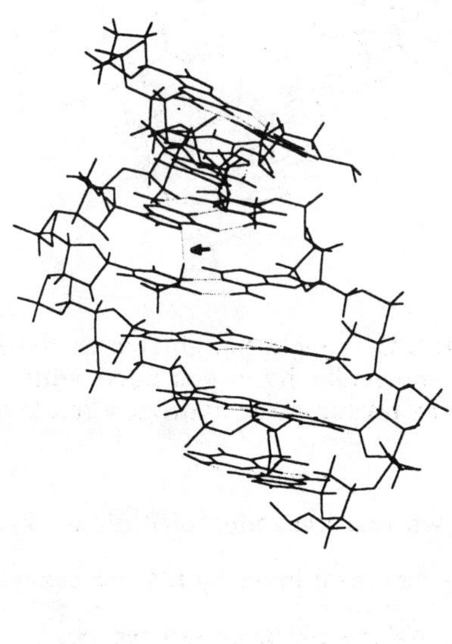

Figure 5. The HMdU.A Watson Crick base pair in which the HMdU residue forms an intermolecular hydrogen bond with G3. The arrow indicates the presumptive hydrogen bond.

Relative to a T methyl group, the hydroxymethyl group is much bulkier and may significantly modify interactions with sequence specific DNA binding proteins. Interbase hydrogen bond formation by HMdU could also obscure other contact points.

Biological studies suggest that base substitution mutations induced by HMdU would arise via the formation of intermediate HMdU.G base pairs during DNA replication and we have therefore investigated the HMdU.G base pair in the same sequence. At physiological pH we observe that both bases are intrahelical and base paired in a classical wobble geometry as for G.T pairs. The hydroxymethyl group again adopts a unique conformation, but is now found on the 3' side of the base and the hydrogen bond is intrabase, as shown in Fig. 6.

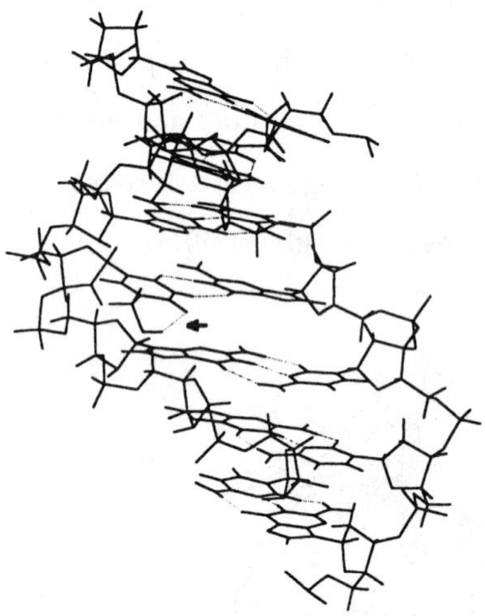

Figure 6. View of the HMdU.G wobble pair in which the hydroxy methyl group forms an intramolecular hydrogen bond with the O4 carbonyl. The arrow indicates the presumptive hydrogen bond.

We have previously shown that ionization of BrdU is likely to account for its mispairing potential (10). We therefore investigated the behavior of the HMdU.G mispair as a function of pH. As the pH is raised we observed spectral changes demonstrating that the HMdU.G base pair ionizes with an apparent pK of 9.7. In contrast none of the systems T.G, U.G nor HMdU.A undergo ionization with pH prior to alkaline denaturation of the helix. Spectra were recorded at sufficiently high pH such that the ionized base pair was the predominant structure. Both bases remain intrahelical and the duplex stays as a B form helix. The HMdU and G bases are now paired in Watson Crick geometry and the hydroxymethyl group has returned to the 5' side of the base and is probably hydrogen bonded to the adjacent G residue. The HMdU residue adopts essentially the same position as that observed for HMdU.A. The ionized Watson Crick configuration of the HMdU.G base pair would be present at a level of roughly 1 in 300 at physiological pH.

HMdU may therefore mispair with G, in Watson Crick geometry as a consequence of ionization, between 1 and 2 orders of magnitude more frequently than T. If the ionized configuration of the HMdU.G mispair is stabilized by the formation of an interbase pair hydrogen bond, mispairing of HMdU may be highly sequence specific. We note that a recent biochemical study which used HMdU in the template for DNA polymerase did not show enhanced mispairing (11). In this sequence the HMdU residue was flanked by two cytosines. The proposed intermolecular hydrogen bond formed between HMdU and the guanine in the 5' direction could not form when HMdU is adjacent to a C residue. Further studies will be necessary to verify these hypotheses.

Oxo Purines: 7,8-dihydro-8-oxoadenine (oxoA) was the first identified major radiation product resulting from the degradation of adenine under the action of gamma rays (12). This product has been improperly named 8-hydroxyadenine. We observed that T is still preferentially inserted opposite this modified base even though it had been shown that the monomer predominantly adopts the *syn* conformation (13). An NMR study of the oxoA.T system showed that the oxoA base was an *anti* conformation and that there were no major modifications relative to a normal B form helix (14). We directly observed the N7H but this was not involved in the hydrogen bonding pattern.

A crystal structure of a deoxydodecamer with oxoA.G base pairs has been reported (15). This showed base pairing G *anti*.oxoA *syn* in which both the N7H and C80 were involved in hydrogen bonding with the base in front.

Patel and coworkers have reported an NMR study of oxoG paired with adenosine (16). They observed that the oxoG base was in a *syn* conformation and that hydrogen bonding occurred between the G keto group - A NH2 and GN7H-AN1. The proposed structure is similar to that observed for the protonated G.A mismatch (17) for which the additional proton is found on the N1 of the adenine.

Ureadeoxyribose: One of the products of radical attack upon T residues is a ureadeoxyribose moiety resulting from splitting of the thymidine ring. While this residue has lost part of its original coding functions it could still base pair with two hydrogen bonds. We have studied the conformation of the ureadeoxyribose opposite a T residue (18) which is a mutagenic situation as the natural base would be an adenine. Initially attack on the T residue will probably give rise to a urea residue with the peptide in the normal *trans* configuration. However, in our study we observed two species in a ratio of ca. 3:2. We identified the major species as the *trans* isomer hydrogen bonded to T by two hydrogen bonds (Fig. 7). For pairing to occur the T residue has to move towards the minor groove relative to an A.T base pair. On the other hand the *cis* isomer is highly unfavorable for the formation of hydrogen bonds with T. In this case both the ur and T residues are extrahelical.

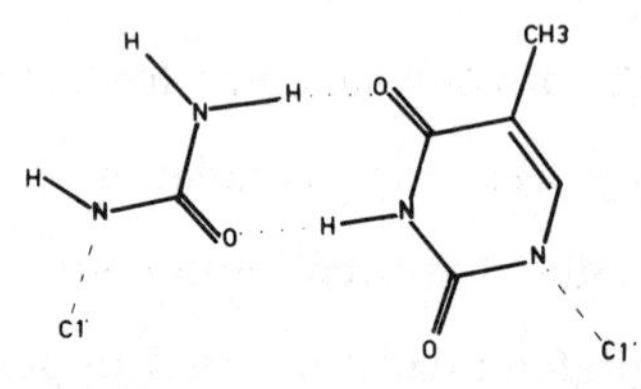

Figure 7. Possible pairing structure for ur.T

In a subsequent study (GVF unpublished results) we were able to initially separate the two isomers in the sequence 5'd(AGGA-CCACG).d(CGTGGurTCCT) where ur stands for the ureadeoxyribose and the dash in the first strand, 9 mer, indicates that there is no base in front of the ur residue when the sequences are aligned. It might be expected that this system would behave like an abasic frameshift with the ur residue extrahelical. We observe for one of the two isomers of the ur that this is so. Surprisingly for the other isomer we find that the ur residue is intrahelical squeezing apart the neighboring bases. Although we could carry out the initial measurements on the separate systems isomerization occurs with time such that both systems give identical spectra, that of a mixture of the two isomers and thus two conformations, after several weeks. To explain the biological activity of the damaged product, it may be necessary to understand the quite different structures adopted by both isomers.

References

1 Ph. Cuniasse, L.C. Sowers, R. Eritja, B. Kaplan, M.F. Goodman, J.A.H. Cognet, M. LeBret, W. Guschlbauer & G. V. Fazakerley (1987) Nucleic Acids Res. 15, 8003-8022.

2. Ph. Cuniasse, L.C. Sowers, B. Kaplan, W. Guschlbauer & G.V. Fazakerley (1990) J. Mol. Biol. 213, 303-314.

3. M.W. Kalnik, C.-N. Chang, A.P. Grollman & D.J. Patel (1988) Biochemistry 27, 924-931.

4. M.W. Kalnik, C.-N. Chang, F. Johnson, A.P. Grollman & D.J. Patel (1989) Biochemistry 28, 3373-3383.

5. J.M. Withka, J.A. Wilde, P.H. Bolton, A. Muzumder & J.A. Gerlt (1991) Biochemistry 30, 9931-9940.

6. Ph. Cuniasse, L.C. Sowers, R. Eritja, B. Kaplan, M.F. Goodman, J.A.H. Cognet, M. Le Bret, W. Guschlbauer & G.V. Fazakerley (1989) Biochemistry, 27, 2018-2026.

7. J.A.H. Cognet, J. Gabarro-Arpa, Ph.Cuniasse, G.V. Fazakerley & M. LeBret (1990) J. Biomol. Struct. Dynam. 7, 1095-1115.

8. S. Mellac, G.V. Fazakerley & L.C. Sowers (1993) Biochemistry, 32, 7779-7786.

9. J.M. Clark, N. Pattabiraman, W. Jarvis & G.P. Beardsley (1987) Biochemistry 26, 5404-5409.

10. L.C. Sowers, M.F. Goodman, R. Eritja, B. Kaplan & G.V. Fazakerley (1989) J. Mol Biol. 205, 437-447.

11. D.D. Levy & G.W. Teebor (1991) Nucleic Acids Res. 15, 9975-9983.

12. J.J. Conlay (1963) Nature 197, 555.

13. N. Marrigai, J. Cadet & R. Téoule (1976) Tetrahedron 32, 2385-2387.

14. W. Guschlbauer, A. Duplaa, A. Guy, R. Téoule & G.V. Fazakerley (1991) Nucleic Acids Res., 19, 1953-1958.

15. G.A. Leonard, A. Guy, T. Brown, R. Téoule & W.N. Hunter (1992) Biochemistry 31, 8415-8420.

16. M. Kouchakdjian, V. Bodepudi, S. Shibutani, M. Eisenberg, F. Johnson, A.P. Grollman & D.J. Patel (1991) Biochemistry 30, 1403-1412.

17. C. Carbonnaux, G.A. van der Marel, J.H. van Boom, W. Guschlbauer & G.V. Fazakerley (1991) Biochemistry, 30, 5449-5458.

18. V. Gervais, A. Guy, R. Téoule & G.V. Fazakerley (1992) Nucleic Acids Res. 20, 6455-6460.

THE RADIATION CHEMISTRY OF DNA OLIGOMERS

Harold C. Box, Elizabeth A. Schroder and Edwin E. Budzinski
Biophysics Department
Roswell Park Cancer Institute
Elm & Carlton Streets
Buffalo, NY 14263 USA

Introduction

Studies of the radiation chemistry of DNA dimers, tetramers and hexamers have turned out to be far more rewarding than we ever anticipated. Radiation-damaged dimers are proving useful as carriers in radioactive postlabeling assays of DNA damage. Tetramers, on the other hand, are preferred for modeling the radiation chemistry of DNA. All four bases can be represented, yet tetramers are still small enough so that the mixture of radiation products is generally separable by chromatography. Hexamers bearing a specific lesion are difficult to isolate from the mix of products. Nevertheless, hexamers are of considerable interest because they may prove to be just long enough so that the modified molecule can be enzymatically ligated onto a longer DNA sequence. The longer sequence may then serve as template for studying the mutational effect of the lesion. Before any of these applications can be exercised, oligomers bearing specific radiation-induced lesions must be isolated and the lesion identified.

We have largely relied on NMR spectroscopy to identify lesions in intact radiation-damaged DNA oligomers. The greatest difficulty is usually the isolation of individual products and the collection of enough material to make identification by NMR feasible. Variations of HPLC have been used to isolate products, initially at the dimer level, subsequently at the tetramer level and more recently at the hexamer level. Table 1 is a compilation of radiation-induced modifications that have been identified in DNA oligomers X-irradiated in solution in the presence of oxygen. Table 2 is a compilation of radiation-induced modifications that have been identified in DNA oligomers X-irradiated in aqueous solution in the absence of oxygen.

Dimers: Nuclease P1 is primarily an endonuclease that is capable of hydrolyzing DNA to the monomer level. There is wide variation, however, in the

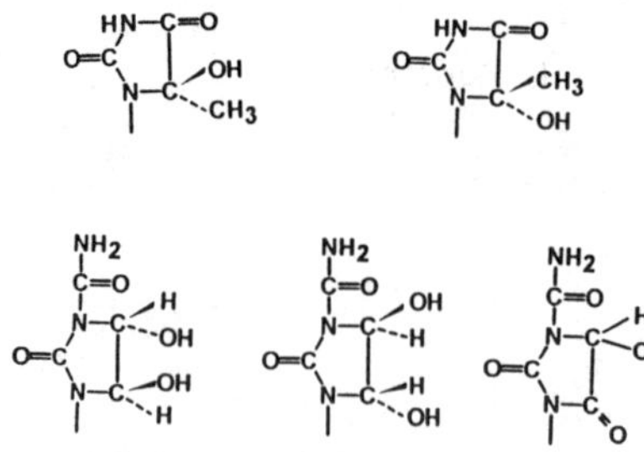	Five membered ring products are formed subsequent to opening of the pyrimidine ring. Formation of both 5-hydroxy-5-methylhydantoin and of 1-carbamoyl-2-oxo-4,5-dihydroxy-imidazolidene products is enhanced by Cu^{++} catalyst (Schroder et al., 1994). Only one of two possible stereoisomers of 3-carbamoy-4-hydroxy hydantoin products was observed.
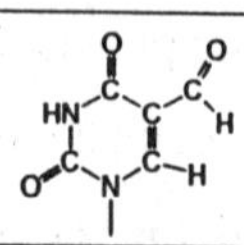	Generally not a major product (Budzinski et al., 1993), this product can also be produced by light irradiation in the presence of a sensitizer (Decarroz et al., 1986).
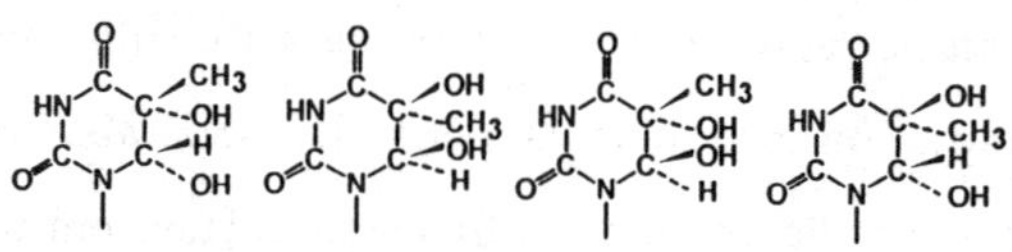	The early work of Cadet et al. (1979) on irradiated thymidine was key to recognizing these lesions in oligomers (Belfi et al., 1986).
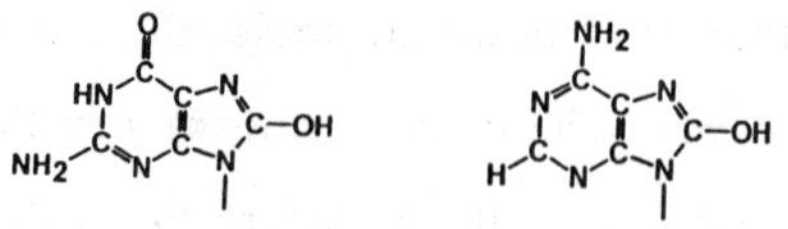	Although both 8-hydroxyguanine and 8-hydroxyadenine lesions are generated in X-irradiated solutions of oligomers, the former is produced more abundantly in oxygenated solutions (Schroder et al., 1994).
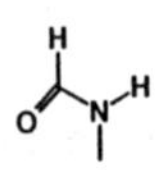	The formamido remnant can be derived from either a thymine or cytosine base and is prominent in irradiated oligomer solutions. It is easily identified in the intact modified oligomer but can escape detection as a modified nucleotide due to weak UV absorption (Budzinski et al., 1994).
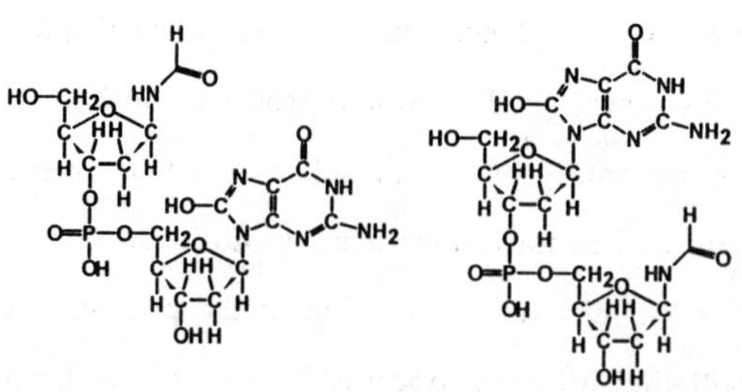	These double base lesions can be formed from GT and GC sequences or TG and CG sequences (Box et al., 1994).

Table 1. Base modifications observed in DNA oligomers X-irradiated in oxygenated aqueous solutions.

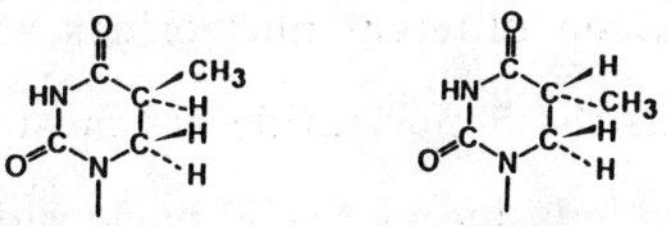	In studies of d(ACGT), the dihydrothymine lesions became predominant when the irradiated solution was purged with nitrogen (Schroder et al., 1994).
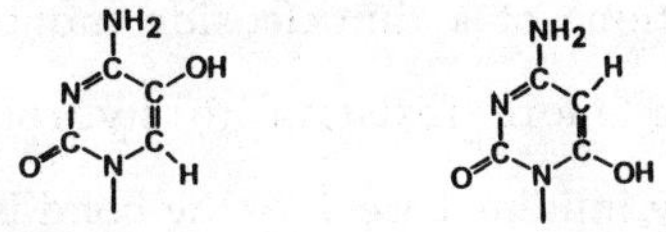	These products were observed by Paul et al. (1987) in irradiated solutions saturated with N_2O.
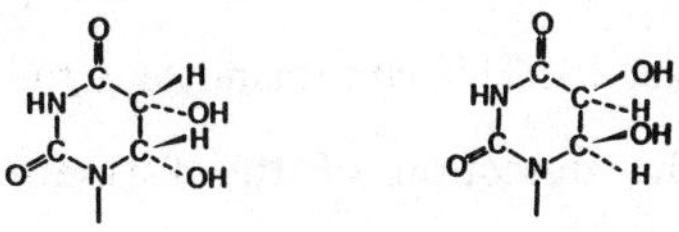	Of four possible stereoisomers only the two <u>cis</u> isomers were observed in studies of irradiated dimers (Paul et al., 1987).
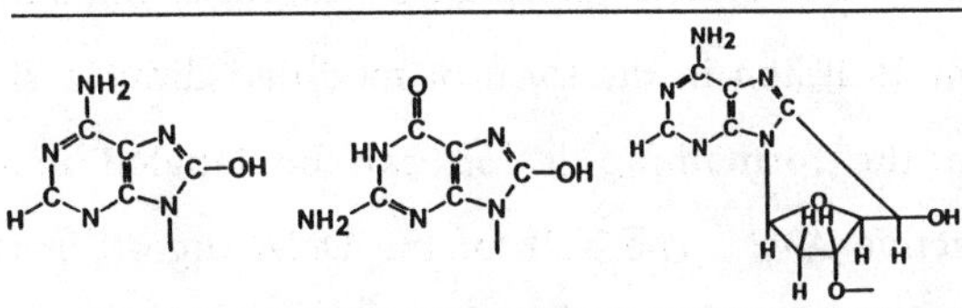	Compared with pyrimidine products the number of purine products is limited (Schroder et al., 1994). Formation of the 8-hydroxyguanine moiety is severely curtailed in the absence of oxygen.
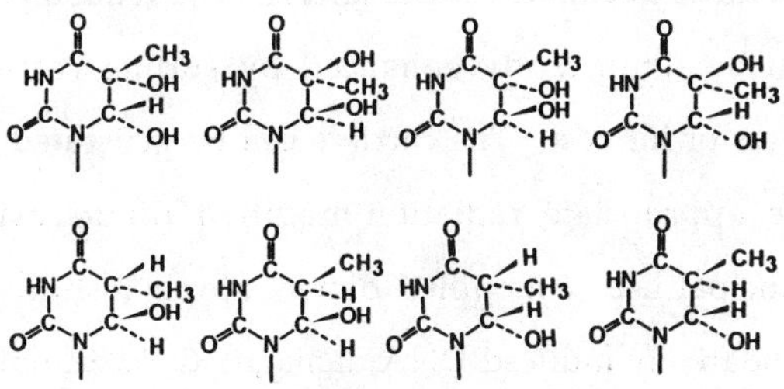	The complete sets of stereoisomers of 5,6-dihydrothymine modifications of dimers have been observed by Belfi et al. (1986) in N_2O saturated solutions. This work relied heavily on the pioneering studies on thymidine by Cadet et al. (1979).
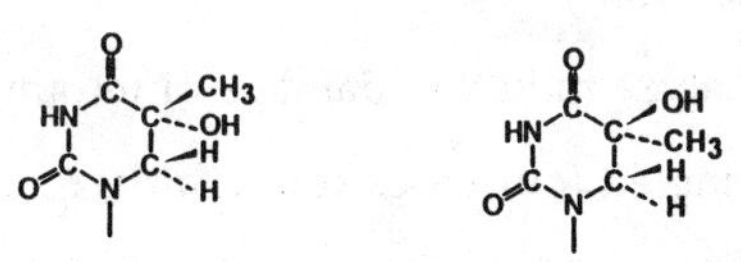	Both stereoisomers were observed in irradiated solutions of tetramer containing glutathione (Paul et al., 1990).
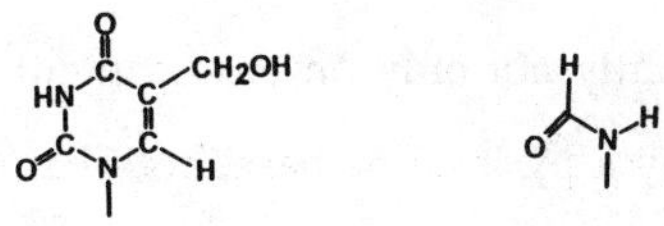	To date the 5-hydroxymethyluracil product has only been observed in oligomers irradiated in the absence of oxygen. Interestingly, however, the lesion has been detected in cellular DNA (Teebor et al., 1984). Conversely, formation of the formamido lesion is severely curtailed in the absence of oxygen (Paul et al., 1987).

Table 2. Base modifications observed in DNA oligomers X-irradiated in oxygen deprived aqueous solutions.

rate at which phosphodiester bonds between different nucleosides are attacked. Phosphodiester bonds having thymidine as the 5' nucleoside are most resistant to hydrolysis, whereas bonds having deoxyadenosine as the 5' nucleoside are most susceptible. Radiation-induced modifications of a dinucleoside monophosphate usually make the phosphodiester bond more resistant to hydrolysis. An outstanding example is the case where a pyrimidine base 5' to the bond is degraded to a formamido remnant, (see Figs. 1 and 2). This phosphodiester bond appears to be completely resistant to hydrolysis by nuclease P1. This circumstance is the basis for a ^{32}P-postlabeling assay developed for the detection of the formamido lesion (Maccubbin et al., 1993). A nuclease P1 (plus acid phosphatase) digest of radiation-damaged DNA leaves the formamido lesion in the form of modified dinucleoside monophosphates. In this form the formamido lesion can be labeled using polynucleotide kinase and radioactive ATP. The bulk of the DNA digest, in the form of mononucleosides, is not a suitable substrate for the kinase. The reliability of the ^{32}P-postlabeling procedure can be significantly enhanced by adding carriers corresponding to the modified dimers of interest. The carriers can be generated by enzymatic phosphorylation of the appropriate radiation-modified dinucleoside monophosphates. Currently the principal use of modified dimers at our institute is as carriers in biological assays of oxidatively-induced DNA damage. Only a minute amount of dimer carrier is required for each assay.

Tetramers: An early goal was to analyze radiation damage in tetramers based on the expectation that a tetramer containing all four bases could fairly represent the chemistry of DNA itself. In the main this is probably true but the most interesting product identification to date belies this notion. Some of the products that are formed by ionizing radiation depend importantly not only on base content but also on base sequence. When guanine is flanked by a pyrimidine base, both bases can be modified with the guanine base hydroxylated at the 8-position and the adjacent pyrimidine base degraded to a formamido remnant (see Fig. 1). In our view the double base lesion is the result of a single free radical initiating event. Arguments in support of this mechanism have been presented elsewhere (Box et al., 1993, Budzinski et al., 1994). It has long been known that the principal lesions produced by photoirradiation of DNA are produced at adjacent pyrimidine bases. Now it appears there is a base sequence dependence for ionizing radiation also.

Hexamers: Progress in analyzing radiation damage in hexamers has been slow. As the length of the oligomer is increased the number of products becomes greater and their chromatographic elution properties less different. However, it has been possible recently to isolate several of the main products from an irradiated solution of hexamers. For example, the formamido and 8-hydroxyguanine modifications of d(CGATCG) have been isolated from irradiated oxygenated solution (Schroder, 1994). The anticipated use of hexamers containing an identified lesion is for the construction of longer DNA sequences that can serve as templates in mutation assays. The aim is to enzymatically ligate the modified hexamer into the longer DNA sequence. This method of constructing templates containing a radiation-induced lesion is less elegant than the approach in which templates of arbitrary sequence are synthesized from protected modified monomers. However, for some radiation-induced lesions it may be difficult or impossible to synthesize the protected modified monomer. In these instances modified hexamers may provide an alternative approach.

Discussion

The major products produced by ionizing radiation in aqueous solutions of short DNA oligomers have largely been identified by means of NMR spectroscopy. The nucleoside modifications identified to date are shown in Figs. 1 and 2. It is of some interest to compare the base modifications listed here with the base modifications detected in irradiated DNA by GC-MS (Fuciarelli et al., 1990). The compilations in Figs. 1 and 2 do not represent two other common forms of DNA radiation damage that are manifested in oligomer studies, namely depyrimidation and depurination. Nor do the compilations take account of strand scissions which are also clearly indicated by oligomer product profiles.

The task ahead is to utilize DNA oligomers containing identified lesions in studies of the biological effects of radiation. The use of dimers as carriers for the sensitive detection of DNA damage caused by radiation or by oxidative stress in biological systems has received some attention (Maccubbin et al., 1993) but the technique has not been fully exploited. The use of longer sequences containing identified lesions for *in vitro* studies of the mutational consequences of radiation-induced lesions is currently under development.

Acknowledgements

This work has been supported by grant CA 46838 from the National Cancer Institute and by a NORCUS fellowship held by Elizabeth Schroder. We are grateful to Mr. John C. Wallace for having synthesized many of the oligomers that made this research possible.

References

Belfi CA, Arakali AV, Paul CR, Box HC (1986) Radiation chemistry of a dinucleoside monophosphate and its sequence isomer. Radiat Res 106:17-30

Box HC, Freund HG, Budzinski EE, Wallace JC Maccubbin, AE (1994) Free radical-induced double base lesions. Radiat Res, in press

Box HC, Budzinski EE, Freund HG, Evans MS, Patrzyc HB, Wallace JC, Maccubbin AE (1993) Vicinal lesions in X-irradiated DNA? Int J Radiat Biol 3:261-263

Budzinski EE, Dawidzik JD, Wallace JC, Freund HG, Box HC, (1994) The radiation chemistry of d(CpGpTpA) in the presence of oxygen. Radiat Res, submitted

Budzinski EE, Maccubbin AE, Freund HG, Wallace JC, Box HC (1993) Characterization of the products of dinucleoside monophosphates d(GpN) irradiated in aqueous solutions. Radiat Res 136:171-177

Cadet J, Ducolomb R, Hruska FE (1979) Proton magnetic resonance studies of 5,6-saturated thymidine derivatives produced by ionizing radiation. Conformational analysis of 6-hydroxylated diastereoisomers. Biochim. Biophys. Acta 563:206-215

Decarroz C, Wagner JR, VanLier JE, Krishna CM, Riesz, P, Cadet J (1986) Sensitized photo-oxidation of thymidine by 2-methyl-1,4-napthoquinone. Characterization of the stable photoproducts. Int J Radiat Res 50:491-505

Fuciarelli AF, Wegher BJ, Blakely WF, Dizdaroglu M (1990) Yields of radiation-induced base products in DNA: Effects of DNA conformation and gassing conditions. Int. J. Radiat. Biol. 58:397-415

Maccubbin AE, Budzinski EE, Patrzyc H, Evans MS, Ersing N, Box, HC (1993) ^{32}P-postlabeling assay for free radical-induced DNA damage: The formamido remnant of thymine. Free Rad Res Comms 1:17-28

Paul CR, Budzinski EE, Maccubbin AE, Wallace JC, Box H.C. (1990) Characterization of radiation-induced damage in d(TpApCpG). Int. J. Radiat. Biol. 58:759-768

Paul CR, Arakali AV, Wallace JC, McReynolds J, Box H.C. (1987) Radiation chemistry of d(CpG) and its sequence isomer in N_2O and O_2 saturated solutions. Radiation Research 112: 464-477

Schroder EA, Budzinski EE, Wallace JC, Zimbrick JD, Box HC (1994) The radiation chemistry of d(ApCpGpT). Int J Radiat Biol, submitted

Teebor GW, Frenkel K, Goldstein MS (1984) Ionizing radiation and tritium transmutation both cause formation of 5-hydroxymethyl-2'-deoxyuridine in cellular DNA. Proc Natl Acad Sci, USA 81:318-321c

Chapter 9

Organic Chemistry Involved in the Analysis of Structure/Function Relationships in Irradiated DNA

MISCODING AND MUTAGENIC PROPERTIES OF 8-OXOGUANINE AND ABASIC SITES: UBIQUITOUS LESIONS IN DAMAGED DNA

Arthur P. Grollman and Masaru Takeshita
SUNY Stony Brook
Department of Pharmacological Sciences
Stony Brook, New York 11794-8651

Introduction

More than twenty oxidatively-damaged bases, including 8-oxoguanine, have been found to occur in genomic DNA (Gajewski et al, 1990). Some of these lesions block DNA replication and are potentially lethal; others generate mutations which can initiate carcinogenesis and promote cellular aging (Ames and Gold, 1991). In this report, we focus attention on the mutagenicity and repair of 8-oxoguanine. Kasai and Nishimura's discovery that hydroxyl radicals react with guanine residues in DNA to form 8-oxoguanine (Kasai and Nishimura, 1984; reviewed in Kasai and Nishimura, 1991), and the development of sensitive methods for the detection and quantitation of this modified base (Gajewski et al., 1990, Floyd, 1990) led to the observation that approximately 1 in 10^5 guanine residues in mammalian DNA are oxidized at the C-8 position (Fraga et al., 1990). The number of 8-oxoguanine residues in genomic DNA is significantly higher than the number of DNA adducts in mice treated with established chemical carcinogens, prompting us to explore the mutagenic potential of this lesion (Shibutani et al., 1991).

Abasic sites in DNA arise spontaneously by a hydrolytic process which can be markedly accelerated by chemical modification of DNA bases with approximately 100,000 abasic sites being generated per cell each day (Lindahl and Nyberg, 1972). Damaged DNA bases are substrates for DNA N-glycosylases; this process produces abasic sites as an intermediate step in DNA repair (Friedberg, 1985). It has been suggested, based largely on indirect evidence, that abasic sites represent a common intermediate in chemical mutagenesis (Loeb and Preston, 1986).

DNA containing 8-oxoguanine and synthetic analogs of the abasic site have been used to investigate the miscoding and mutagenic potential of these ubiquitous lesions. Studies in our laboratory were facilitated by the development of solid state synthetic methods by which these lesions could be introduced at defined positions in DNA. In this paper, we review studies in which 8-oxoguanine and abasic sites have been used in model systems to explore various early events in the replication of selectively damaged DNA.

Studies on 8-oxoguanine

The mutagenic potential of 8-oxoguanine is reflected in its miscoding properties. Primer extension reactions, catalyzed by various DNA polymerases, have been used to show that dAMP and dCMP are readily incorporated opposite 8-oxodG on site-specifically modified DNA templates (Shibutani et al., 1991). These in vitro studies reveal striking differences between various DNA polymerases with respect to the relative amount of dAMP incorporated opposite 8-oxodG (Table 1). For example, the ratio of dCMP/dAMP incorporated opposite 8-oxodG by DNA polymerase α is 1:200 while the comparable value for DNA polymerase β is 4:1 (Shibutani et al., 1991). Differences between repair and replicative DNA polymerases were subsequently shown to play an important role in the cellular repair of 8-oxoguanine in DNA (Michaels et al., 1992; Tchou and Grollman, 1993).

Table 1 Relative incorporation of nucleotides opposite 8-oxodG by DNA polymerases *in vitro*[a]

DNA Polymerase	Incorporation C:A
pol α	1 : 200
pol δ	1 : 5
pol III	1 : 8
pol β	4 : 1
pol I	7 : 1
pol γ	4 : 1

[a]Data from Shibutani *et al.* (1991) and unpublished experiments.

Structure-function relationships--The structure of duplex DNA in which 8-oxodG is positioned opposite dA was established by 2D-NMR studies (Kouchakdjian et al., 1991) and later confirmed by x-ray crystallographic methods (McAuley-Hecht et al., 1994). When positioned as a single lesion in a B-DNA duplex, 8-oxoguanine assumes the *syn* conformation, forming a Hoogstein base pair with dA (Kouchakdjian et al., 1991; McAuley-Hecht et al., 1994). The dimensions of the 8-

oxodG:dA base pair resemble Watson-Crick geometry, shielding the incoming base at the replication fork and permitting efficient extension of the nascent chain from the 3' primer terminus (Shibutani et al., 1991). In this respect, 8-oxoguanine differs from the structurally-related lesion, formamidopyrimidine (Fapy), which effectively blocks chain extension (O'Connor et al., 1988).

Based on a structure-function analysis (Tchou et al., 1994, Grollman and Moriya, 1993), a mechanism has been proposed for the glycosylase function of Fpg protein in which 8-oxoguanine is initially protonated at the 6-oxo position. Based on analogy to the mechanism of action proposed for T4 endonuclease V (Schrock and Lloyd, 1991), and recent data showing that the Schiff's base formed between such enzymes and their substrates can be trapped (Tchou and Grollman, unpublished data), we propose that the amino function on the enzyme is involved with a nucleophilic attack at C1' of 8-oxodG. Thus, recognition and repair of DNA substrates by Fpg protein involve several functional groups in the enzyme and minimally two specifically positioned carbonyl functions in the major groove of DNA (Grollman et al., 1994).

Mutagenesis studies-- Plasmid vectors, modified site-specifically with 8-oxodG, were used to establish the mutational frequency and specificity of this lesion in *E. coli* (Wood et al., 1990; Moriya et al., 1991; Cheng et al., 1992). A single strand plasmid, containing origins of replication from *E. coli* and SV-40, was used for experiments in simian kidney (COS-7) cells (Moriya, 1993). Following replication in the host cell, mutant plasmids were analyzed by oligonucleotide hybridization and DNA sequencing techniques. The predominant mutations observed were G:C→T:A transversions, targeted to the site of the lesion. Results of these experiments are consistent with primer-extension studies (Shibutani et al., 1991). Mutational frequencies for 8-oxodG in *E. coli* and COS cells were <3%, revealing the modified base to be weakly mutagenic (Moriya, 1993).

DNA repair--Mutagenic DNA lesions are potentially deleterious; several enzymes are involved in their repair. 8-Oxoguanine is excised from DNA by Fpg protein, a DNA N-glycosylase (Tchou et al., 1991) shown previously (Chetsanga and Lindahl, 1979) to remove Fapy lesions from DNA. Fpg protein has a slightly higher affinity for DNA duplexes containing 8-oxodG than for duplexes containing Fapy (Tchou et al., 1994). These observations, together with genetic experiments (Michaels et al., 1992) and the demonstration that Fapy is a blocking lesion (O'Connor et al., 1988), suggest that 8-oxoguanine is the "physiological" target for Fpg protein.

Structure function analysis reveals that 8-oxodG is removed efficiently from DNA by Fpg protein when this oxidized base is paired with dC, whereas DNA containing an 8-oxodG:dA pair is a poor substrate for the enzyme (Tchou et al., 1991, 1994). Another DNA glycosylase, adenine DNA glycosylase, was found to be involved in repairing the mutagenic intermediate (Michaels et al., 1993). Adenine DNA glycosylase had previously been isolated, based on its activity on DNA containing G:A mismatches (Au et al., 1989). It seems likely that the primary physiological function of this enzyme involves repair of oxidatively damaged DNA (Tchou and Grollman, 1993; Michaels and Miller, 1992).

8-OxodGTP can be incorporated into DNA during DNA synthesis (Maki and Sekiguchi, 1992). 8-OxodGTP pairs to dA and dC with equal efficiency (Cheng et al., 1992; Maki and Sekiguchi, 1992); failure to repair 8-oxoguanine when the lesion is introduced into the primer strand, results in A:T→C:G mutations. The presence of an 8-oxoGTPase in cells serves to "sanitize" the nucleotide triphosphate pool, thus preventing a mutational event (Maki and Sekiguchi, 1992).

The ubiquitous presence of reactive oxygen species in nature (reviewed by Frenkel, 1992) ensures that 8-oxodG is present in the DNA of all living organisms. Mutator strains of *E. coli* have been used to define the role of DNA repair enzymes in protecting cells against the mutagenic effects of 8-oxoguanine (Tchou and Grollman, 1993 Michaels and Miller, 1992). MutT strains of *E. coli* are unable to hydrolyze 8-oxoGTP (Maki and Sekiguchi, 1992) and are characterized by a high frequency of A:T→C:G transversions. MutM strains are deficient in Fpg protein and accumulate 8-oxodG (Bessho et al., 1992); these strains are associated with a high frequency of G:C→T:A transversions (Michaels et al., 1991). MutY strains are deficient in adenine DNA glycosylase and also have a high frequency of G:C→T:A mutations. MutM/MutY double mutations act synergistically, suggesting that the products of MutM and MutY combine to protect *E. coli* from the mutagenic effects of 8-oxodG (Michaels et al., 1992).

In summary, reactive oxygen species are generated continuously in cells; the complex repair system shown in Fig. 1 helps protect *E. coli* from the deleterious effects of endogenous and exogenous oxidative damage, represented in this case by 8-oxoguanine. Reviews of this subject have been published by Tchou and Grollman (1993), Michaels and Miller (1992), and Grollman and Moriya (1993). Mammalian counterparts of MutM, MutY, and MutT gene products have been identified (Michaels et al., 1991; Chung et al., 1991; Yeh et al., 1991; Mo et al., 1992); thus, it appears likely that an analogous error avoidance pathway operates in mammalian cells.

Studies on abasic sites

Miscoding properties--During DNA synthesis, dAMP is preferentially inserted opposite abasic sites in DNA. This phenomenon, generally associated with non-instructional lesions, has been observed both *in vitro* (Boiteux and Laval, 1982; Sagher and Strauss, 1983; Randall et al., 1987; Takeshita et al., 1987) and *in vivo* (Shaaper et al., 1983; Lawrence et al., 1990) and has come to be known as the "A rule" (Strauss, 1991). Synthetic abasic sites lacking a deoxyribose moiety (Fig. 2) preferentially incorporate dAMP opposite the lesion, indicating that hydrogen bonding at the lesion site does not play a major role in the selection of dAMP. Thus, preferential insertion of dAMP opposite abasic sites appears to be an intrinsic property of (all) DNA polymerases.

Structure-function relationships--The natural abasic site in DNA consists of a 2' deoxyribose residue linked to neighboring bases through 3' and 5' phosphodiester bonds. In aqueous solution, this structure is represented by an equilibrium mixture of tautomers consisting primarily of the α and β anomers of deoxyribofuranose (dR) and accompanied by a small amount of the partially hydrated (open chain) aldehydo form. Structures of this type are subject to β-elimination, leading to strand scission of DNA.

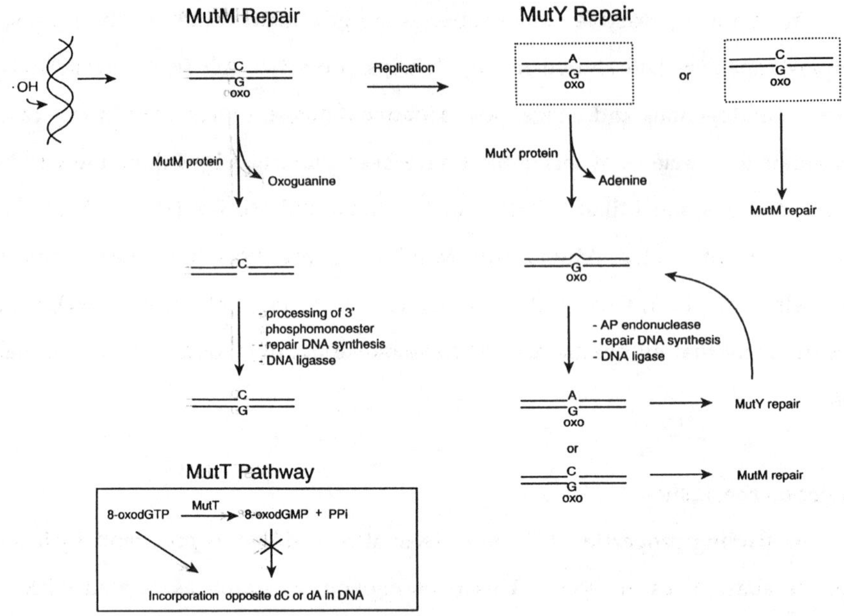

Figure 1. Schematic diagram illustrating how MutM, MutY, and MutT proteins reduce accumulation of 8-oxoguanine residues in DNA. 8-Oxoguanine is introduced into DNA by oxidative damage (unboxed pathway) or by incorporation of 8-oxodGTP during DNA biosynthesis (boxed pathway). MutM (8-oxoguanine DNA glycosylase) and MutY (adenine DNA glycosylase) act on oxidatively damaged DNA to minimize accumulation of 8-oxoguanine. MutT protein (8-oxodGTPase) cleanses the dNTP pool. 8-Oxoguanine is excised from DNA by the DNA-glycosylase and AP lyase activities of MutM protein. The single nucleotide gap is processed by DNA repair synthesis. If 8-oxoguanine is removed during repair, dAMP or dCMP is incorporated opposite the lesion. DNA containing 8-oxodG:dA is repaired by MutY protein, removing adenine. The MutY protein is temporarily bound to DNA, protecting 8-oxoguanine from subsequent attack by MutM protein, thereby preventing loss of one base of information and a double-strand break. Repair DNA synthesis directs incorporation of dAMP and dCMP opposite 8-oxodG. The 8-oxodG:dA pair is repaired through the MutY pathway while the preferentially formed 8-oxodG:dC pair can be repaired through the MutM pathway.

The synthetic abasic sites shown in Fig. 2 were designed as stable analogs of the natural abasic site which could be incorporated into DNA, using automated DNA synthesis technology. In duplex DNA, synthetic abasic sites are substrates for both AP endonucleases (Takeshita et al., 1991, Takeuchi et al., 1994) and in single strand DNA templates, for primer extension reactions catalyzed by DNA polymerases (Takeshita et al., 1987, Randall et al., 1987). The tetrahydrofuran moiety (F) is an analog of the predominant (cyclic) form of the natural abasic site. The propanyl moiety (P) represents an anucleosidic site with a three-carbon phosphodiester backbone. The analog designated E contains a phosphodiester backbone shortened by one carbon atom at the abasic site. Thus, acyclic structures P and E (Fig. 2) resemble the deoxyribitol moiety, generated by reduction of natural abasic sites and used in previous studies of abasic sites (Sagher and Strauss, 1983).

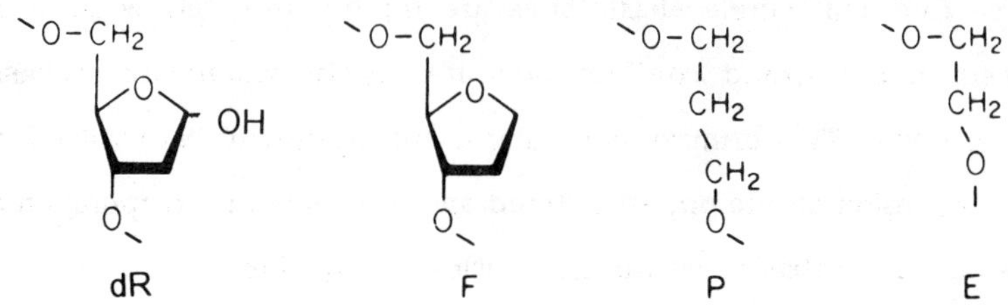

Figure 2. Structures of natural abasic sites (dR) and tetrahydrofuran (F), propanyl (P), and ethanyl (E) analogs.

Mutagenesis studies--Under physiological conditions, isolated abasic sites are rapidly repaired and it is doubtful whether such lesions contribute significantly to spontaneous or chemically-induced mutagenesis in wild-type cells. The most convincing evidence for mutagenesis at abasic sites derives from recent studies in *E. coli* and yeast which show that cells deficient in AP endonuclease have a slightly higher rate of spontaneous mutagenesis (Kunz et al., 1994). The mutagenic potential of abasic sites also has been tested by site-specific mutagenesis experiments in which a single strand shuttle plasmid vector containing a single abasic site is introduced into *E. coli* and mammalian cells. Using such a model system, Sarasin and his co-workers reported that the "A rule" is not observed in simian kidney (COS) cells (Gentil et al., 1992). Klinedinst and Drinkwater (1992) also observed random incorporation of dNMPs opposite abasic sites in mammalian cells. Lawrence et al. (1990) reports that dAMP is inserted opposite abasic sites in E. coli while Kunz et al. (1994) observed an increase of G→C transversions in AP endonuclease deficient strains of *S. Cerevisiae.* Using a duplex vector, Takeshita and his colleagues find that dAMP is preferentially incorporated opposite synthetic abasic sites in mammalian cells (Takeshita and Eisenberg, 1994). Further studies will be required to resolve these apparent discrepancies.

Bleomycin and neocarzinostatin cleave duplex DNA, generating discrete bistrand lesions composed of an abasic site and single strand gap positioned on opposite strands. Mutagenic properties of these radiomimetic agents have been attributed to the abasic site moiety (Povirk and Goldberg, 1985; Bennett et al., 1993). Using synthetic abasic sites to explore mutagenesis at bistrand lesions in duplex DNA, we find that single abasic sites are rapidly repaired while abasic sites positioned in a bistrand configuration are highly mutagenic (Takeshita and Eisenberg, 1994). Two primary molecular events appear to be involved, reflecting repair of the lesion on the opposite strand and concomitant incorporation of dAMP opposite one of the abasic sites during translesional synthesis.

Abasic sites are also associated with frameshift deletion mutations (Kunkel, 1984). An *in vitro* system, developed by Shibutani (1993), was used to quantify frameshift deletions and base substitutions formed during DNA synthesis in vitro. Oligodeoxynucleotides, modified site-specifically with synthetic abasic sites, were used as templates in primer extension reactions catalyzed by various DNA polymerases (Shibutani and Grollman, 1993). The influence of DNA sequence context on frameshift mutagenesis was determined by modifying systematically the bases flanking the lesion. The frequency of nucleotide insertion opposite abasic sites and of chain extension from the 3' primer terminus were established by steady state kinetic analysis. The ability of an abasic site to generate one-base and two-base frameshift deletions appears to be determined primarily by two parameters: the nature of the base inserted opposite the abasic site with respect to the sequence context in which the lesion is embedded, and the overall rate of translesional DNA synthesis. Misinsertion of bases opposite the lesion precedes misalignment of the template-primer. Extending on earlier studies (Kunkel and Soni, 1988; Boosalis et al., 1989), a general model for frameshift deletion mutagenesis shown in Fig. 3 has been proposed (Shibutani and Grollman, 1993). Depending on the nature of the base located 5' to the abasic site and the degree of block created by the lesion, abasic sites in DNA should promote incorporation of dAMP opposite the lesion, leading either to insertion of a new base (often T) in the template strand or to deletion of the lesion, leading to a frameshift mutation. Anucleosidic abasic sites are more deletion prone than cyclic analogs of deoxyribose.

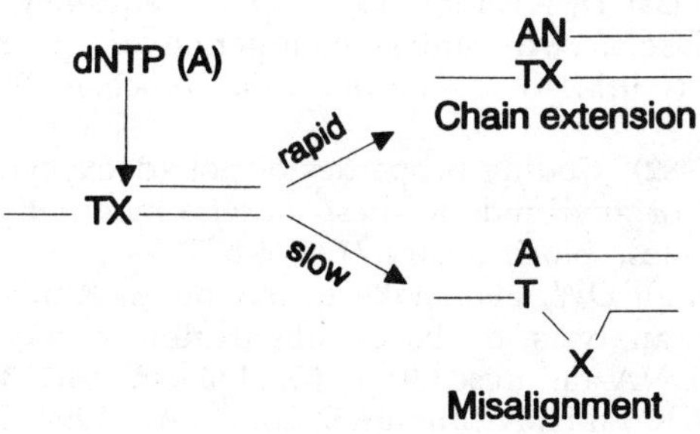

Figure 3. A model showing two pathways by which abasic sites could lead to mutations in DNA. Base insertion, followed by chain extension, is favored when translesional synthesis is relatively rapid; deletions, facilitated by misalignment of the template, occurs when DNA synthesis is slow or blocked at the replication fork.

In summary, single abasic sites in duplex DNA are rapidly repaired; however, in a bistrand configuration, as might be present in clustered lesions produced by ionizing radiation (Ward, 1993) or radiomimetic agents, the same lesions are highly mutagenic. Abasic sites block progression of DNA polymerase during DNA synthesis; however, translesional synthesis past abasic sites is observed with dAMP being preferentially incorporated opposite the lesion. Base substitution mutations may occur at abasic sites when a base is inserted opposite the lesion; however, depending on the nature of the base 5' to the lesion, frameshift deletions may be generated by the mechanism shown in Fig. 3.

Acknowledgements

The authors acknowledge the major contributions of their colleagues to the research described in this paper, especially C.-N. Chang, F. Johnson, M. Michaels, J. Miller, M. Moriya, S. Nishimura, D. Patel, S. Shibutani, and J. Tchou.

References

Ames BN, Gold LS (1991) Endogenous mutations and the causes of aging and cancer. Mutation Res 250:3-16

Au KG, Clark S, Miller JH, Modrich P. (1989) *Escherichia coli* mutY gene encodes an adenine glycosylase active on GA mispairs. Proc Natl Acad Sci USA 86:8877-8881

Bennett RAO, Swerdlow PS, Povirk LF (1993) Spontaneous cleavage of bleomycin-induced abasic sites in chromatin and their mutagenicity in mammalian shuttle vectors. Biochemistry 32:3188-3195

Bessho T, Tano K, Kasai H, Nishimura S. (1992) Deficiency of 8-hydroxyguanine DNA endonuclease activity and accumulation of the 8-hydroxyguanine in mutator mutant (mutM) of *Escherichia coli.* Biochem Biophys Res Commun 188:372-378

Boiteux S, Laval J (1982) Coding properties of poly(deoxycytidylic acid) templates containing uracil or apyrimidinic sites: *In vitro* modulation of mutagenesis by DNA repair enzymes. Biochemistry 21:6746-6751

Boosalis MS, Mosbaugh DW, Hamatake R, Sugino A, Kunkel TA, Goodman MF (1989) Kinetic analysis of base substitution mutagenesis by transient misalignment of DNA and miscoding. J Biol Chem 264:11360-11366

Cheng KC, Cahill DS, Kasai H, Nishimura S, Loeb L.A. (1992) 8-Hydroxyguanine, an abundant form of oxidative DNA damage, causes G→T and A→C substitutions. J Biol Chem 267:166-172

Chetsanga CJ, Lindahl T (1979) Release of 7-methylguanine residues whose imidazole rings have been opened from damaged DNA by a DNA glycosylase from *Escherichia coli.* Nucleic Acids Res. 6:3673-3684

Chung MH, Kim HS, Ohtsuka E, Kasai H, Yamamoto F, Nishimura S. An endonuclease activity of human polymorphonuclear nutrophils removes 8-hydroxyguanine residues from DNA. Biochem Biophys Res Commun 178:1472-1478

Floyd RA The role of 8-hydroxyguanine in carcinogenesis. Carcinogenesis 11:1447-1450

Fraga CG, Shigenaga MK, Park J-W, Degan P, Ames, BN (1990) Oxidative damage to DNA during aging: 8-hydroxy-2'-deoxyguanosine in rat organ DNA and urine. Proc. Natl. Acad. Sci. USA 87:4533-4537

Frenkel K (1992) Carcinogen-mediated oxidant formation and oxidative DNA damage. Pharmac Ther 53:127-166

Friedberg EC (1985) DNA Repair, WH Freeman and Company, New York

Gajewski E, Rao G, Nackerdein Z, Dizdaroglu M. (1992) Modification of DNA bases in mammalian chromatin by radiation-generated free radicals. Biochemistry 49:7876-7882

Gentil A, Cabral-Neto JB, Mariage-Samson R, Margot A, Imbach JL, Rayner B, Sarasin A (1992) Mutagenicity of a unique apurinic/apyrimidinic site in mammalian cells. J Mol Biol 227:981-984

Grollman AP, Moriya M (1993) Mutagenesis by 8-oxoguanine: an enemy within. Trends in Genetics 9:246-249

Grollman AP, Johnson F, Tchou J, Eisenberg M (1994) Recognition and repair of 8-oxoguanine and formamidopyrimidine lesions in DNA. In: DNA Damage: Effects on DNA Structure and Protein Recognition, Wallace S, VanHouten B, Wah Kow E (Eds.), Annals of the NY Academy of Sciences Vol. 726, pp. 208-214.

Kasai H and Nishimura S. (1984) Hydroxylation of deoxyguanosine at the C-8 position by ascorbic acid and other reducing agents. Nucleic Acids Res 12:2137-2145.

Kasai H and Nishimura S. (1991) Formation of 8-hydroxydeoxyguanosine in DNA by oxygen radicals and its biological significance. In: Oxidative Stress: Oxidants and Antioxidants (Sies H, Ed.) Academic Press, NY, pp. 99-116

Klinedinst DK, Drinkwater NR (1992) Mutagenesis by apurinic sites in normal and ataxia telangectasia human lymphoblastoid cells. Mol Carcinogen 6:32-42

Kouchakdjian M, Bodepudi V, Shibutani S, Eisenberg M, Johnson F, Grollman AP, Patel DJ (1991) NMR structural studies of the ionizing radiation adduct 7-hydro-8-oxodeoxyguanosine (8-oxo-7H-dG) opposite deoxyadenosine in a DNA duplex. 8-Oxo-7H-dG(syn).dA(anti) alignment at lesion site. Biochemistry 30:1403-1412

Kunkel TA, Soni A (1988) Mutagenesis by transient misalignment. J Biol Chem 263:14784-14789

Kunkel TA (1984) Mutational specificity of duprination. Proc Natl Acad Sci USA 81:1494-1498

Kunz BA, Henson ES, Roche H, Ramotar D, Nunoshiba T, Demple B (1994) Specificity of the mutator caused by deletion of the yeast structural gene (APN1) for the major apurinic endonuclease. Proc Natl Acad Sci USA 91:8165-8169

Lawrence CW, Borden A, Banerjee SK, LeClerc JE (1990) Mutation frequency and spectrum resulting from a single abasic site in a single-stranded vector. Nucl Acids Res 18:2153-2157

Lindahl T, Nyberg B. (1972) Rate of depurination of native deoxyribonucleic acid. Biochemistry 11:3610-3617

Loeb LA, Preston BD (1986) Mutagenesis by apurinic/apyrimidinic sites. Ann. Rev. Genet. 20:201-230

Maki H and Sekiguchi M (1992) MutT protein specifically hydrolyses a potent mutagenic substrate for DNA synthesis. Nature 355:273-275

McAuley-Hecht KE, Leonard GA, Gibson NJ, Thomson JB, Watson WP, Hunter WN, Brown T (1994) Crystal structure of a DNA duplex containing 8-hydroxydeoxyguanine-adenine base pairs. Biochemistry 33:10266-10270

Michaels ML, Pham Y, Cruz C, Miller JH (1991) MutM, a protein that prevents GCÆTA transversions, is formamidopyrimidine-DNA glycosylase. Nucleic Acids Res 19:3629-3632

Michaels ML, Cruz, C, Grollman, AP, Miller, JH (1992) Evidence that MutY and MutM combine to prevent mutations by an oxidatively damaged form of guanine in DNA. Proc Natl Acad Sci USA 89:7022-7025

Michaels ML, Miller JH (1992) The GO system protects organisms from the mutagenic effect of the spontaneous lesion 8-hydroxyguanine (7,8-dihydro-8-oxoguanine). J. Bacteriology 174:6321-6325

Michaels, ML, Tchou J, Grollman AP, Miller JH (1993) A repair system for 8-oxo-7,8-dihydrodeoxyguanine (8-hydroxyguanine). Biochemistry 31:10964-10968

Mo JY, Maki H, Sekiguchi M (1992) Hydrolytic elimination of a mutagenic nucleotide, 8-oxodGTP, by human 18-kilodalton protein: Sanitization of nucleotide pool. Proc Natl Acad Sci USA 89:11021-11025

Moriya M, Ou C, Bodepudi V, Johnson F, Takeshita M, Grollman AP (1991) Site specific mutagenesis using a gapped duplex vector: a study of translesion synthesis past 8-oxodeoxyguanosine in E. coli. Mutation Res 254:281-288

Moriya M. (1993) Single strand shuttle phagemid for mutagenesis studies in mammalian cells: 8-Oxoguanine in DNA induces targeted G:C→T:A transversions in simian kidney cells. Proc Natl Acad Sci USA 90:1122-1126

O'Connor TR, Boiteux S, Laval J (1988) Ring-opened 7-methylguanine residues in DNA are a block to in vitro DNA synthesis. Nucleic Acids Res 16:5879-5894

Povirk LF and Goldberg IH (1985) Endonuclease resistant apyrimidinic sites formed by neocarzinostatin at cytosine residue in DNA: Evidence for a possible role in mutagenesis. Proc Natl Acad Sci USA 82:3182-3186

Randall SK, Eritja R, Kaplan BE, Petruska J, Goodman MF (1987) Nucleotide insertion kinetics opposite abasic lesions in DNA. J Biol Chem 262:6864-6870

Sagher D, Strauss B (1983) Insertion of nucleotides opposite apurinic/apyrimidinic sites in deoxyribonucleic acid during in vitro synthesis: Uniqueness of adenine nucleotides. Biochemistry 22:4518-4526

Schrock RD III, Lloyd RS (1993) Site-directed mutagenesis of the NH2 terminus of T4 endonuclease V: the position of the a-NH2 moiety affects catalytic activity. J Biol Chem 268:880-886.

Shaaper RM, Kunkel TA, Loeb LA (1983) Infidelity of DNA synthesis associated with bypass of apurinic sites. Proc Natl Acad Sci USA 80:487-491

Shibutani S, Takeshita M, Grollman AP (1991) Insertion of specific bases during DNA synthesis past the oxidation-damaged base 8-oxodG. Nature 349:431-434

Shibutani S, Grollman AP (1993) On the mechanism of frameshift (deletion) mutagenesis *in vitro* . J Biol Chem 268:11703-11710

Strauss BS (1991) The "A Rule" of mutagenic specificity: a consequence of DNA polymerase bypass. BioEssays 13:79-84

Takeshita M. Chang C-N, Johnson F, Will S, Grollman AP (1987) Oligodeoxynucleotides containing synthetic abasic sites. Model substrates for DNA polymerases and apurinic/apyrimidinic endonucleases. J Biol Chem 262:10171-10179

Takeshita M, Eisenberg W (1994) Mechanism of mutation on DNA templates containing synthetic abasic sites: study with a double strand vector. Nucleic Acids Res 22:1897-1902

Takeuchi M, Lillis R, Demple B, Takeshita M (1994) Interactions of *Escherichia coli* endonuclease IV and exonuclease III with abasic sites in DNA. J Biol Chem 269:21907-21914.

Tchou J, Kasai H, Shibutani S, Chung M-H, Laval J, Grollman AP, Nishimura S (1991) 8-Oxoguanine (8-hydroxyguanine) DNA glycosylase and its substrate specificity. Proc Natl Acad Sci USA 88:4690-4694

Tchou J, Grollman, AP (1993) Repair of DNA containing the oxidatively-damaged base, 8-oxoguanine. Mutation Res 299:277-287

Tchou J, Bodepudi V, Shibutani, S, Antoshechkin I, Miller J, Grollman AP, Johnson F (1994) Substrate specificity of Fpg protein. J. Biol. Chem. 269:15318-15324

Ward, J.F. (1993) DNA damage produced by ionizing radiation in mammalian cells: identities, mechanisms of formation and repairability. Prog in Nucleic Acid Res and Molec Biol Vol. 35, pp. 95-125.

Wood ML, Dizdaroglu M, Gajewski, E, Essigmann JM (1990) Mechanistic studies of ionizing radiation and oxidative mutagenesis: genetic effects of a single 8-hydroxyguanine (7-hydro-8-oxoguanine) residue inserted at a unique site in a viral genome. Biochemistry 29:7024-7032

Yeh YC, Chang DY, Musin, J, Lu AL (1991) Two nicking enzyme systems specific for mismatch-containing DNA in nuclear extracts from human cells. J Biol Chem 266:6480-6484

HYDROXYL AND ONE-ELECTRON OXIDATION MEDIATED REACTIONS OF THE PURINE BASES OF DNA AND MODEL COMPOUNDS: MECHANISTIC ASPECTS

J. Cadet, M. Berger, G.W. Buchko, P.C. Joshi, B. Morin, S. Raoul and J.-L. Ravanat
CEA/DRFMC/SESAM/LAN
F-38054 Grenoble, Cedex 9, France

Introduction

A large body of information is now available on the early events involved in the radiation-induced decomposition of DNA and related compounds in both aqueous solutions and in the solid state (for comprehensive reviews see von Sonntag, 1987; Steenken 1989; Becker & Sevilla, 1993; Chatterjee & Holley, 1993; O'Neill & Fielden, 1993). However, despite many efforts devoted to the identification of radiation-induced DNA modifications in the past two decades (for recent reviews, see Wagner et al; Dizdaroglu, 1992; Cadet, 1994) there is still a paucity of data on the final decomposition products which arise from the fate of initially generated DNA purine radicals, at least, in aqueous aerated solutions. One major exception deals with the characterization of 7,8-dihydro-8-oxo-2'-deoxyguanosine [6] (8-oxodG) within isolated and nuclear DNA upon exposure to ionizing radiation (Fuciarelli et al., 1990; Mori et al., 1993). It should be noted that the biological significance of 8-oxodG 6 as inferred from comprehensive DNA repair and mutagenic studies has been recently assessed (for recent reviews, see Tchou et al., 1994; Grollman & Moriya, 1993). The present survey focuses on recent mechanistic aspects of the decomposition of the guanine moiety of DNA and model compounds as the result of hydroxyl radical and one-electron oxidation reactions in aerated aqueous solutions. In addition, a model system aimed at explaining the radiation-induced formation of DNA-protein crosslinks in aerated aqueous solution is presented.

Results and Discussion

The indirect effects of ionizing radiation were investigated by exposing diluted aqueous solutions of the substrates (nucleosides, dinucleoside monophosphates, DNA) to hydroxyl radicals generated through gamma radiolysis. Photosensitizers which operate mostly through type I mechanism (electron transfer reaction) such as benzophenone and riboflavin (Cadet & Vigny, 1990), were used to produce the guanine radical cation, one of the main reactive intermediates resulting from the interaction of ionizing radiation with DNA (Symons, 1987; Sevilla et al., 1991).

Hydroxyl radical and one-electron oxidation reactions of the guanine moiety of DNA model compounds.

In contrast to observations made for 2'-deoxyadenosine (Berger and Cadet, 1985; Cadet and Berger, 1985) the hydroxylation the imidazole ring of 2'-deoxyguanosine [1] at C8 by OH radicals, which give rise to 8-oxodG **6** in aerated aqueous solution is only a minor process (Berger et al., 1993), although the yield of the radical precursor is close to 25% (Steenken, 1989). Instead the two major oxidation products of the purine moiety of **1** which result from both its reaction with OH· radical and the transformation of the guanine radical cation (one-electron oxidation) have been isolated and identified as 2,2-diamino-4-[(2-deoxy-β-D-erythro-pentofuranosyl)amino]-5(2H)-oxazolone **(6)** and its precursor 2-amino-5-[(2-deoxy-β-D-erythro-pentofuranosyl)amino]-4H-imidazol-4-one [5] (Cadet et al., 1994a; 1994b). The mechanism of their formation may be rationalized in terms of transient formation of the oxidizing guanilyl radical **3** as depicted in Figures 1 & 2. The hydroxyl radical adduct at C(4) **3** dehydrates in neutral aqueous solution (O'Neill, 1983) with a rate constant of k = 5 x 10^3 sec^{-1} (Candeias and Steenken, 1991), yielding **4** which can exist in several tautomeric forms (Steenken, 1989). Interestingly, the neutral guanylyl radical **4** may also be generated through the quantitative deprotonation of the cation radical **2** as shown from pulse radiolysis and laser photolysis experiments (Candelas and Steenken, 1992). The resulting neutral radical **4** or rather the C5 centered radical tautomer **8** then follows a complicated decomposition pathway involving, in the early steps, the addition of a molecule of O_2 and H_2O to generate the transient peroxyl radical **9** (Fig. 2). Opening of the pyrimidine ring at the C5-C6 bond following either the formation of the hydroperoxide **10** or through the β-scission of the related oxyl radical gives rise to **11**. After the loss of a formamide residue, rearrangement leads to the formation of the

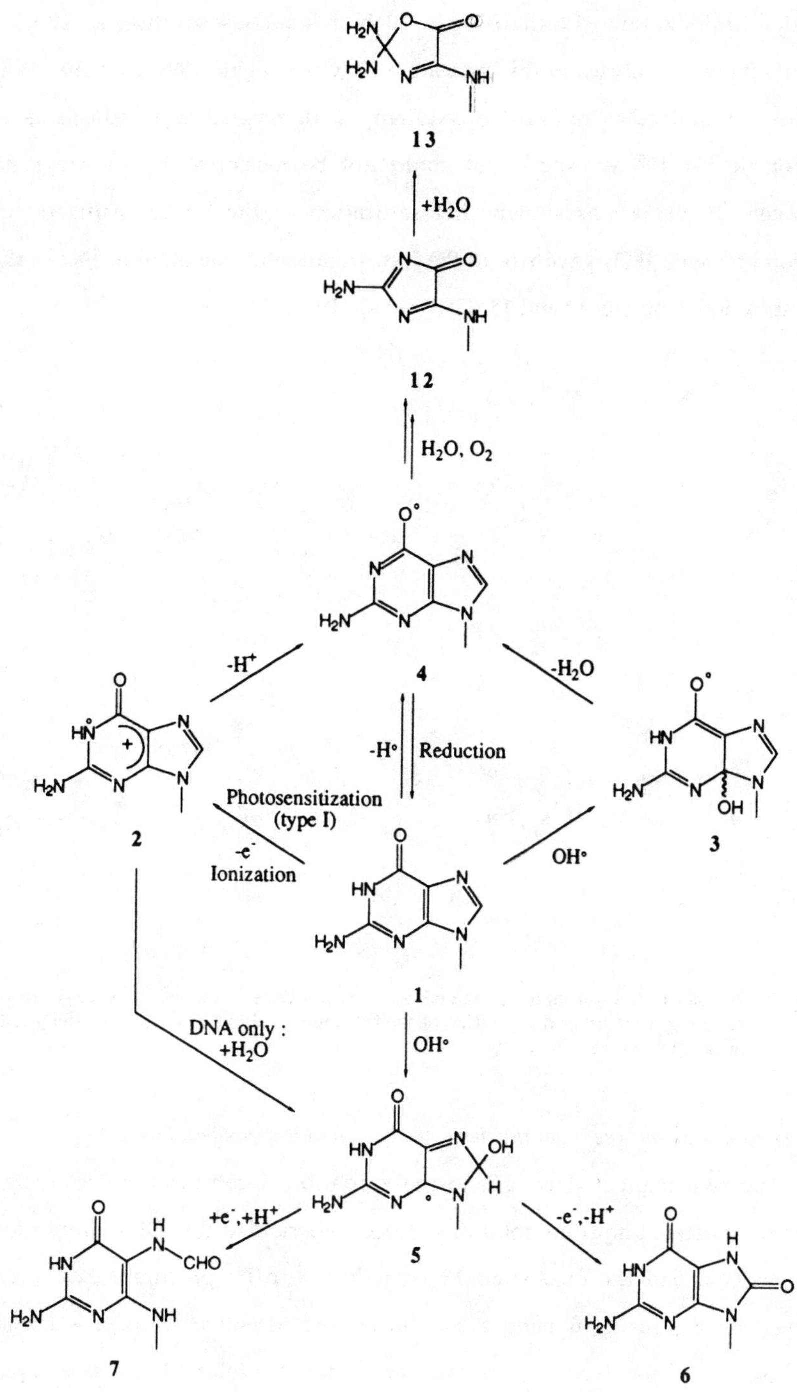

Fig. 1. Main radical oxidation pathways of the guanine moiety within nucleosides and DNA.

unstable imidazolone **12** (half-life = 10 h in aqueous solution at 20°C) which quantitatively hydrolyses to the oxazolone **13** (Cadet et al., 1994a; 1994b). While the addition of molecular oxygen to oxidizing 2'-deoxyguanosyl radicals is a slow reaction (k < < 10^7 M^{-1} sec^{-1} that could not be measured by pulsed radiolysis (Steenken, 1989), it was shown that saturation of the photosensitized aqueous solution of **1** with $^{18}O_2$ gave rise to the incorporation of one atom of ^{18}O in the final decomposition products **12** and **13** (Cadet et al., 1994a).

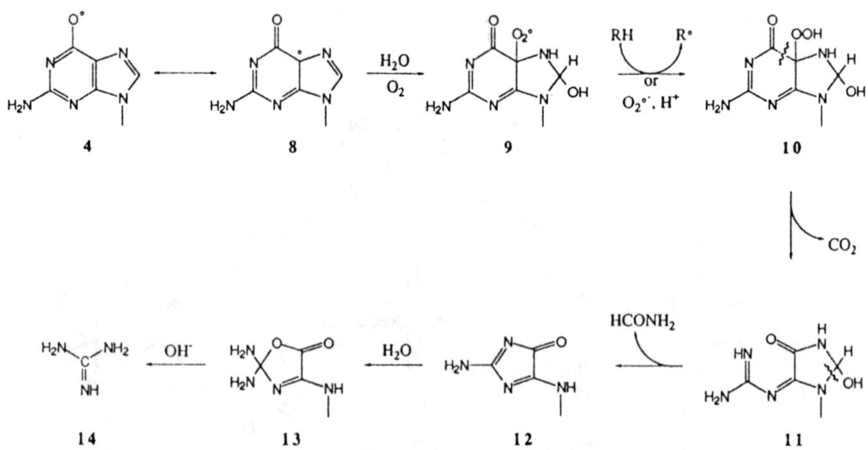

Fig. 2. Transformation reactions in aerated aqueous solutions of the oxidizing oxyl radical **3** resulting from either dehydration of the OH radical adduct at C(4) or the dehydration of the guanine radical cation.

Radical oxidation of the guanine base within double-stranded DNA.

The two main 2'-deoxyguanosine oxidation decomposition products upon exposure of aerated aqueous solutions of double-stranded (ds) DNA were identified as 8-oxodG **6** and the oxazolone **13** (together with its precursor **12**). This was achieved by measuring **6** using the accurate and sensitive HPLC-electrochemical detection assay (for a recent review, see Cadet & Weinfeld, 1993) whereas the formation of the imidazolone **12** and oxazolone **13** was determined by a fluorescent method involving the alkali-mediated release of guanidine (Fig. 2). It is interesting to note the dramatic increase in the OH-induced formation of 8-oxodG in ds DNA with respect to its yield when the free nucleoside **1** is used as the substrate. This parallels the observation of a significant increase in the yield of **6** at the expense of **12** and **13** when reducing agents such as Fe^{2+} and ascorbic acid are present in the gamma-irradiated aerated aqueous solutions of **1** (Berger et al., 1993). However, further studies are required to better assess the origin of this drastic difference in the quantitative distribution of the oxidation products of the guanine moiety with ds DNA and the free nucleoside. Even more striking is the significant yield of **6** through the riboflavin-mediated photosensitized formation of the guanine radical cation **2** within ds DNA (Kasai et al., 1992). This may be explained by a competitive hydration reaction of **2**, leading to the formation of intermediate **5** which may also be generated through OH radical addition at the C8 position. It should be noted that the α and β pyranosidic isomers of the formamidopyrimidine (Fapy) nucleoside **7**, which are the main radiation-induced decomposition products of **1** in oxygen-free aqueous solutions (Berger & Cadet, 1985) are not produced, at least, in detectable amounts when oxygen is present. Work is currently in progress to determine whether the Fapy derivative **7** is formed within ds DNA by either OH addition or one-electron oxidation reaction in aerated aqueous solutions.

Radical oxidation formation of DNA-protein crosslinks.

There is no available structural information on the radiation-induced formation of DNA-protein cross-links in aerated aqueous solutions. In this respect, the involvement of a nucleophilic addition of a water molecule in the formation of the imidazolone 12 from the oxyl radical **4** represents an interesting model system for investigating the mechanism of the generation of DNA-protein cross-links under radical oxidative conditions. The key observation which provided strong support to the involvement of nucleophilic reactions in the radical oxidative decomposition of the guanine moiety dealt with the competitive intramolecular cyclization of the nucleoside radical derived from **4**. The reaction gives rise to the anydronucleoside 15 (Buchko et al., 1992). This received further confirmation from the isolation and characterization of the (2R*) and (2S*) diastereoisomers of 1-(2-

deoxy-β-D-erythro-pentofuranosyl)2-methoxy-4,5-imidazolinedione **16** upon benzophenone photosensitization of **1** in an aerated solution of (1:1) water and methanol (Morin & Cadet, 1994). Another interesting example of a nucleophilic reaction involving radical **4** is given by the efficient formation of the anhydronucleoside **17** through intramolecular cyclization of the 5-amino group to the C8 position of 5'-amino-2',5'dideoxyguanosine (Fig. 3).

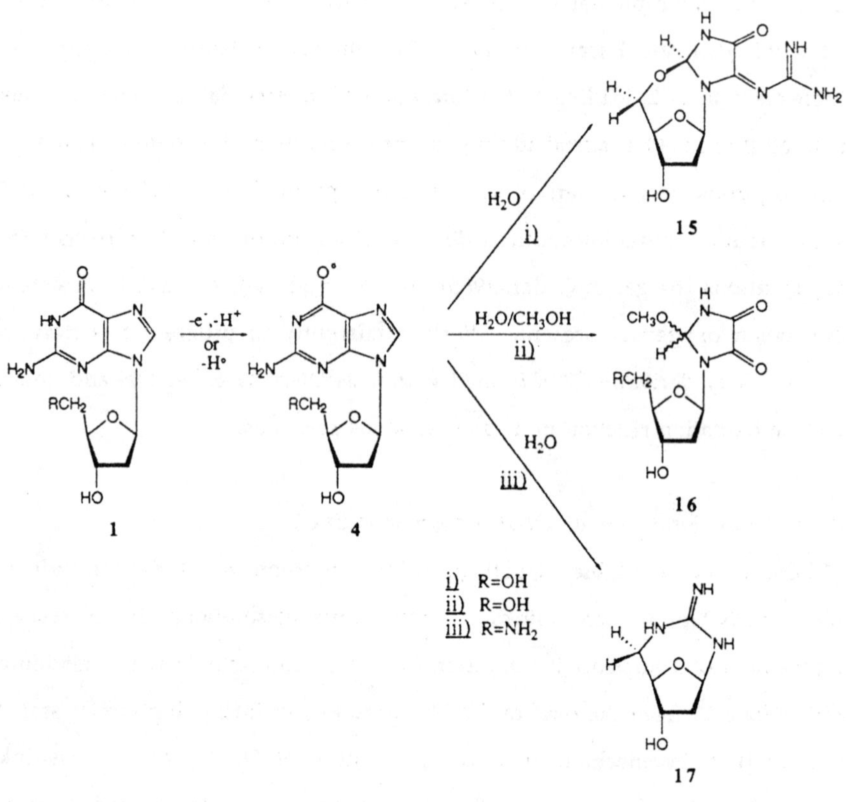

Fig. 3. Nucleophilic reactions of the oxidizing oxyl radical **3** involving hydroxyl and amino groups.

Conclusions

This short review clearly shows the complexity of the radical oxidation reactions involved in the radiation-induced decomposition of the guanine base of DNA and model compounds. In particular, the radical reactions appear to strongly depend on structural and environmental factors. It is interesting to note that the product distribution for the purine bases is much simpler than for the pyrimidines bases. In addition, at least, in ds DNA, similar decomposition pathways are induced by both OH radical and one electron oxidation reactions. Current extension of the present work should provide a better understanding of the mechanism of radiation-induced formation of DNA-protein crosslinks.

References

Becker D, Sevilla D (1993) The chemical consequences of radiation damage to DNA. Adv Radiat Biol 17: 121-180

Berger M, Cadet J (1985) Isolation and characterization of the radiation-induced degradation products of 2'-deoxyguanosine in oxygen-free aqueous solutions. Z. Naturforsch 40b: 1519-1531

Berger M, de Hazen M, Nejjari A, Fournier J, Guignard J, Pezerat H, Cadet J, (1993) Radical oxidation reactions of the purine moiety of 2'-deoxyribonucleosides and DNA by iron-containing minerals. Carcinogenesis 14: 41-46

Buchko GW, Cadet J, Ravanat JL, Labataille P (1993) Isolation and characterization of a new product produced by ionizing irradiation and type I photosensitization of 2'-deoxyguanosine in oxygen-saturated aqueous solution: (2S)-2,5'-anhydro-l-(2-deoxy-β-D-erythro-pentofuranosyl)-5guanidinylidene-2-hydroxy-4-oxoimidazol-idine. Int J Radiat Biol 63: 669-676

Cadet J, Berger M (1985) Radiation-induced decomposition of the purine bases within DNA and related model compounds. Int J Radiat Biol 47: 127-143

Cadet J, Vigny P (1990) The photochemistry of nucleic acids. Bioorganic Photochemistry, Vol. 1. Morrison H (ed) Wiley and Sons New York pp 1-272

Cadet J, Weinfeld M. (1993) Detecting DNA damage. Anal Chem 65: 675A-682A

Cadet J, (1994) DNA damage caused by oxidation, deamination, ultraviolet radiation and photoexcited psoralens. DNA adducts: Identification and biological significance. IARC Scientific Publications, No. 125 Hemminki K, Dipple A, Shuker DEG, Kadlubar FF, Segerbdck D, Bartsch H (eds.) International Agency for Research on Cancer Lyon pp 245-276

Cadet J, Berger M, Buchko GW, Joshi PC, Raoul S, Ravanat JL (1994a) 2,2-Diamino-4-[(3,5-di-O-acetyl-2-deoxy-β-D-erythro-pentofuranosyl)amino]5-(2H)-oxazolone: A novel and predominant radical oxidation product of 3',5'-di-O-acetyl-2'-deoxyguanosine. J Am Chem Soc 116: 7403-7404

Cadet J, Berger M, Morin B, Ravanat JL, Raoul S, Buchko GW, Weinfeld M (1994b) Photooxidation reactions of the guanine moiety of DNA and nucleosides. The Spectrum 7: 21-24

Candelas LP, Steenken S (1991) Transformation reactions of two isomeric OH-adducts of 2'-deoxyguanosine. The early effects of radiation on DNA Fielden EM, O'Neill M (eds.) NATO ASI Series Vol 54 pp 265-266

Candeias LP, Steenken S (1992) Ionization of purine nucleosides and nucleotides and their components by 193-nm laser photolysis in aqueous solution: model studies for oxidative damage of DNA. J Am Chem Soc 114: 699-704

Chatterjee A, Holley WR (1993) Computer simulation of initial events in the biochemical mechanisms of DNA damage. Adv Radiat Biol 17: 181-226

Dizdaroglu M (1992) Measurement of radiation-induced damage to DNA at the molecular level. Int J Radiat Biol 61: 175-183

Fuciarelli AF, Wegher BJ, Blakely WF, Dizdaroglu M (1990) Yields of radiation-induced base products in DNA: effects of DNA conformation and gassing conditions Int J Radiat Biol 58: 397-415

Grollman AP, Moriya M (1993) Mutagenesis by 8-oxoguanine: an enemy within. Trends Genet 9: 246-249

Kasai H, Yamaizumi Z, Berger M, Cadet J (1992) Photosensitized formation of 7,8-dihydro-8-oxo-2'-deoxyguanosine (8-hydroxy-2'-deoxyguanosine) in DNA by riboflavin: a non singlet oxygen mediated reaction. J Am Chem Soc 114: 9692-9694

Mori T, Hori Y, Dizdaroglu M (1993) DNA base damage generated *in vivo* in hepatic chromatin of mice upon whole body irradiation. Int J Radiat Biol 64: 645-650

Motin B, Cadet, J (1994) Benzophenone photosensitisation of 2'-deoxyguanosine: Characterization of the 2R and 2S diastereoisomers of 1-(2-deoxy-β-D-erythro pentofuranosyl)-2-methoxy-4,5-imidazolidinedione. A model system for the investigation of photosensitized formation of DNA-protein crosslinks. Photochem Photobiol 60: 102-109

O'Neill (1983) Pulse radiolytic study of the interaction of thiols and ascorbate with OH adducts of DGMP and dG: implications for DNA repair processes. Radiat Res 96: 198-210

O'Neill P, Fielden EM (1993) Primary free radical processes in DNA. Adv Radiat Biol 17: 53-120

Sevilla MD, Becker D, Yan M, Summerfield SR (1991) Relative abundances of primary ion radicals in gamma-irradiated DNA: cytosine vs thymine anions and guanine vs adenine cations. J Phys Chem 95: 3409-3415

Symons MCR (1987) Application of electron spin resonance spectroscopy to the study of the effects of ionising radiation on DNA and DNA complexes. J Chem Soc Faraday Trans I 83: 1-11

Tchou J, Bodepudi V, Shibutani S, Antoshechkin I, Miller J, Grollman AP, Johnson F (1994) Substrate specificity of Fpg protein. J Biol Chem 269:15318-15324

von Sonntag C (1987). The Chemical Basis of Radiation Biology, Taylor and Francis, London.

Wagner JR, van Lier JE, Decarroz C, Berger, M, Cadet, J (1990) Photodynamic methods for oxy radical-induced DNA damage. Methods Enzymol 186:502-511

THE USE OF PHOTOSENSITIZERS TO SELECTIVELY GENERATE RADIATION-INDUCED GUANINE DECOMPOSITION PRODUCTS

G.W. Buchko and M. Weinfeld
Department of Radiobiology
Cross Cancer Institute
11560 University Ave.
Edmonton, Alberta, Canada

M. Berger, J. Cadet and S. Raoul
Laboratoire des Lesions des Acides Nucléiques
CEA/DRFMC/SESAM
Centre d'Etudes Nucléaires de Grenoble
France

Introduction

Free radicals generated *in vivo* through biological processes, including photosensitization, may lead to oxidative damage to cellular DNA (Hélène, 1978; Cadet & Vigny, 1990). Such photosensitizer-mediated damage to DNA occurs through two principal competitive mechanisms, labelled type I and type II, depending on the process by which an excited photosensitizer decays from its triplet state back to the ground state (Foote, 1976). Type I mechanisms involve initial electron or hydrogen transfer by the excited photosensitizer to, or from, the substrate (DNA) to generate free radicals. Type II mechanisms involve the initial generation of singlet oxygen by the excited photosensitizer which subsequently reacts with the substrate to generate unstable peroxidic intermediates. Although in both cases the primary target in DNA is the guanine base (Ennever & Speck, 1981; Cadet et al., 1986; Cadet & Vigny, 1990), the two mechanisms generate a different set of photoproducts.

Another source of free radical damage to DNA is the hydroxyl radical, •OH, which can be generated through biological processes (Britigan et al., 1987) and the radiolysis of water (Greenstock, 1981; Ward, 1981; von Sonntag, 1987). Unlike photosensitizer mediated DNA damage, OH• radicals show relatively little bias towards any of the bases as indicated by the variety of base- and sugar-derived products that have been identified (for reviews see Téoule & Cadet, 1978; von Sonntag, 1987; Téoule, 1987; Cadet & Vigny, 1990; Dizdaroglu, 1992). However, the HPLC damage profile obtained upon exposure of 3',5'-di-O-acetyl-2'-deoxyguanosine to OH radicals in aerated aqueous solution is similar to that obtained by exposing the same substrate to photoexcited sensitizers, such as benzophenone and riboflavin, which act primarily through a type I mechanism (Cadet et al., 1994a). Although initiated by different mechanisms, similar reaction intermediates are proposed to explain the generation of the same products (Cadet et al., 1994a,b). Thus, the use of the excited photosensitizers represents a facile method for selectively obtaining radiation-induced guanine lesions in DNA.

The lesions 2-amino-5-[(3,5-di-O-acetyl-2-deoxy-β-D-*erythro*-pentofuran-osyl)-amino]-4H-imidazol-4-one and its major decomposition product, 2,2-diamino-4-[(3,5-di-O-acetyl-2-deoxy-β-D-*erythro*-pentofuranosyl)amino]-5(2H)-oxazolone, have previously been identified and characterized following the exposure of 3',5'-di-O-acetyl-2'-deoxyguanosine to OH radicals in aerated aqueous solution (Cadet et al., 1994a). In this report we more fully characterize 2-amino-5-[(2-deoxy-β-D-*erythro*-pentofuranosyl)amino]-4H-imidazol-4-one (dIZ) and its major decomposition product, 2,2-diamino-4-[(2-deoxy-β-D-*erythro*-pentofuranosyl)amino]-5(2H)-oxazolone (dZ) (Figure 1), obtained from the γ-radiolysis of the non-acetyled nucleoside, 2'-deoxyguanosine (dG), in aerated aqueous solution.

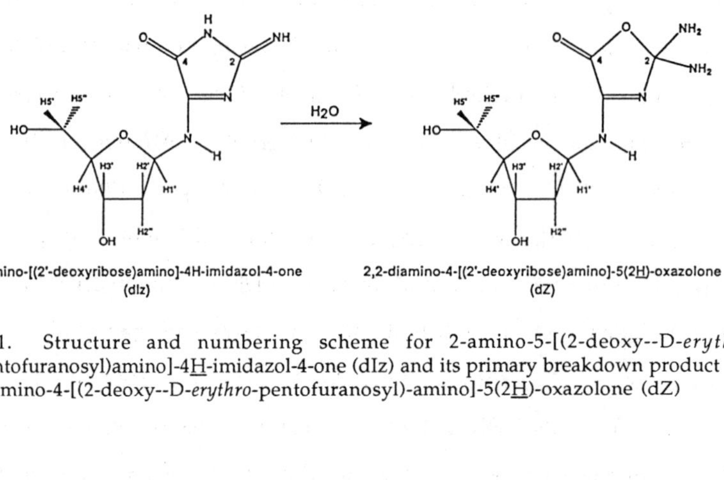

Figure 1. Structure and numbering scheme for 2-amino-5-[(2-deoxy--D-*erythro*-pentofuranosyl)amino]-4H-imidazol-4-one (dIz) and its primary breakdown product 2,2-diamino-4-[(2-deoxy--D-*erythro*-pentofuranosyl)-amino]-5(2H)-oxazolone (dZ)

Methods

Photosensitization reactions : A 10 mL solution of 7.5 mM dG, saturated with riboflavin, was prepared in a 20 mL beaker and irradiated predominately at 365 nm with a Brickman Chromato-vue Model CC-20 ultra-violet lamp at a distance of 5 cm. A dark blue glass filter was used to remove visible light. After 2 and 24 h, 80-μL aliquots were removed for high performance liquid chromatography (HPLC) analysis. NMR quantities of dIz were prepared by removing the water by rotary evaporation (<30°C) and resuspending the residue in the minimum volume of water for HPLC purification.

γ-Irradiation: A 50 mL aqueous solution of 7.5 mM dG was irradiated (200 Gy) using a ^{60}Co source at a dose rate of 100 Gy/min. The water was removed by rotary evaporation (<30°C) and the residue resuspended in the minimum volume of water for HPLC analysis.

Chromatography of photoproducts: Samples were injected onto a 5 µm Hypersil (1.0 cm x 25 cm) semipreparative octadecylsilyl column (Phenomenix). After eluting the column for 1 min with 100% H_2O, a linear gradient of methanol was introduced at a rate of change of 0.5%/min and at a flow rate of 1.2 mL/min. The HPLC elution profile was monitored by UV absorbance at 260 nm. The major bands were collected, pooled, and concentrated under reduced pressure by rotary evaporation (<30°C). Note that the use of ammonium acetate as the aqueous phase should be avoided because dIz completely decomposes during rotary evaporation or lyophilization.

Proton and carbon nuclear magnetic resonance (NMR) spectrometry.; Samples were lyophilized twice from 99.8% D_2O prior to resuspension in 400 µL of NMR buffer (pD = 7.0) consisting of 0.1 M sodium chloride, 0.01 M sodium phosphate, and 0.001 M ethylenediaminotetraacetic acid. A trace of *tert*-butanol was added to serve as an internal NMR reference (assigned a chemical shift of 1.231 ppm on the DSS scale). Conventional homonuclear decoupling, nuclear Overhauser enhancement and two-dimensional coherence (Aue et al., 1976) experiments were carried out for proton spectral assignments. The H5' was assigned downfield of H5" according to Remin and Shugar (1974) whereas the H2'/H2" assignments were based on coupling constant arguments (Wood et al., 1974, Davies & Danyluk, 1974). To verify the assignments and to obtain accurate coupling constants, all the spectra were simulated using the Bruker PANIC program.

A proton decoupled ^{13}C spectrum for dZ was obtained at 50.3 MHz (Bruker AC200); a trace of dioxane was used as the internal reference (assigned a chemical shift of 67.86 ppm relative to TMS). The carbon resonances of dZ were assigned on the basis of previous assignments for dG (Jones et al., 1970; Buchko et al., 1993) and continuous wave single frequency proton decoupling experiments.

pH Stability of dIz; A 300 mL solution containing 2.0 A_{260} units of dIz was incubated at 37°C in 10 mM phosphate buffers of pH 4.8, 7.4 and 10.4. A 30 mL aliquot was removed at various time intervals and analyzed immediately by HPLC.

Results

HPLC profile of the photoproducts; Fig. 2A is an HPLC chromatogram of the riboflavin-mediated photosensitized oxidation products of dG after a 2 h exposure to light. The major product elutes at 28 min and has been shown to contain dIz. Fig. 2B is the HPLC chromatogram of the riboflavin-mediated photosensitized oxidation products of dG after a 24 h exposure to light. All of the dG has been consumed and the major band elutes at 28 min. While all the other products have not yet been fully characterized, the HPLC band which elutes at 16 min has been identified as dZ. The product which elutes at 26 min has previously been shown to contain (2S)-2,5'-anhydro-1-(2'-deoxy)-5-guanidinylidene-2-hydroxy-4-oxoimidazolidine (d<G) (Buchko et al., 1993). Incubation of the products in Fig. 2B in the dark and in aqueous solution for 24 h at 37°C sees a decrease in the band which elutes at 28 min (dIz) and a quantitative increase in the band which elutes at 16 min (dZ).

The HPLC chromatogram of the damage profile observed following γ-irradiation of dG in aerated aqueous solution (not shown) also contains a major band which elutes at 28 min and decomposes primarily to a product with a retention time of 16 min. Spectral analysis (NMR, mass spectrometry, UV, IR) of the latter two products showed them to be identical to the products isolated from riboflavin-mediated oxidation of dG.

Carbon NMR: The ^{13}C NMR spectrum of dZ contains only 8 carbon resonances. The missing dZ carbon resonances, relative to dG, can be attributed to base carbons after assigning all the dZ furanosyl resonances. C1': 83.21, C2': 39.75, C3': 72.63, C4': 87.44, C5': 63.07, Base carbons: 158.38, 164.08, 167.23.

Proton NMR; Table 1 lists the proton chemical shifts and coupling constants for dG, and dZ. The H8 resonance observed in the spectrum of dG is absent in the spectra of dZ indicating that the guanine ring has been modified in the nucleoside. On the other hand, the chemical shifts and coupling constants for the sugar resonances of dZ do not differ dramatically from dG suggesting that the furanose

Table 1. Proton chemical shifts (σ, ppm from DSS) and coupling constants (J, Hz) for dG and dZ obtained at 23°C in 0.1 M NaCl.

σ	dG	dZ	J	dG	dZ
8	7.96	---	1'2'	6.9	6.9
1'	6.27	5.77	1'2"	6.6	6.5
2'	2.76	2.26	2'2"	-14.1	-14.0
2"	2.49	2.28	2'3'	6.7	6.2
3'	4.60	4.41	2"3'	3.5	3.2
4'	4.11	3.96	3'4'	3.2	2.9
5'	3.79	3.67	4'5'	3.5	4.2
5"	3.74	3.62	4'5"	4.7	5.4
			5'5"	-12.5	-12.2

ring has not been altered in this photoproduct.

pH Stability of dIz: Fig. 3 is a plot of the decomposition of dIz as a function of time at various pHs at 37°C. Decreasing the pH from 7.4 to 4.8 has little effect on the rate of dIz decomposition. On the other hand, at pH 10.4, less then 5% of dIz remains after 3 h. Note that these rates of decomposition are temperature dependent, increasing with increasing temperature (data not shown).

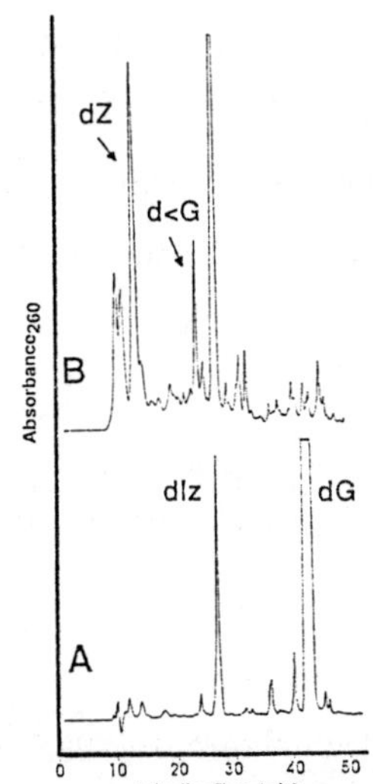

Figure 2. Reverse-phase HPLC chromatograms of the products of 2 h (A) and 24 h (B) riboflavin-mediated photosensitization of dG. UV detection at 260 nm.

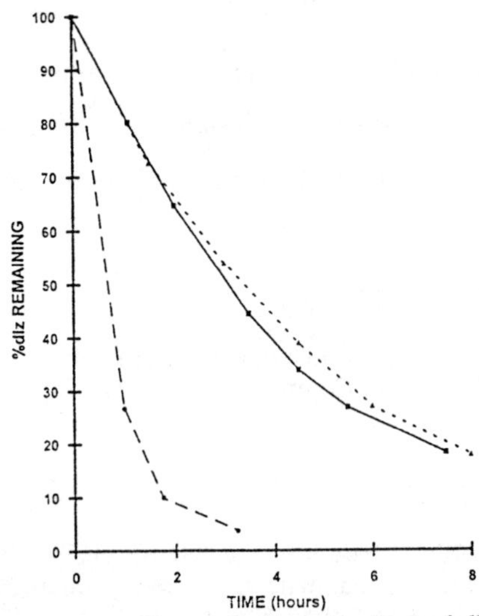

Figure 3. Plot of the decomposition of dIz as a function of time at pH 4.8 (dotted), 7.4 (solid) and 10.4 (dashed).

Discussion

We have shown that the major radiation induced decomposition product of dG in aerated aqueous solution is the same major product of type I mediated photosensitization of dG in aerated aqueous solution. While the presence of an unstable photooxidation product of guanine has been reported previously following γ-irradiation or photosensitization of dG and 3',5'-diacetyldG (Ivanchenko et al., 1977; Cadet et al., 1986), only recently has it been possible to fully characterize the initial lesion and its primary degradation product. One of the reasons for this difficulty is the lability of dIz in aqueous solutions containing various salts typically used to elute these products from HPLC columns; a process aggrevated by heating the eluant to concentrate the product. Although dIz is initially detected by UV-absorbance on the HPLC chromatogram, attempts to remove the solvent leads to its decomposition with the type and ratio of various products depending on the buffer used. This problem is minimized by using unbuffered water as the aqueous phase and gently removing the water by rotary evaporation (<30°C) or freeze-drying. Once pure product is obtained the interpretation of spectroscopic data is simplified.

The lability of dIz in aqueous solution will make detection of this lesion difficult within cellular DNA. Instead, it may be possible to detect the major decomposition product of dIz using mass spectral (Cadet et al., 1994c; Cadet & Weinfeld, 1993) or post-labelling protocols (Weinfeld & Soderlind, 1991; Buchko & Weinfeld, 1993). Efforts are in progress to prepare markers for such analyses.

Acknowledgments

We acknowledge the support of NCI Canada, the Alberta Cancer Board, and the European Commission CHCM program (contract no. CHRX CT 93 0275).

References

Aue WP, Bartholdi E, Ernst RR (1976) Two-dimensional spectroscopy. Application to nuclear magnetic resonance. J Chem Phys 64:2229-2246.

Britigan BE, Cohen MS, Rosen GM, (1987) Detection of the production of oxygen-centered free radicals by human neutrophils using spin trapping techniques: a critical perspective. J Leukocyte Biol 41:349-362.

Buchko GW, Cadet J, Ravanat JL, Labataille P (1993) Isolation and characterization of a new product produced by ionizing irradiation and type I photosensitization of 2'-deoxyguanosine in oxygen-saturated aqueous solution: (2S)-2,5'-anhydro-1-(2'-deoxy-b-D-*erythro*-pentofuranosyl)-5-guanidinylidene-2-hydroxy-4-oxoimidazolidine. Int J Rad Biol 63:669-676.

Buchko GW, Weinfeld M (1993) Influence of nitrogen, oxygen, and nitroimidazole radiosensitizers on DNA damage induced by ionizing radiation. Biochemistry 32:2186-2193.

Cadet J, Berger M, Decarroz C, Wagner JR, van Lier JE, Ginot Y, Vigny P (1986)

Photosensitized reactions of nucleic acids. Biochimie 68:813-834.
Cadet J, Vigny P (1990) The photochemistry of nucleic acids. Bioorganic Photochemistry, Vol 1, Morrison H, (ed) John Wiley & Sons, Inc., N.Y. pp 1-272.
Cadet J, Berger M, Buchko GW, Joshi PC, Raoul S, Ravanat JL (1994a) 2,2-Diamino-4-[(3,5-di-O-acetyl-2-deoxy-β-D-erythro-pentofuranosyl)amino]-5-(2H)-oxazolone: A novel and predominant radical oxidation product of 3',5'-di-O-acetyl-2'-deoxyguanosine. J Amer Chem Soc 116:7403-7404.
Cadet J, Berger M, Buchko GW, Joshi PC, Morin B, Raoul S, Ravanat JL (1994b) Hydroxyl and one-electron oxidation mediated reactions of the purine bases of DNA and model compounds: Mechanistic aspects. (This series).
Cadet J, Ravanat JL, Buchko GW, Yeo HC, Ames BN (1994c) Singlet oxygen DNA damage: Chromatographic and mass spectrometry analysis of damage products. Methods in Enzymology: Oxygen Radicals in Biological Systems, Part D, Vol. 234, Parker L (ed) pp 79-88.
Cadet J, Weinfeld M (1993) Detecting DNA damage. Anal Chem 65:675A-682A.
Davies DB, Danyluk SS (1974) Nuclear magnetic resonance studies of 5'-ribo- and deoxyribonucleotide structures in solution. Biochemistry 13:4417-4434.
Dizdaroglu M (1992) Measurement of radiation-induced damage to DNA at the molecular level. Int J Rad Biol 61:175-183.
Ennever JF, Speck WT (1981) Photodynamic reaction of riboflavin and deoxyguanosine. Pediatric Res 15:956-958.
Foote CS (1976) Photosensitized oxidation and singlet oxygen: Consequences in biological systems. Free Radicals in Biology. Prior WA (ed), Academic Press, N.Y., pp 85-133.
Greenstock CL (1981) Redox processes in radiation biology and cancer. Radiat Res 86:196-211.
Hélène C (1987) Excited states and photochemical reactions in DNA, DNA-photosensitizer, and DNA-protein complexes. A review. Photobiochem Photobiophy Suppl: 3-22.
Ivanchenko VA, Budowsky EI, Simukova NA, Vul'fson NS, Tishchenko AI, Askerov DB (1977) The photochemistry of purine components of nucleic acids. II. Photolysis of deoxyguanosine. Nucleic Acids Res 4:955-968.
Jones AJ, Grant DM, Winkley MW, Robins RK (1970) Carbon-13 magnetic resonance. XVII. Pyrimidine and purine nucleosides. J Am Chem Soc 92:4079-4087.
Remin M, Shugar D (1972) Conformation of the exocyclic 5'-CH_2OH in nucleosides and nucleotides in aqueous solution from specific assignments of the $H_{5'}$ and $H_{5''}$ signals in the NMR spectra. Biochem Biophys Res Comm 48:636-642.
Téoule R (1987) Radiation-induced DNA damage and repair. Int J Rad Biol 51:573-589.
Téoule R, Cadet J (1978) Radiation-induced degradation of the base components in DNA and related substances. Effects of Ionizing in DNA Huttermann J, Kohnlein W, Téoule R, Bertinchamps A (eds) Berlin: Springer-Verlag pp 171-202.
von Sonntag C (1987) The Chemical Basis of Radiation Biology Taylor & Francis, N.Y.
Ward JF (1981) Some biochemical consequences of the spatial distribution of ionizing radiation-produced free radicals, Radiat Res 86:185-195.
Weinfeld M, Soderlind KM (1991) ^{32}P-Postlabelling detection of radiation induced DNA damage: Indentification and estimation of thymine glycols and phosphoglycolate termini. Biochemistry 30:1091-1097.
Wood DJ, Hruska FE, Ogilvie KK (1974) Proton magnetic resonance studies of 2'-deoxythymidine, its 3'- and 5'-monophosphates and 2'-deoxythymidylyl-(3', 5')-2'-deoxythymidine in aqueous solution. Can J Chem 52:3353-3366.

Chapter 10

Application of Computer Modeling to Structure/Function Analysis of Radiation Damage to DNA

DISTRIBUTION OF IONS AROUND THYMINE DIMER CONTAINING DNA: A POSSIBLE RECOGNITION ELEMENT FOR ENDONUCLEASE V

Roman Osman and Ning Luo
Department of Physiology and Biophysics
Mount Sinai School of Medicine
New York, NY 10029 USA

Karol Miaskiewicz and John Miller
Pacific Northwest Laboratory
Richland, WA 99352 USA

Introduction

The molecular link between sunlight exposure and skin cancer can be traced to the formation of cyclobutane pyrimidine dimers together with (6-4) photoadducts of pyrimidines in DNA upon exposure to UV radiation (Hutchinson 1987; Cadet and Vigny 1990). The mutagenicity of these lesions is frequently explained by miscoding during DNA replication due to perturbations of base-pairing interactions. However, the mutagenicity of UV photoproducts depends of their sequence context (Brash et al. 1987), suggesting that more global structural changes in DNA contribute to mutation induction.

One of the most effective protections against the deleterious effects of cyclobutane pyrimidine dimers is the wide range of repair of this lesion by different enzymatic pathways. DNA photolyases (Sancar 1990) catalyze photocleavage of the C-C bonds linking the adjacent thymines of the dimer, thereby restoring the pyrimidines to their native state. Cyclobutane pyrimidine dimers are also removed by base-excision. The most extensively studied enzyme of this type is the bacteriophage T4 endonuclease V. This enzyme hydrolyses the glycosyl bond of the 5' pyrimidine in the dimer and also exhibits an apurinic/apyrimidinic endonuclease activity proceeding through the β-elimination of the 3'-phosphate of an abasic site (Dodson and Lloyd 1989). The enzyme has also been shown to bind to DNA duplex non-specifically and to scan it in a one-dimensional diffusion along the DNA until it encounters a thymine dimer. Upon encounter with the thymine dimer the enzyme binds to DNA with higher affinity and engages in the enzymatic excision of the damage. The nonspecific binding is sensitive to ionic strength: below 50mM NaCl the enzyme performs a one-dimensional scanning, whereas above that concentration the enzyme activity is controlled by a three dimensional diffusion (Gruskin and Lloyd 1986; Dowd and Lloyd 1990). This fundamental property of rapid one-dimensional scanning along DNA (Von Hippel and Berg 1989; Dowd and Lloyd 1990; von Hippel 1994) is shared by many DNA binding proteins. Thus, the understanding of the underlying mechanism of such a process is of major importance.

In spite of the high resolution X-ray crystal structure of endonuclease V (Morikawa et al. 1992) and numerous site directed mutagenesis studies (Dowd and Lloyd 1989; Lloyd and Augustine 1989; Augustine et al. 1991; Dodson et al. 1991; Doi et al. 1992; Dodson et al. 1993), the mechanism by which it recognizes the pyrimidine dimer lesion in DNA is still obscure. NMR studies performed on DNA oligonucleotides containing *cis,syn*-pyrimidine dimers showed that the distortions caused by the lesion are surprisingly small (Kemmink et al. 1987; Kemmink et al. 1987; Kemmink et al. 1987; Taylor and ODay 1990; Lee et al. 1994). Nevertheless, endonuclease V is able to recognize this change with very high fidelity. Since distortions of DNA structure are usually associated with changes in the electrostatic shielding by counterions and in solvation of DNA it is possible that the enzyme is acutely sensitive to this property, as is demonstrated by the dependence of the scanning ability on ionic strength.

This paper presents the results of a 200 ps molecular dynamics simulation on the dodecamer $d(CGCGAATTCGCG)_2$ containing a *cis,syn*-cyclobutane thymine dimer, explicit water and counterions. The averaged structure calculated from the simulation shows good agreement with the available NMR data. The distribution of counterions around the damaged DNA is different from that around a non damaged DNA and suggests a possible mechanism of damage recognition by the enzyme.

Methods

A *cis-syn*-thymine cyclobutane dimer was incorporated into the double helical dodecamer $d(CGCGAATTCGCG)_2$ with a canonical B-DNA initial geometry. To neutralize the phosphates, 22 Na^+ ions were placed around the DNA and the system was surrounded by a box of water with periodic boundary conditions. The complete system of DNA, water and counterions contained 6435 atoms.

The thymine dimer was built inside DNA during the initial energy optimization of the system. First, the positions of counterions and water were optimized with the DNA fixed in the initial conformation. This was followed by a stepwise minimization of the whole system with a restraining force constant of 100 kcal/mole on the distances between the thymines. The restraints were removed in the final stage of minimization and the covalent bonds between the thymines were defined by changing the atom types and the atomic charges.

Minimizations were followed by ten 0.2ps molecular dynamics 'heating' steps in which the temperature was gradually raised from 0K to 300K. This was followed by a 1.0ps run at 300K assigning initial velocities from a Maxwell-Boltzmann distribution. A 200ps molecular dynamics simulation was performed with constant temperature and volume. The SHAKE constraint was applied to all bonds and a 0.001ps time step was used. A non bonded cutoff distance of 10Å was used and the matrix of non bonded interactions was updated every 20 steps.

The resulting trajectories were used for structure analysis and for calculations of counterion density distribution.

Results and Discussion

The simulation can be divided into three distinct regions. The initial 20ps can be considered as an equilibration period followed by additional 80ps during which the structure of the DNA evolved with time. The RMS deviation of backbone dihedral angles and helical parameters from their starting values shows a gradual increase between 20 and 80ps. The third period, 100 - 200ps, is characterized by fluctuations about an apparently stable structure judged by the values of structural and energetic parameters.

Several structural parameters, e.g., the *base pair axis* displacement and inclination, are very good indicators of the global character of DNA structure, which appears to be close to a canonical B-DNA. On the other hand, values of the tip angle for the T7-A18 base pair indicate that the position of the 5′ thymine of the dimer is slightly distorted within the helix. The *axis* parameters defined by relative arrangements of base pairs also show the distortions of the helical axis induced by the thymine dimer. Axis inclination at the A6-T19/T7-A18 step as well as axis tip and axis displacement at the T7-A18/T8-A17 step are significantly increased, causing a slight bending of the helical axis at the lesion.

Backbone torsional angles and fluctuations of the sugar puckering phase angle ϕ cover the range of conformations characteristic for A-DNA and B-DNA without extending into the high energy structures. The thymine dimer does not change the pattern of backbone angles in a significant way. The only change is in χ, the glycosyl bond dihedral at the 5' thymine of the dimer, which shows a *syn* instead of an *anti* conformation. Also, the coupling of the thymines produces a more rigid deoxyribose ring. The puckering of the 5' thymine is mostly a B-DNA type whereas the 3' thymine is more A-DNA type. This dynamic behavior of the DNA backbone is consistent with that observed in simulations performed on the native sequence (Swaminathan et al. 1991; Miaskiewicz et al. 1993).

Effects of the thymine dimer on *intra base pair* parameters, that describe the geometry of base pairing, show an increase in buckle and in propeller twist at T7-A18. The *inter base pair* parameters, that describe base stacking, show an increase in tilt angle for the A6-T19/T7-A18 step (more positive) and a decrease for the T7-A18/T8-A17 step (more negative). Most Watson-Crick hydrogen bonds are preserved in the dodecamer during the simulation except for the T7-A18 base pair that contains the 5′ thymine of the dimer. It is substantially stretched and bent suggesting a weakening of this bond. The other hydrogen bond in the T7-A18 base pair does not show significant distortion. The major DNA groove width does not show changes attributable to the presence of lesion, but the minor groove seems to be widened by 2-3Å in the vicinity of the dimer.

NMR studies give strong evidence that distortions to DNA structure due to cyclobutane pyrimidine dimer formation are small. The results of our molecular dynamics simulation on a DNA dodecamer containing a cyclobutane thymine dimer are in line with these NMR studies in that they predict only small structural perturbations close to the site of the lesion. Our results are also consistent with NMR spectra of thymine dimers in the dinucleotide d(TpT) (Kim and Alderfer 1992; Kim et al. 1993), which suggest significant deviations of the glycosyl torsional angle from an *anti* conformation and varying preferences in sugar puckering. The dynamic simulations also suggest an increased rigidity of the deoxyribose moieties in the dimer relative to native DNA nucleotides in the dodecamer.

The major structural changes predicted by our calculations are not in the backbone but rather in the position, base-pairing and base stacking geometries of the damaged bases within the helix. This finding is consistent with 2D-NOE spectra of an octamer duplex containing a thymine dimer (Kemmink et al. 1987) from which it was deduced that the 5′ thymine was more reoriented with respect to helical axis than the 3′ thymine. This effect is clearly seen from our results. As a consequence of this reorientation, our calculations predict that one of the hydrogen bonds of the 5′ thymine with complementary adenine is significantly weakened. This result is in full agreement with observations based on NMR spectra of imino protons that the hydrogen bonds of base pairs in a thymine dimer are weakened but not broken (Kemmink et al. 1987; Lee et al. 1994).

In a recent study using synthetic oligonucleotides, the minor groove of DNA containing a thymine dimer was implicated as the binding site for the repair enzyme T4 endonuclease V (Iwai et al. 1994). Our calculations show that the minor groove is widened by 2-3Å in the vicinity of the thymine dimer. This raises the question whether the subtle structural changes in the DNA are a sufficient recognition element identified by the enzyme. In view of the dependence of the enzyme-DNA interaction on the ionic strength of the environment it would be reasonable to expect that the distribution of counterions around the damaged DNA may be sufficiently different from the distribution around normal DNA. We have analyzed the distribution of counterions from the present simulation to assess the effect of DNA distortion of counterion distribution and compared it to the counterion distribution in a similar simulation of a normal DNA that does not contain a thymine dimer.

The contour density plots shown in Figure 1 represent three types of distributions that were calculated from the trajectories of the simulations. The increasing dark colors in the plots depict increasing densities of counterions as a function of a radial distance from the DNA (horizontal displacement) and as a

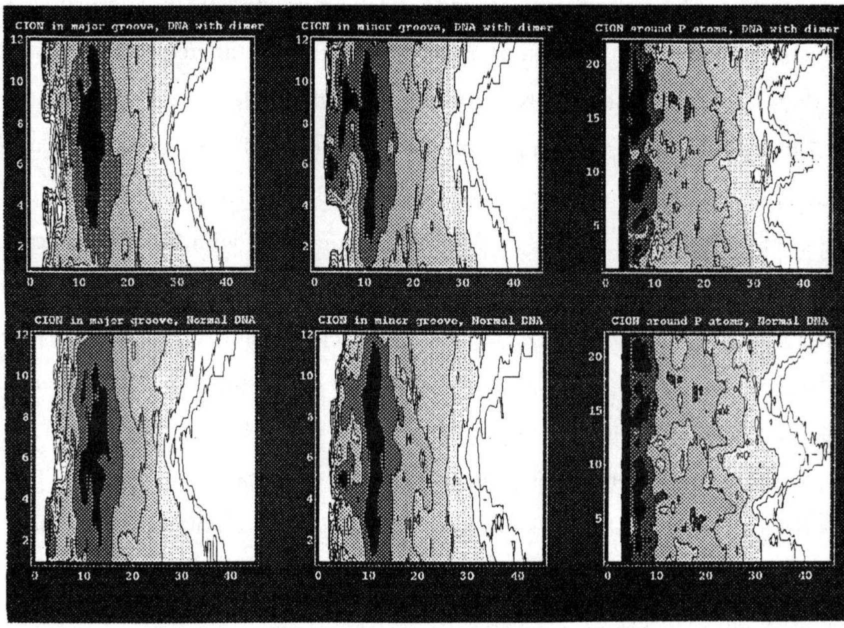

Figure 1. Contour density plots of the distribution of ions in major groove (left), minor groove (center), and along the phosphates (right).

function of the position of the base pairs along the sequence (vertical displacement). The pair of plots on the left show the distribution of counterions in the major groove of the damaged DNA (top) and normal DNA (bottom). The density of counterions increases as the distance from the bottom of the minor groove increases and reaches a maximum near 14 Å. Moving further away from the DNA the density is reduced reflecting the fact that a considerable number of counterions is condensed on the DNA. It is clear that the differences between the distributions in the damaged and normal DNAs is small. The only region in which the densities of counterions differ is in the vicinity of T7 of the DNA with thymine dimer. Thus, the area near the thymine dimer in the major groove of the damaged DNA has a lower density than the corresponding place in normal DNA.

The center pair of plots shows the distribution of counterions in the minor groove of damaged DNA (top) and normal DNA (bottom). The radial distribution of counterions in the minor groove peaks around 10 Å from the bottom showing a closer packing of ions than in the major groove. The distribution in the damaged DNA shows an interesting polarization of counterions which is not observed in the normal DNA. There is a clear depletion of counterions near the C3 and a concomitant increase near C9. In the normal DNA the distribution is more uniform. Thus, the depletion of counterions on the 5′ side of the thymine dimer may act as an electrostatic trigger for the enzyme to change its affinity to the particular sequence that contains the lesion. A reduction in the density of counterions will reduce the required energy to displace the counterions and consequently will increase the affinity of the protein to DNA.

A comparison of the densities around the phosphates, shown in the pair of plots on the right, demonstrates that the difference between the distributions is very small. This would suggest that the phosphates and the counterions condensed around these negatively charged groups may not be important for the recognition of the damage. It is noteworthy that the changes in counterion distribution should also have an effect on the structure of water around the damaged site. These studies are now in progress.

Conclusions

Structural analysis of the results of molecular dynamics simulations on the dodecamer $d(CGCGAATTCGCG)_2$ with a *cis,syn* cyclobutane dimer at T7-T8 reveals several distortions that are easily attributable to lesion. These perturbations of DNA structure are largest for the 5′ thymine of the dimer which exhibits a *syn* conformation about the glycosyl bond, extensive reorientation of the base relative to the helical axis, and weakening, but not breaking, of the hydrogen bond to complementary adenine. Significant distortions of the 3′ thymine in the dimer were not detected in our simulations. Base stacking is perturbed within the dimer and between the linked thymines and their nearest neighbors. The dimer increases the minor groove width by 2-3Å but significant bending or kinking was not detected at sight of the lesion. Our calculations agree with NMR results in that they predict only small distortions of DNA structure localized at the site of the lesion. This level of agreement suggests that molecular dynamics simulations can provide reliable data on the structure of damaged DNA sequences for studies of pyrimidine dimer recognition by repair enzymes.

The subtle changes in DNA structure produce a significant redistribution of counterions. The area near the T7-T8 in the major groove is slightly depleted from counterions but a large change in the density of counterions is observed in the minor groove. The area 5′ to the damage is strongly depleted from counterions suggesting an electrostatic mechanism for the recognition of the damage by the enzyme.

Acknowledgment

This work was supported in part by the Office of Health and Environmental Research of the US Department of Energy under grant DE-FG02-88ER60675 and contract DE-AC06-76RLO1830.

References

Augustine ML, Hamilton RW, Dodson ML, Lloyd RS (1991). Oligonucleotide site directed mutagenesis of all histidine residues within the T4 endonuclease V gene: role in enzyme-nontarget DNA binding. Biochemistry **30**(32): 8052-9.

Brash DE, Seetharam S, Kraemer KH, Seidman MM, Bredberg A (1987). Photoproduct frequency is not the major determinant of UV base substitution hot spots or cold spots in human cells. Proc. Natl. Acad. Sci. U.S.A. **84**: 3782-3786.

Cadet J, Vigny P, Eds. (1990). Bioorganic Photochemistry. New York, John Wiley & Sons.

Dodson ML, Lloyd RS (1989). Structure-function studies of the T4 endonuclease V repair enzyme. Mutat. Res. **218**(2): 49-65.

Dodson ML, Prince MA, Anderson WF, Lloyd RS (1991). Site-directed deletion mutagenesis within the T4 endonuclease V gene: dispensable sequences within putative loop regions. Mutat. Res. **255**(1): 19-29.

Dodson ML, Schrock RD, Lloyd RS (1993). Evidence for an imino intermediate in the T4 endonuclease V reaction. Biochemistry **32**(32): 8284-90.

Doi T, Recktenwald A, Karaki Y, Kikuchi M, Morikawa K, Ikehara M, Inaoka T, Hori N, Ohtsuka E (1992). Role of the basic amino acid cluster and Glu-23 in pyrimidine dimer glycosylase activity of T4 endonuclease V. Proc. Natl. Acad. Sci. U.S.A. **89**(20): 9420-4.

Dowd DR, Lloyd RS (1989). Site-directed mutagenesis of the T4 endonuclease V gene: the role of arginine-3 in the target search. Biochemistry **28**(22): 8699-705.

Dowd DR, Lloyd RS (1990). Biological significance of facilitated diffusion in protein-DNA interactions. Applications to T4 endonuclease V-initiated DNA repair. J. Biol. Chem. **265**(6): 3424-31.

Gruskin EA, Lloyd RS (1986). The DNA scanning mechanism of T4 endonuclease V. Effect of NaCl concentration on processive nicking activity. J. Biol. Chem. **261**(21): 9607-13.

Hutchinson F (1987). A review of some topics concerning mutagenesis by ultraviolet light. Photochem. Photobiol. **45**: 897-903.

Iwai S, Maeda M, Shimada Y, Hori N, Murata T, Morioka H, Ohtsuka E (1994). Endonuclease V from bacteriophage T4 interacts with its substrate in the minor groove. Biochemistry **33**: 5581-5588.

Kemmink J, Boelens R, Kaptein R (1987). Two-dimensional proton NMR study of two cyclobutane type photodimers of thymidylyl-(3'-> 5')-thymidine. Eur. Biophys. J. **14**(5): 293-9.

Kemmink J, Boelens R, Koning T, Van der Marel GA, Van Boom JH, Kaptein R (1987). Proton NMR study of the exchangeable protons of the duplex d(GCGTTGCG)•d(CGCAACGC) containing a thymine photodimer. Nucleic Acids Res. **15**(11): 4645-53.

Kemmink J, Boelens R, Koning TMG, Kaptein R, Van der Marel GA, Van Boom JH (1987). Conformational changes in the oligonucleotide duplex d(GCGTTGCG)•d(CGCAACGC) induced by formation of a cis-syn thymine dimer . A two-dimensional NMR study. Eur. J. Biochem. **162**(1): 37-43.

Kim J-K, Alderfer JL (1992). Conformational variations of the cis-syn cyclobutane-type photodimer in DNA and RNA. J. Biomol. Struct. Dynamics **9**: 705-718.

Kim JK, Wallace JC, Alderfer JL (1993). Substituent effects on the puckering mode of the cyclobutane ring and the glycosyl bond of cis-syn photodimers. Biopolymers **33**(4): 713-21.

Lee BJ, Sakashita H, Ohkubo T, Ikehara M, Doi T, Morikawa K, Kyogoku Y, Osafune T, Iwai S, Ohtsuka E (1994). Nuclear magnetic resonance study of the interaction of T4 endonuclease V with DNA. Biochemistry **33**: 57-64.

Lloyd RS, Augustine ML (1989). Site-directed mutagenesis of the T4 endonuclease V gene: mutations which enhance enzyme specific activity at low salt concentrations. Proteins **6**(2): 128-38.

Miaskiewicz K, Osman R, Weinstein H (1993). Molecular dynamics simulation of the hydrated d(CGCGAATTCGCG)$_2$ dodecamer. J. Am. Chem. Soc. **115**: 1515-1526.

Morikawa K, Matsumoto O, Tsujimoto M, Katayanagi K, Ariyoshi M, Doi T, Ikehara M, Inaoka T, Ohtsuka E (1992). X-ray structure of T4 endonuclease V: an excision repair enzyme specific for a pyrimidine dimer. Science **256**(5056): 523-526.

Sancar GB (1990). DNA photolyases: physical properties, action mechanism, and roles in dark repair. Mutat. Res. **236**: 147-160.

Swaminathan S, Ravishanker G, Beveridge DL (1991). Molecular dynamics of B-DNA including water and counterions: a 140-ps trajectory for d(CGCGAATTCGCG) based on the GROMOS force field. J. Am. Chem. Soc. **113**(13): 5027-40.

Taylor JS, ODay CL (1990). Cis-syn thymine dimers are not absolute blocks to replication by DNA polymerase I of Escherichia coli in vitro. Biochemistry **29**(6): 1624-32.

von Hippel PH (1994). Protein-DNA Recognition: New Perspectives and Underlying Themes. Science **263**: 769-770.

von Hippel PH, Berg OG (1989). Facilitated target location in biological systems. J. Biol. Chem. **264**(2): 675-8.

PREDICTION OF STRUCTURAL PERTURBATIONS CAUSED BY URACIL, DIHYDROURACIL AND CYTOSINE HYDRATE IN A DNA DECAMER DUPLEX FROM THE ADENOVIRUS E2A PROMOTER BY MOLECULAR MODELING

M. G. Cooney[1] and P. W. Doetsch[1,2]
Departments of [1]Biochemistry and [2]Radiation Oncology
Emory University School of Medicine
Atlanta, GA 30322 USA

Introduction

Ionizing radiation and oxidizing agents produce a variety of base damage products in the genomic DNA of all living organisms (Doetsch et al., 1990). These modifications include ring saturation products, *e.g.* thymine glycol (Basu et al., 1989; Clark et al., 1987); 5,6-dihydrothymine and 5,6-dihydrouracil (Dizdaroglu et al., 1993), and cytosine glycol; exocyclic derivatives such as uracil from the deamination of cytosine (Frederico et al., 1993) and 5-hydroxymethyluracil (Mellac et al., 1993); and purine oxidation products such as 8-oxoguanine (Kouchakdjian et al., 1991; Poltev et al., 1993). Prokaryotic and eukaryotic cells possess a variety of DNA repair N-glycosylases to remove these damages, thereby initiating base excision repair. The redoxyendonucleases constitute one of the most versatile and widespread classes of such base excision repair enzymes, of which the best and most extensively characterized example is *E. coli* endonuclease III (Dizdaroglu et al., 1993).

Saturated pyrimidine derivatives, but not normal pyrimidines, are well known to be redoxyendonuclease substrates, suggesting that this class of enzymes is sensitive to loss of base aromaticity and/or local disruptions in the damaged nucleotide unit (Doetsch et al., 1990). Two such lesions, 5,6-dihydrouracil (DHU) and *R*-6-hydroxy-5,6-dihydrocytosine (CytHyd), are formed by electron and hydroxyl radical attack, respectively, on cytosine. DHU results from anoxic deamination and reduction of cytosine, while CytHyd arises from oxidative hydration of the 5-6 double bond of cytosine. Cytosine also can deaminate to uracil; if left unrepaired, either a G:U or G:DHU wobble type mispair will lead to a C-->T transition mutation by directing incorporation of A on the complementary strand. The biological relevance of the structural changes that uracil, dihydrouracil and cytosine hydrate may induce stems from the fact that all of these lesions are derived from cytosine *in vivo* and are potentially premutagenic.

To gain further insight into the structural basis for DNA damage recognition by redoxyendonucleases, we have carried out computer-assisted molecular modeling of single-site pyrimidine damages in a 10-base-pair DNA duplex that forms the activating transcription factor (ATF) binding site within the adenovirus E2A promoter (Borden, 1993). Molecular modeling predictions are valuable in that they give an estimate of the stability, conformations and possibly the biological lifetime of the altered residue in the damaged DNA.

Molecular modeling studies carried out on short duplex DNA in the absence of experimental data include dynamics simulations by Mrigank and Kothekar (1991), Ravishanker et al. (1989), Swaminathan et al. (1991) and Miaskewicz et al. (1993). Additionally, 8-oxoguanine (Kouchakdjian et al., 1991) and abasic (AP) sites (Kalnik et al., 1989) have been analyzed in short DNA molecules by NMR and computer modeling. Although there are no previous modeling studies on dihydrouracil or dihydrouridine in nucleic acids, the crystal structure of a dihydrouridine dimer was solved 22 years ago (Suck et al., 1972).

The sequences of the strands modeled in this study are 5'-d(CCGXGTYTTC)-3' (E2APYR), and 5'-d(GAAGACGCGG)-3' (E2APUR) (see Figure 1). We based the choice of this DNA sequence on biological relevance and the availability of 2-dimensional NMR data on the undamaged sequence (Borden, 1993). Since this sequence corresponds to the ATF binding site, a base damage introduced into it could diminish or prevent transcription factor binding as well as initiate repair enzyme binding, making it biologically interesting from two aspects.

The residue at position X in E2APYR was chosen as cytosine, uracil, DHU, or CytHyd. With residue X as cytosine, residue Y was set as cytosine or dihydrouracil so that the latter could be examined within two different sequence contexts, between two purines or two pyrimidines; the duplex scheme is diagrammed in Figure 1.

Figure 1

E2APYR = 5'-d(C_1 C_2 G_3 X_4 G_5 T_6 Y_7 T_8 T_9 C_{10})-3'
E2APUR = 3'-d($G_{20}G_{19}C_{18}G_{17}C_{16}A_{15}G_{14}A_{13}A_{12}G_{11}$)-5'

Materials and Methods

All computations were performed on an Iris 4D computer with Silicon Graphics and SPARC workstations using the program Sybyl (Tripos Associates, St Louis, MO). The DNA sequences were first preset in a B-type helical conformation stored in a database provided with Sybyl. The fourth base (X) on the strand E2APYR was modeled as cytosine and then modified to uracil, DHU, and CytHyd. The seventh base (Y) was similarly set as cytosine or DHU. Nonplanar analogs (DHU and CytHyd) were thus modeled as planar high-energy structures and allowed to settle into low-energy puckered configurations during minimization. The AMBER force field (Weiner et al., 1986) was used throughout; to reduce computation time, parameter modifications were made as proposed by Veal and Wilson (1991) for nucleic acid modeling without explicit inclusion of solvent molecules or counterions. The distance dependent dielectric expression used was $\varepsilon = 4r_{ij}$, simulating an aqueous solvent layer, and the phosphate groups were given a net charge of −0.2 electrostatic charge units (e.s.u.) each to simulate one Na^+ counterion per phosphate. Other minor charge modifications were made after Veal and Wilson (1991). Torsion force constants were maintained at their Kollman values.

Energy minimizations were carried out first with the steepest-descent or conjugate gradient algorithms for 5 to 10 steps to relieve high-energy contacts, followed by the Powell algorithm, a modified Newton-Raphson method, until the RMS value of the energy gradient was 0.05 kcal/(mol·Å) or less.

Molecular dynamics calculations for 50 picoseconds were carried out on the minimized conformations of each of the five molecules, with a time step of 2 fsec at a temperature of 300K; conformations were saved every 0.25 psec. Overall B-type helical geometry was preserved by applying light distance range constraints (force constant = 5 kcal/mol·$Å^2$) to all base pairs except the one with the damaged base, with hydrogen bonding groups constrained to remain 2.5 Å to 3.5 Å apart. The SHAKE algorithm was applied to all bonds.

Results and Discussion

Energy Minimizations. Figure 2 shows stereo views of three of the energy-minimized ATF duplexes, undamaged (2A) and with dihydrouracil as the fourth (2B) and seventh (2C) residues on the E2APYR strand, while Figure 3 shows the G17·X4 base pairs in more detail. In Figure 2, the bases X and Y are displayed, marked with an arrow, so that their position in the major groove is apparent. From Figure 2B the DHU atoms O4, C5 and the two hydrogens on C5 are seen to protrude more into the major groove than their cytosine counterparts N4, C5 and H5. The calculations predict a similar wobble pairing for uracil (Figure 3C). Both DHU and CytHyd (Figs. 3B and 3D) have puckered base rings, but CytHyd is more distended at C5 and C6 than DHU (Fig. 3D), and its C5 and C6 substituents cause slight vertical distortion along the pyrimidine strand.

From Figure 2 it is apparent that the DHU substitution causes only local disruption of the helix. To quantitate the structural features shown in Figure 2, Table 1 shows torsion angles indicative of such features, *i.e.* the sugar pucker parameter ν_2 and backbone torsions α, γ, δ and ζ, for this residue and the two flanking guanosine units. (For an explanation of the torsion angles and atom numbering system see the text by W. Saenger (1984).) The hydrogen bonding groups on the DHU (O2 and N3) are displaced into the major groove of the helix to allow for wobble pairing. This prediction is consistent with the wobble dG·dT pair found by Hare et al. (1986) from NMR (NOESY) and distance geometry studies. The sugar pucker of the DHU residue torsion angle ν_2 (C1'-C2'-C3'-C4'), which shifts from -10.1^o to $+9.5^o$, indicates a partial S- to N-type transition (Table 1, ν_2(X4) row). The duplex with X=U shows similar wobble hydrogen bonding and an N-type pucker where ν_2 has shifted to $+14.5^o$.

Figure 3 shows close-ups of the base pairs G with C (A), DHU (B), U (C) and CytHyd (D); for U and DHU the wobble hydrogen bonding distances, *i.e.* O6 of G to N3 of U or DHU, are all approximately 3 Å, suggesting that any pyrimidine derivative that can base pair like uracil or thymine should engage in wobble pairing. The nonaromatic DHU base adopts a puckered conformation; the N3-C4-C5-C6 torsion is similar in magnitude but opposite in sign to the crystal structure value reported by Suck et al., (1972). In contrast, the base CytHyd maintains normal Watson-Crick pairing with G (Figure 3D), although the base ring is puckered. Although not based on direct structural measurements, our DHU and U results are consistent with the previous G·T results of Hare et al., (1986).

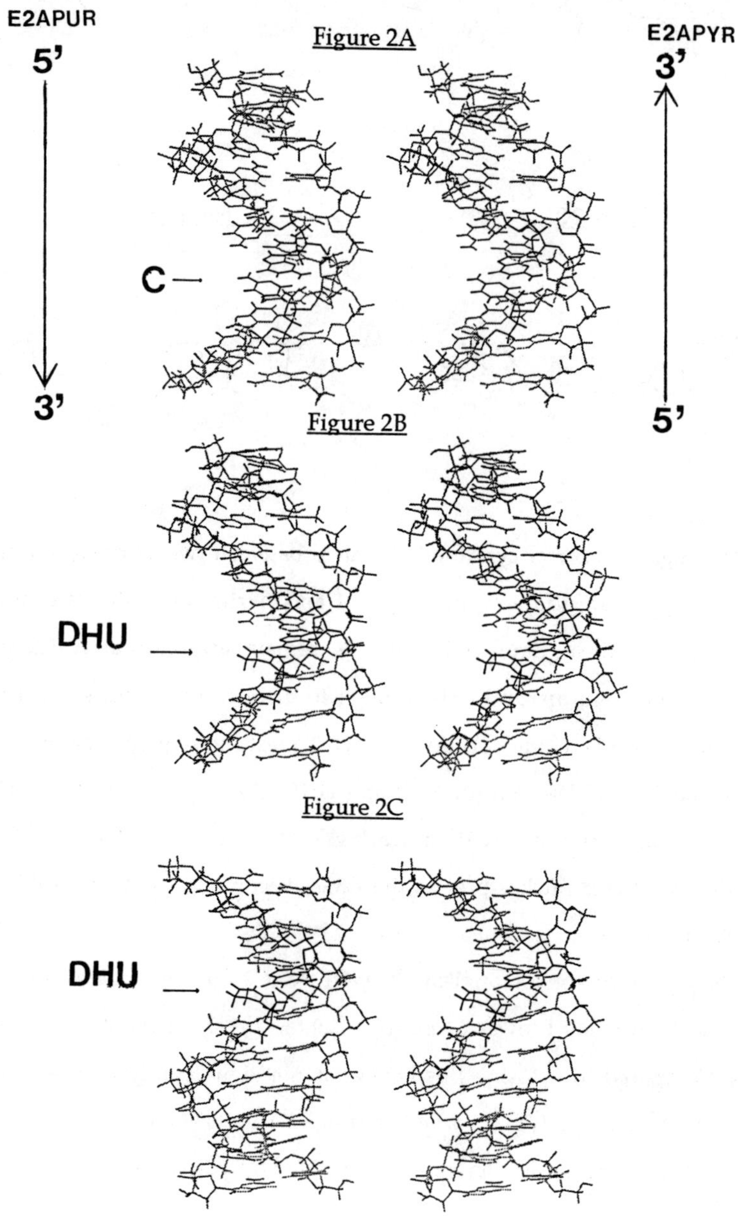

Figure 2A

Figure 2B

Figure 2C

Figure 3A

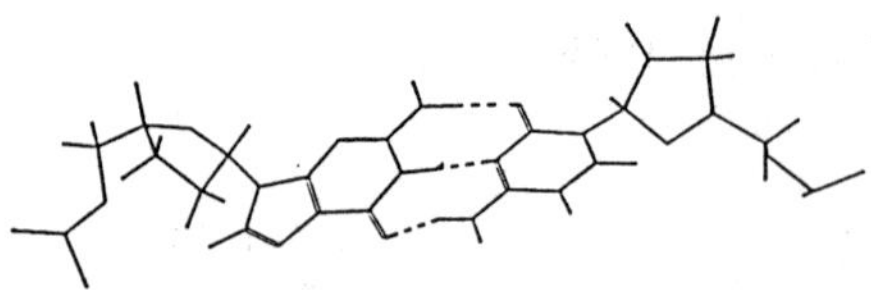

Figure 3B

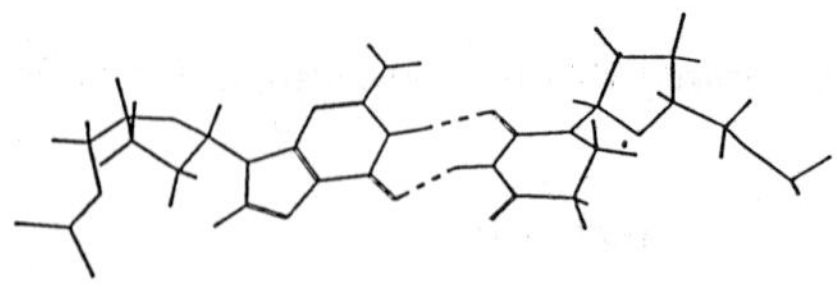

Figure 3C

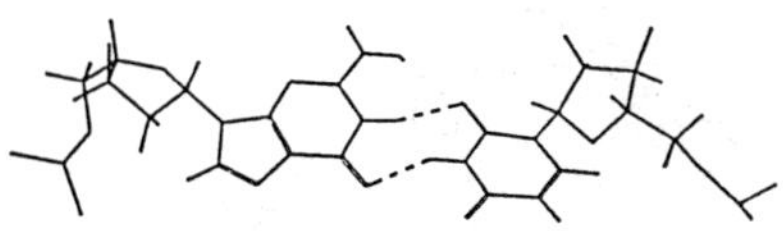

Figure 3D

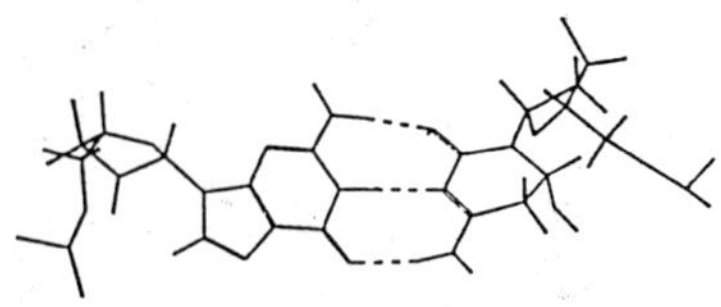

The sugar pucker of cytidine 16, located between two purines, adopts the C3'-*endo* or N-type configuration, although C4 and pyrimidines that are not between two purines on the undamaged molecule remain approximately C2'-*endo*. This minor conformational disruption is difficult to discern from Figure 2 but evident in Tables 1 and 2 (see Dynamics section below). To test the sequence dependence of the conformational shift observed for X=U and DHU, the duplex with Y=DHU was also energy-minimized (Figure 2C), indicating that even when DHU is flanked by pyrimidines its sugar pucker prefers the C3'-*endo* configuration (see Table 1, entries for T6, Y7 and T8).

The glycosidic torsion angles χ (O4'-C1'-N1-C2 for pyrimidines, O4'-C1'-N9-C4 for purines) tend to shift toward *trans* by -12° to -15°, for pyrimidines between two purines. Compared to cytosine this change is not significantly enhanced for U and DHU; however, for CytHyd a marked shift of ~27° to -88° (closer to *gauche* minus) is seen, probably owing to the steric intrusion of the 6-hydroxyl into the inter-base pair space.

Table 1
Energy-Minimized Glycosidic Base Pair Angles*, Interatomic Distances* and Sugar Pucker and Backbone Torsions* in Vicinity of Lesions for E2APYR:E2APUR Duplexes with X and Y Base Substitutions

Parameter	**Residue X**				**Parameter**	**Residue Y**	
	C	**U**	**DHU**	**CytHyd**		**C**	**DHU**
ζ(G3)	−96.2	−92.9	−91.1	−100.1	ζ(T6)	−99.5	−91.6
α(X4)	−65.1	−67.1	−62.6	−63.3	α(Y7)	−62.9	−66.4
γ(X4)	+62.1	+60.9	+63.8	+55.9	γ(Y7)	+61.1	+61.2
δ(X4)	+103.7	+83.4	+85.5	+136.9	δ(Y7)	+126.4	+78.3
ζ(X4)	−91.0	−86.6	−85.8	−132.5	ζ(Y7)	−94.9	−79.6
α(G5)	−61.5	−59.7	−61.9	−65.5	α(T8)	−65.4	−60.9
Glyc. Angle**	49.6	41.8	40.5	51.4	Glyc. Angle **	50.3	39.4
ν_2(G3)	−25.9	−26.8	−24.4	−32.3	ν_2(T6)	−25.8	−24.5
ν_2(X4)	−10.1	+14.5	+9.5	−35.7	ν_2(Y7)	−27.2	+24.6
ν_2(G5)	−29.6	−33.3	−32.6	−32.4	ν_2(T8)	−21.3	−32.8
χ(G3)	−119.3	−116.5	−121.3	−114.3	χ(T6)	−123.5	−124.7
χ(X4)	−132.9	−140.2	−133.6	−87.6	χ(Y7)	−116.3	−138.3
χ(G5)	−113.9	−110.1	−111.3	−116.4	χ(T8)	−128.8	−105.8
r(G17N2-X4O2)#	2.96	3.41	3.61	2.95	r(G14N2-Y7O2)	2.98	3.85
r(G17O6-X4N3)	3.53	2.92	2.91	3.56	r(G14O6-Y7N3)	3.56	2.92

*All angle measurements are in degrees and distances in Angstroms.
**Glycosidic angle is defined as angle N9(G)-C1′(G)-C1′(X) and provides a measure of overall base pair geometry.
Distance between the indicated atoms.

Molecular Dynamics. To examine in greater detail the conformational preferences of the damaged residues, we have carried out 50 psec of molecular dynamics (MD) simulation on the energy-minimized structures at a temperature of 300K. To preserve overall helical geometry, weak harmonic distance range constraints (2.5 to 3.5 Å, k=5 kcal/mol·$Å^2$) were applied to the N-N and N-O distances in the hydrogen bonding groups for all base pairs except G17·X4. Table 2 shows the mean, maximum, minimum, and standard deviation of some critical torsion angles and interatomic distances during the MD simulations. In keeping with the deviations found by energy minimization, the sugar puckers of U and DHU at position X show increased flexibility and average values closer to N-type compared to other residues except for cytidines on the E2APUR strand; cytosine hydrate shows similarly enhanced flexibility.

Table 2
Statistical Analysis of Torsion Angles* and Interatomic Distances* of Interest during 50 psec of Molecular Dynamics at 300K

X = C	ν_2(G3)	ν_2(X4)	ν_2(G5)	ν_2(C16)	ν_2(G17)
Mean	−22.5	−5.16	−27.4	−1.35	−25.2
Std. Dev.	17.0	20.8	8.20	23.3	12.4
Max.	+36.2	+39.4	+0.65	+41.8	+20.9
Min.	−46.5	−41.4	−44.3	−37.9	−43.1
X = U					
Mean	−29.1	+11.2	−31.2	−0.47	−29.7
Std. Dev.	7.52	17.6	6.57	21.8	7.95
Max.	−3.43	+46.7	−6.50	+40.3	+13.4
Min.	−47.3	−47.1	−45.2	−44.5	−44.0
X = DHU					
Mean	−25.0	+10.3	−32.0	0.0	−30.0
Std. Dev.	15.9	17.2	6.43	23.3	7.71
Max.	+37.1	+40.0	−9.86	+40.3	+5.26
Min.	−47.1	−40.6	−46.6	−43.5	−48.0
X = CytHyd					
Mean	−30.8	−13.9	−27.4	−10.2	−19.7
Std. Dev.	9.27	21.6	10.9	21.5	18.9
Max.	+18.1	+36.5	+27.2	+48.0	+42.0
Min.	−46.1	−43.8	−45.0	−45.6	−42.3

X=DHU	r(G17N1-DHUO2)#	**Y=DHU** ν_2(G14)	ν_2(T6)	ν_2(Y7)
Mean	2.9	−27.7	−22.3	−0.15
Std. Dev.	0.1	10.3	15.6	21.9
Max.	3.3	+19.6	+35.1	+42.4
Min.	2.7	−43.6	−42.0	−44.1

*All angle values are in degrees and distances in Angstroms.
#Distance between the indicated atoms.

Figure 4 shows traces of the sugar pucker torsions for G17 and X4, in the molecules where X=C and DHU, as a function of time, indicating relative preference for particular conformations. In the X = DHU duplex, the ν_2 sugar pucker torsion is considerably more flexible and closer to N-type over the entire 50 psec for DHU than for G17. Similar behavior (not shown) is seen for the same residues (G17 and X4) in the U4 duplex, suggesting that wobble pairing increases the flexibility of the damaged pyrimidine. Pyrimidines C4, C16 and C18 additionally show considerably more flexibility than purines G3, G5 and G17 in both the damaged and undamaged molecules (see Fig. 1 for numbering scheme), suggesting a possible stacking-related effect for pyrimidines located between purines.

Figure 4.

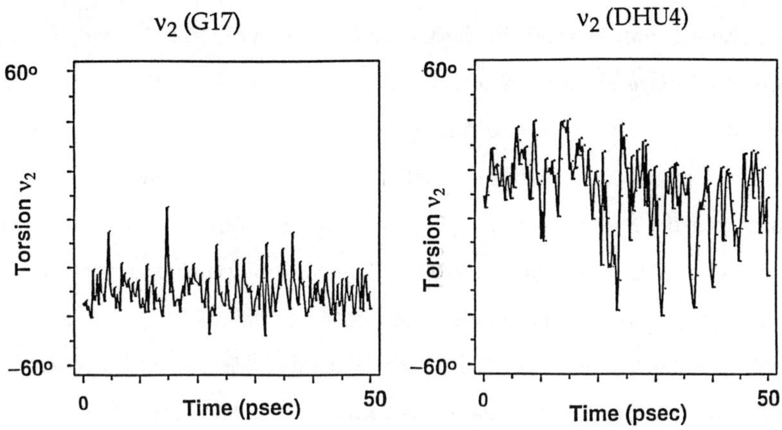

Together these minimization and dynamics results suggest that the effects of a deamination-induced lesion are local, being restricted essentially to the damaged residue itself. A slight shift of pyrimidines toward N-type sugar puckers has been noted from NMR-derived solution structures (Chuprina et al., 1993); however, we currently have no explanation for the greater variability in sugar pucker of the C specifically in G-C-G tracts; stacking interactions could account for this effectively more favorable N-type configuration. The ν_2 torsion for Y=DHU showed a somewhat greater shift toward N-type (C3'-endo) for DHU7 than for DHU4, confirming our hypothesis that wobble pairing contributes substantially to such shifts in geometry. Dynamics simulations, however, show this torsion (DHU7) to have a mean of $-0.15°$, closer to S-type than for DHU4, but with maximum, minimum and standard deviation similar to DHU4. The important conclusion from the DHU7 modeling is that a DHU lesion alone (U was not tested), regardless of its placement in the helix, is probably sufficient to perturb the sugar pucker and thus the local backbone conformation.

Comparison with NMR of undamaged duplex. K. Borden (1993) has reported two-dimensional NMR (NOESY, COSY and TOCSY) data on the undamaged version of the duplex that we are modeling, suggesting a B-type helix with A-type sugar puckers, with phase angles P consistently in the range 35°-90°, indicative of C3'-endo-type puckers. In contrast, our modeling studies on the undamaged sequence show P values more in the C2'-endo range (100°-140°), with the notable exception of C16 (see Figure 1) which is calculated to be closer to C3'-endo with a P value of approximately 80°. These discrepancies may reflect limitations on modeling in the absence of explicit aqueous solvent and sodium counterions.

Base pairing/hydrogen bonding. Upon energy minimization the duplexes with either U or DHU exhibited G17-X4 or G14-X7 base pairing consistent with the well-characterized "wobble" type pair (Crick, 1966; Romaniuk et al., 1979). Several N-O and N-N distances that indicate either Watson-Crick or wobble pairing were monitored for their preferred value in the course of the MD simulations. As seen in Table 2, the distance N1(G17)-O2(DHU4) remains near 3.0 Å throughout with a very small standard deviation of 0.1 Å, strongly supporting the hypothesis that the G and DHU maintain wobble pairing at 27°C.

Potential energies of the optimized B-conformations. The total potential energy ranking of the various lesions is C4<CytHyd4<DHU4<U4<DHU7. As expected, the undamaged duplex has the lowest minimized energy of the five molecules studied, but overall energy differences are small, with only ~11 kcal/mol difference between lowest and highest (DHU7); the duplexes containing wobble pairs are slightly higher in energy than the two containing Watson-Crick pairs, owing largely to hydrogen bond energies. The energy cost of placing DHU in position 7 rather than 4 is difficult to explain, but may be due to differences in stacking interactions.

When molecular dynamics is performed on this structure at 300K, an A-to-B transition occurs after about 50 psec. Sugar pucker torsion angle variations suggest that the transition takes place at about 10 ps and is relatively, though not completely, stable for the remainder of the simulation. This result suggests that, although the E2A duplex can adopt an A type local minimum, this sequence prefers the B-form as corroborated by the data of Borden (1993). Dynamics simulation run on the DHU4 duplex reveals a similar result.

Conclusions

Our results support the hypothesis that a nonplanar, nonaromatic pyrimidine detivative will cause some structural disruption, but in the case of dihydrouracil this does not appear to be very large. Wobble pairing, causing some extrusion of the pair into the major groove, seems to be the main feature of the DHU substitution, but this situation is also calculated to occur for uracil. Cytosine hydrate maintains Watson-Crick pairing with guanine, but the nonplanar ring and the 6-OH group disrupt the glycosidic χ and backbone ζ torsions. The NMR data available (Borden, 1993) on the undamaged duplex are consistent with a fairly wide range of sugar puckers and suggest that while the overall helical conformation is B-type, some puckers are especially flexible in this particular sequence. Our modeling results appear to confirm this flexibility at least for cytidines 16 and 18. Additionally, our dynamics results show a marked preference for the B helical type.

Repair enzymes such as E. coli endonuclease III recognize thymine glycol and a variety of other lesions; altered electronic structure of the dihydropyrimidine or local geometric perturbations may play a key role in this recognition process. Our modeling studies, however, predict sugar-backbone disruptions to be independent of whether or not the uracil base ring is hydrogenated, and it is known that endo III fails to recognize and remove normal uracil from DNA (Doetsch et al., 1990). Together, these results support the notion that loss of aromaticity and a concomitant change in electronic charge distribution are critical to recognition by base excision repair enzymes.

Acknowledgements

This work was supported by NIH grants CA42607 and CA55896.

References

Basu, A.K., Loechler, E.L., Leadon, S.A. and Essigmann, J.M. (1989). Genetic effects of thymine glycol: Site-specific mutagenesis and molecular modeling studies. *Proc. Natl. Acad. Sci. USA* **86**, 7677-7681.

Borden, K.L.B. (1993). The Activating Transcription Factor Region within the E2A Promoter Exists in a Novel Conformation. *Biochemistry* **32**, 6506-6514.

Chuprina, V.P., Nerdal, W., Sletten, E., Poltev, V.I. and Fedoroff O.Y. (1993). Base Dependence of B-DNA Sugar Conformation in Solution and in the Solid State. *Journal of Biomolecular Structure & Dynamics* **11** (3), 671-683.

Clark, J.M., Pattabiraman, N., Jarvis, W., and Beardsley, G.P. (1987). Modeling and Molecular Mechanical Studies of the *cis*-Thymine Glycol Radiation Damage Lesion in DNA. *Biochemistry* **26,** 5404-5409.

Crick, F.H.C. (1966). Codon-anticodon pairing: The wobble hypothesis. *J. Mol. Biol.* **19**, 548-555.

Dickerson, R.E. *et al.* (1989). Definitions and Nomenclature of Nucleic Acid Structural Parameters. *J. Mol. Biol.* **205**, 787-791.

Dizdaroglu, M., Laval, J. and Boiteux, S. (1993). Substrate Specificity of the*Escherichia coli* Endonuclease III: Excision of Thymine- and Cytosine-Derived Lesions in DNA Produced by Radiation-Generated Free Radicals. *Biochemistry* **32**, 12105-12111.

Doetsch, P.W. and Cunningham, R.P. (1990). The enzymology of apurinic/apyrimidinic endonucleases. *Mutation Research* **236**, 173-201.

Doetsch, P.W., Hamilton, K.K., Rapkin, L., Okenquist, S.A. and Lenz, J. (1990). Base Excision Repair Enzymes from Eukaryotic Sources. In *Ionizing Radiation Damage to DNA: Molecular Aspects*, @1990 Wiley-Liss, Inc., pp.109-125.

Frederico, L.A., Kunkel, T.A. and Shaw, B.R. (1993). Cytosine Deamination in Mismatched Base Pairs. *Biochemistry* **32**, 6523-6530.

Hare, D., Shapiro, L. and Patel, D.J. (1986). Wobble dG·dT Pairing in Right-Handed DNA: Solution Conformation of the d(C-G-T-G-A-A-T-T-C-G-C-G) Duplex Deduced from Distance Geometry Analysis of Nuclear Overhauser Effect Spectra. *Biochemistry* **25,** 7445-7456.

Kalnik, M.W., Chang, C.-N., Johnson, F., Grollman, A.P. and Patel, D.J. (1989). NMR Studies of Abasic Sites in DNA Duplexes: Deoxyadenosine Stacks into the Helix Opposite Acyclic Lesions. *Biochemistry* **28**, 3373-3383.

Kouchakdjian, M., Bodepudi, V., Shibutani. S., Eisenberg, M., Johnson, F., Grollman, A.P. and Patel, D.J. (1991). NMR Structural Studies of the Ionizing Radiation Adduct 7-Hydro-8-oxodeoxyguanosine (8-oxo-7H-dG) opposite Deoxy-adenosine in a DNA Duplex. 8-Oxo-7H-dG(syn).dA(anti) Alignment at Lesion Site. *Biochemistry* **30**, 1403-1412.

Kuo, C.-F., McRee, D.E., Fisher, C.L., O'Handley, S.F., Cunningham, R.P., and Tainer, J.A. (1992). Atomic Structure of the DNA Repair [4Fe-4S] Enzyme Endonuclease III. *Science* **258**, 434-440.

Laughton, C.A. and Neidle, S. (1992). Prediction of the structure of the $Y^{+}\cdot R^{-}\cdot R^{+}$-type DNA triple helix by molecular modelling. *Nucleic Acids Research* **20** (24), 6535-6541.

Melamede, R.J., Hatahet, Z., Kow, Y.W., Ide, H. and Wallace, S.S. (1994). Isolation and Characterization of Endonuclease VIII from *Escherichia coli. Biochemistry* **33**, 1255-1264.

Mellac. S., Fazakerley, G.V., and Sowers, L.C. (1993). Structures of Base Pairs with 5-(Hydroxymethyl)-2'-deoxyuridine in DNA Determined by NMR Spectroscopy. *Biochemistry* **32**, 7779-7786.

Miaskiewicz, K., Miller, J. and Osman, R. (1994). Energetic basis for structural preferences in 5,6-hydroxy-5,6-dihydropyrimidines: products of ionizing and ultraviolet radiation action on DNA bases. *Biochimica et Biophysica Acta* **1218**, 283-291.

Miaskiewicz, K., Osman, R. amd Weinstein, H. (1993). Molecular Dynamics Simulation of the Hydrated d(CGCGAATTCGCG)$_2$ Dodecamer. *J. Am. Chem. Soc.* **115**, 1526-1537.

Mrigank and Kothekar, V. (1991). 100 ps Molecular Dynamic Simulation of d(TATCACC)$_2$. *Journal of Biomolecular Structure & Dynamics* **8** (6), 1147-1167.

Poltev, V.I., Smirnov, S.L., Issarafutdinova, O.V. and Lavery, R. (1993). Conformations of DNA Duplexes Containing 8-Oxoguanine. *Journal of Biomolecular Structure & Dynamics* **11** (2), 293-301.

Ravishanker, G., Swaminathan, S., Beveridge, D.L., Lavery, R. and Sklenar, H. (1989). Conformational and Helicoidal Analysis of 30 PS of Molecular Dynamics on the d(CGCGAATTCGCG) Double Helix: "Curves", Dials and Windows. *Journal of Biomolecular Structure & Dynamics* **6** (4), 669-699.

Romaniuk, P., Hughes, D.W., Gregorie, R.J., Neilson, T. & Bell, R.A. (1979). Contribution of a G:U base pair to the stability of a short RNA helix. *J. Chem. Soc. Chem. Commun.*, 559-560.

Saenger, W. *Principles of Nucleic Acid Structure.* New York: Springer-Verlag, 1984.

Sekharudu, C.Y., Yathindra, N. and Sundaralingam, M. (1993). Molecular Dynamics Investigations of DNA Triple Helical Models: Unique Features of the Watson Crick Duplex. *Journal of Biomolecular Structure & Dynamics* **11** (2), 225-244.

Suck, D., Saenger, W., and Zechmeister, K. (1972). Molecular and Crystal Structure of the tRNA Minor Constituent Dihydrouridine. *Acta Cryst.* B**28**, 596-605.

Swaminathan, S., Ravishanker, G. and Beveridge, D.L. (1991). Molecular Dynamics of B-DNA Including Water and Counterions: A 140-ps Trajectory for d(CGCGAATTCGCG) Based on the GROMOS Force Field. *J. Am. Chem. Soc.* **113** (13), 5027-5040.

Veal, J.M. and Wilson, D.W. (1991). Modeling of Nucleic Acid Complexes with Cationic Ligands: A Specialized Molecular Mechanics Force Field and Its Application. *Journal of Biomolecular Structure & Dynamics* **8** (6), 1119-1145.

Weiner, S.J., Kollman, P.A., Nguyen, D.T. and Case, D.A. (1986). An All Atom Force Field for Simulations of Proteins and Nucleic Acids. *Journal of Computational Chemistry* **7** (2), 230-252.

CATALYTIC ACTIVITY OF NUCLEASE P1: EXPERIMENT AND THEORY

J. M. Falcone, M. Shibata and H. C. Box
Roswell Park Cancer Institute
Buffalo, NY 14263

J. H. Miller
Pacific Northwest Laboratory
Richland WA 99352

Introduction

Nuclease P1 from *Penicillium citrinum* is a zinc dependent glyco-enzyme that recognizes single stranded DNA and RNA as substrates and hydrolyzes the phosphate ester bond. As a single chain of 270 amino acid residues, the enzyme has a molecular mass of about 36 kilodaltons (Maekawa et al. 1991). Although it exhibits 3′-phosphatase activity, P1 is primarily an endonuclease that is capable of hydrolyzing single stranded DNA completely to the level of mononucleosides. Nuclease P1 seems to recognize particular conformations of the phosphodiester backbone and shows significant variation in the rate of hydrolytic activity depending upon which nucleosides are coupled by the phosphodiester bond. The efficiency of nuclease P1 in hydrolyzing the phosphodiester bonds of a substrate can be altered by modifications to one of the substrate bases induced by ionizing radiation or oxidative stress. Measurements have been made of the effect of several radiation induced lesions on the catalytic rate of nuclease P1. A model of the structure of the enzyme has been constructed in order to better understand the binding and activity of this enzyme on various ssDNA substrates.

Biochemical Materials and Methods

Normal dinucleoside monophosphates were purchased from Sigma and nuclease P1 was purchased from Boehringer Mannheim. Modified dinucleoside monophosphates were isolated by reverse phase HPLC. Dinucleoside monophosphates (6 nmol) were solubilized in 100 μL of reaction buffer (60 mM Na acetate, pH 5.3, containing 42 μM Zn acetate). Hydrolysis was initiated by addition of 50 ng of P1 nuclease in 2 μL of buffer. The mixture was incubated at 37° C for the duration of the reaction which was terminated with the addition of 15 μL of 1 M TRIS. Decrease of the integrated area of absorption of reactants and increase of the integrated area of products were monitored by HPLC with detection at UV 254 nm.

Kinetic Measurements

Rates of hydrolysis by nuclease P1 for equimolar mixtures of the 16 possible dinucleoside monophosphates differ by more than two orders of magnitude. Box et al. (1993) showed that these differences correlated with the 5′ nucleoside. Furthermore, the rate of hydrolysis is reduced if the 5′ nucleoside is damaged by ionizing radiation or oxidative stress, (Box et al. 1993; Weinfeld et al. 1992). Results of more recent studies shown in Table I support the hypothesis that the 5′ nucleoside is the principal binding locus between enzyme and substrate; however, they indicate that the 3′ nucleoside also influences hydrolytic activity. Substrate d(T^FpA), which

d(TpG)	d(TpC)	d(TpA)	d(TpT)	d(TpT^F)[a]	d(T^FpT)	d(T^GpA)[b]	d(TpG^{OH})[c]
112	90	70	40	60	0	3	325

a) T^F = Formamido remnant of Thymine

b) T^G = R-5,6-Dihydroxy-5,6-Dihydrothymine (Thymine Glycol)

c) G^{OH} = 8-OxoGuanine

Table I: Catalytic Constant (moles substrate hydrolyzed/mole enzyme) for P1 hydrolysis of the phosphodiester bond in dinucleoside monophosphates.

nuclease P1 was unable to cleave even after several hours of digestion, is a competitive inhibitor of the enzyme probably due to substrate binding at the 3′ nucleoside only.

Structural Data

The crystal structure of nuclease P1 has been solved by Volbeda et al. (1991) at 2.8 A resolution. Three water molecules and three zinc ions are present in the active site. One water molecule bridges Zn1 and Zn3. The Zn2 ion is coordinated by two water molecules, His126, His149 and Asp153. This configuration is structurally conserved in the active sites of several zinc enzymes (Valee and Auld 1990) and suggests that Zn2 is directly involved in catalysis. All three zinc ions are pentacoordinated with quasi-trigonal bipyramidal geometry.

The active site is characterized by two hydrophobic binding regions separated by about 20 Å, with the three zinc ions at the bottom of the binding cleft. One of these regions, which is characterized by nucleic-acid base stacking between Tyr144 and Tyr155, is believed to be the 5′ nucleotide binding site. The other binding region, which is characterized by base stacking between Val132 and Phe61, has been identified as the binding site for the 3′ nucleotides, which Volbeda et al. (1991) proposed as the most likely substrate binding site based on crystal soaking experiments with dAp(S)A, an uncleavable phosphorothioate substrate analog. We have undertaken modeling studies to gain more insight into the DNA binding domain and catalytic activity of nuclease P1.

Protein Model

Currently, the only structural information available for nuclease P1 comes from the published stereodiagrams of the Cα trace and the active-site residues (Volbeda et al. 1991). These diagrams were digitized and 3-dimensional cartesian coordinates derived from them were used to create a coordinate file in the standard Protein Data Base (PDB) format. Given the Cα coordinates and the amino acid sequence, MAX-SPROUT (Holm and Sander 1991) was used to generate coordinates for the whole protein. MAX-SPROUT utilizes an algorithm that scans a database of known crystal structures for fragments that locally fit the Cα coordinates according to distance criteria. The fragments are joined to create the full backbone. Sidechains are added from a database of frequently occurring rotamers. The reconstructed coordinates for P1 nuclease were further refined using the SYBYL program (TRIPOS Associates, Inc.). Peptide bonds were added at 12 gaps in the protein backbone that could not be sealed by MAX-SPROUT and the disulfide bridges reported in the crystal structure between Cys80-Cys85, and Cys72-Cys217 (Volbeda et al. 1991) were also added. Finally, the Simplex algorithm in the energy minimization utility of SYBYL was used to optimize bond lengths within the structure.

Active-Site Charge Distribution

Table II shows calculated geometries and atomic charges for zinc ions coordinated to a single water molecule or an amino-acid residue analog. Histidine and aspartic acid residues were abbreviated to methylimidizole and acetate, respectively. Quantum chemical calculations were carried out in the Modified Intermediate Neglect of Differential Overlap (MINDO) approximation on each zinc complex and residual atomic charges were determined by fitting the electrostatic potential at the optimal geometry. Molecular mechanics optimization was then performed with the AMBER force field using the default charges (Weiner et al. 1984) or the charges derived from our MINDO calculations. In the latter case, residual positive charge on Zn is reduced and the residual negative charge on the coordinating atom increases relative to the default AMBER charges. If the fragment containing the atom binding to Zn is large (i.e. an amino acid rather than a water molecule) the net electrostatic interaction is reduced and the complex expands.

Source	Zn Charge	WAT(O) Charge	Zn-WAT(O) Bond
AMBER	2.0000	-0.8340	2.143
MINDO	1.7649	-1.1120	1.896
Source	Zn-Charge	ASP(Oe1) Charge	Zn-ASP(Oe1) Bond
AMBER	2.0000	-0.7210	1.793
MINDO	1.4679	-0.7592	2.076
Source	Zn-Charge	HIS(N) Charge	Zn-HIS(N) Bond
AMBER	2.0000	-0.2084	1.870
MINDO	1.3784	-0.3024	2.065

Table II: Charges (esu) and Bond Lengths (Å) in Simple Zinc Ligands

Our results for Zn coordinated to simple ligands (Table II) suggested that the default AMBER charges would not be adequate to describe electrostatic interactions in the active site of P1 nuclease. For this reason, the catalytic site was separated into the two molecular areas for quantum investigation of residual atomic charges. Area 1 contained Zn2, His126, His149, Asp153, Wat1 and Wat2 while Area 2 contained Zn1,

Zn3,Trp1, His6, Asp45, His116, Asp120 and Wat3. Coordinates of atoms in the two areas were extracted from the reconstructed coordinate set of the entire protein. Some bond lengths and angles were adjusted based on similar work for the active site of carboxypeptidase A (Banci et al. 1992). MINDO calculations were carried out on the two areas and residual atomic charges were calculated by fitting the eletrostatic potential of each area at its optimal geometry, which did not differ greatly from that extracted from the reconstructed coordinate set. The geometry and residual charges of Area 1 derived by this semi-empirical quantum-chemical approach are shown in Figure 1.

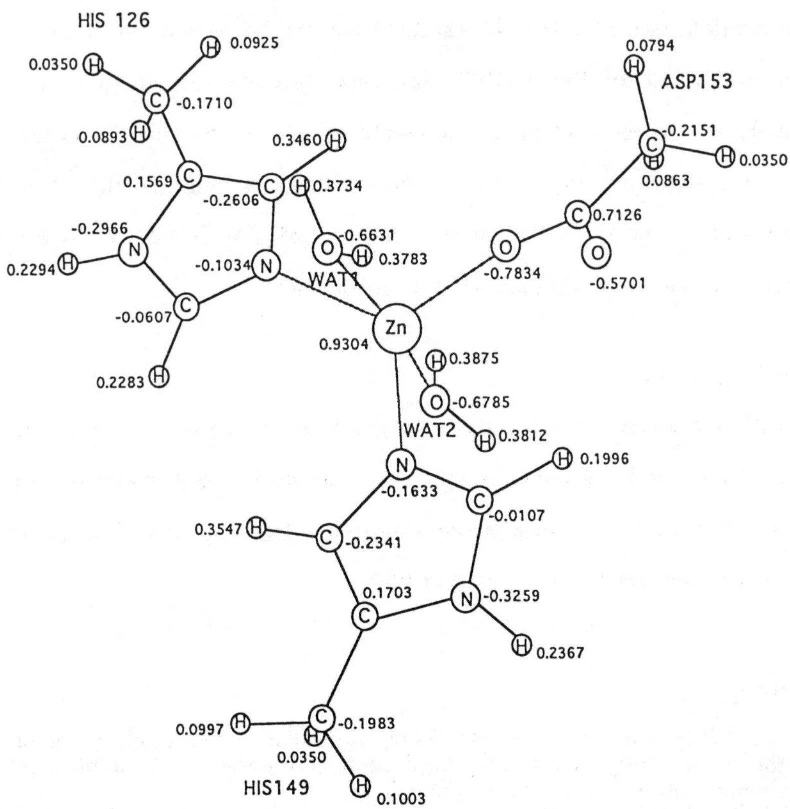

Figure 1: Semi-empirical quantum results for the geometry and atomic charges of the portion of the active site of P1 nuclease most important to its catalytic activity.

Future Work

All of the components have been prepared for refinement of our model of P1 nuclease by molecular dynamics (MD) simulation. The reconstructed atomic coordinates, with small modifications to accommodate the MINDO optimum geometries of areas 1 and 2, provide a starting structure for these simulations. The AMBER all-atom force field (Weiner et al. 1984) augmented by the MINDO residual charges of the catalytic site and parameters for bonded interactions with Zn (Merz et.al,1991) provide the information needed to approximate the multi-dimensional Born-Oppenheimer potential-energy surface. For the enzyme in the absence of substrate, Na^+ and Cl^- counterions are added to neutralize any residues with net charge. Finally, a layer of water will be added to the protein/counterion solute and several hundred picoseconds of MD simulations will be carried out in the constant-temperature mode of the AMBER software package (Pearlman et al. 1991). Eventually, we expect to compare the results of this simulation with experimental atomic-level structural data. In the mean time, we will develop models for complexes of P1 nuclease with native and modified dinucleotide monophosphates in an effort to gain more insight from our kinetic data.

Acknowledgements

JMF gratefully acknowledges support by a graduate fellowship from Associated Western Universities Northwest. This work was partially supported by the Office of Health and Environmental Research (OHER) of the U. S. Department of Energy under contract DE-AC06-76RLO 1830.

References

Banci L, S Schroder and PA Kollman (1992) Molecular dynamics characterization of the active cavity of carboxypeptidase A and some of its inhibitor adducts. Proteins: Struc. Func. Gen. 13:288-305

Box HC, EE Budzinski, MS Evans, JB French and AE Maccubbin (1993) The differential lysis of phosphoester bonds by nuclease P1. Biochim. Biophys. Acta 1161:291-294

Box HC, AE Maccubbin, EE Budzinski, H Patrzyc, MS Evans and N Ersing (1993) ^{32}P-Postlabeling assay for free radical induced DNA damage: the formamido

remnant of thymine. Free. Rad. Res. Comms. 18:17-28
Holm L and C Sander (1991) Database algorithm for generating protein backbone and sidechain coordinates from a Cα trace. Application to model building and detection of coordinate errors. J. Mol. Biol. 218:183-194
Hoops SC, KW Anderson and KM Merz (1991) Force field design for metalloproteins. J. Am. Chem. Soc. 113:8262-8270
Maekawa K, S Tsunasawa, G Dibo and F Sakiyama (1991) Primary structure of nuclease P1 from *Penicillium citrimum*. Eur. J. Biochem. 200:651-661
Pearlman, DA, DA Case, JC Caldwell, GL Seibel, U Chandra Singh, P Weiner and PA Kollman (1991) *AMBER4.0* (University of California, San Francisco)
Vallee BL and DS Auld (1990) Active site zinc ligands and activated water of zinc enzymes. Proc. Natl. Acad. Sci. USA 87:220-224
Volbeda A, A Lahm, F Sakiyama and D Suck (1991) Crystal structure of *Penicillium citrinum* P1 nuclease at 2.8 A resolution. EMBO J. 10:1607-1618
Weiner, SJ, PA Kollman, DA Case, U Chandra Singh, C Ghio, G Alagona, S Profeta and P Weiner (1984) A new force field for molecular mechanical simulation of nucleic acids and proteins. J. Am. Chem. Soc. 106:765-784
Weinfeld M, K-J Soderlind and GW Buchko (1993) Influence of nucleic acid base aromaticity on substrate reactivity with enzymes acting on single-stranded DNA. Nucleic Acids Res. 21: 621-626

Chapter 11

Use of Radiation Modifiers as Mechanistic Probes

USE OF CHEMICAL AGENT RADIATION MODIFIERS AS PROBES FOR MECHANISMS OF RADIATION DAMAGE

Kathryn D. Held
Department of Radiation Oncology
Massachusetts General Hospital
Harvard Medical School
Boston, MA 02114

Introduction

One of the challenges facing radiation biology and chemistry is to understand which of the many types of DNA lesions caused by ionizing radiation are relevant for the ultimate biological effects of the radiation. One approach to studying this relationship is a radiation chemical one, i.e., to use chemical agents that modify radiation damage and assess whether any given agent modifies a relevant biological endpoint, such as cell killing, mutation induction or oncogenic transformation, in the same manner as it modifies an early DNA damage such as strand breaks, DNA-protein crosslinks or selected types of base damage. In classic studies using this radiation chemical approach, e.g., those of Roots and Okada (1972) or Chapman *et al.* (1973), chemical agents, specifically •OH scavengers such as alcohols and thiols, were used to demonstrate the importance of •OH radicals as a species primarily responsible for induction of DNA strand breaks and killing in irradiated mammalian cells.

In this overview, I will use three types of chemical agents - oxygen, thiols and copper - to illustrate how changes in the chemical environment around DNA at the time of irradiation, and hence changes in the free radical reactions occurring in DNA, can be related across a large number of endpoints from early chemical changes in DNA to late expressions of damage at the cellular level. Study of such relationships may help lead to an understanding of which types of DNA damage are relevant for ultimate biological outcomes. My choice of chemical agents for discussion obviously reflects my research interests, as does my limiting the discussion of studies with intact cells to mammalian cells, and not including bacterial cell work. My goal is to highlight more recent data and to emphasize some areas where research is needed at the chemical level to help explain the biological observations.

Obviously, this is a somewhat simplistic approach as it does not allow for the actions of enzymatic repair processes, the importance of chromatin structure for damage induction, the importance of chromatin structure for damage assay, etc. These topics are discussed in other sections of this volume.

Oxygen as a modifying agent

Oxygen was one of the first chemical agents to be identified as a radiation modifier and remains one of the most effective agents known, as well as the most thoroughly studied. As summarized in Table I, oxygen has been studied as a modifier of radiation damage in many test systems. This table is far from inclusive, but is meant to be illustrative of the types of studies that have been performed, and to serve as a guide for comparisons of results obtained with different endpoints.

It is somewhat difficult to extrapolate from the results of oxygen modification of damage in purified DNA irradiated in solution to studies of DNA in cells. Generally, irradiation of purified DNA in solution in oxygen, as compared to hypoxia, does not increase the yield of single strand breaks or the loss of biological activity of DNA measured in bacterial transfection assays and may even decrease the damage slightly (e.g., Blok and Loman, 1973; Held *et al.*1978). This is clearly a very different result from the commonly observed 3-fold enhancement of cell killing when mammalian cells are irradiated in air. The results of assays for breaks and loss of biological activity of purified DNA change, however, when thiols are added to the DNA solutions during irradiation, and the resultant OERs can vary from about unity to 25, depending on the thiol, its concentration and the test system (e.g., Held *et al.*1984). This has been reviewed previously (Held, 1986).

Interesting new data show that the strand breaks generated by irradiation of DNA in soultion under nitrogen usually have 3'-phosphate termini, but irradiation under oxic conditions generates both 3'-phosphoglycolate and 3'-phosphate end groups in a ratio of about 1.6 (Buchko and Weinfeld, 1993). The implications of this observed change for biological endpoints is unclear.

Table I: Oxygen as a Modifier of Radiation Damage

Endpoint	Damage in oxygen compared to hypoxia?	References
Purified DNA in solution		
Single strand breaks	Unchanged (or decreased slightly)	(Blok and Loman, 1973; Siddiqi and Bothe, 1987; Spotheim-Maurizot *et al.*1991; Prise *et al.*1993)
Types of end groups	Changed	(Buchko and Weinfeld, 1993)
Double strand breaks	Unchanged or increased (OER 1.8)	(Siddiqi and Bothe, 1987; Spotheim-Maurizot *et al.*1991; Prise *et al.*1993)
Base damage (GC-MS)	Some products increased; some decreased	(Fuciarelli *et al.*1990)
Biological activity	Decreased slightly	(Blok and Loman, 1973; Held *et al.*1978)
Mutations	Increased	(Retel *et al.*1993)
	Spectrum changed	(Retel *et al.*1993)
Mammalian chromatin/isolated nuclei		
Single strand breaks	Increased (OER 2-3)	(Edgren, 1991)
Double strand breaks	Unchanged or increased (OER ~ 2.5)	(Held and Hopcia, 1993; Michael *et al.*1991; Prise *et al.*1992b)
Base damage (GC-MS)	Some products increased; some decreased	(Gajewski *et al.*1990)
DNA-protein crosslinks	Inhibited (except for Thy-Tyr)	(Nackerdien *et al.*1991)
Mammalian cells - chemical endpoints		
Single strand breaks	Increased (OER 2-4; usually ~ 4)	(Meyn *et al.*1987; Prise *et al.*1992a)
Double strand breaks	Increased (OER 2-4; usually ~ 3)	(Radford, 1986b; Prise *et al.*1992a; Whitaker and McMillan, 1992; Radford, 1985)
Base damage (ESS)	Unchanged	(Hentosh, 1988; Radford, 1986a)
Base damage (GC-MS)	???	
DNA-protein crosslinks	Decreased	Fornace and Little, 1977; Meyn *et al.*1987; Xue *et al.*1988)
Mammalian cells - biological endpoints		
Survival	Increased (OER ~ 3)	(Hopcia *et al.*1991;Sapora *et al.*1991; Borsa *et al.*1984; Prise *et al.*1992a; Meyn *et al.*1987; Whitaker and McMillan, 1992; Radford, 1986b; Radford, 1985)
Chromosome aberrations	Increased (OER ~ 3)	(Deschner and Gray, 1959; Hawes *et al.*1966)
Mutagenesis	Increased (OER 3-4)	(Sapora *et al.*1991; Denault and Liber, 1993)
	Spectrum changed slightly	(Denault and Liber, 1993)
Neoplastic transformation	Increased (OER ~ 3)	(Borsa *et al.*1984)
Apoptosis	Increases (OER ~ 3)	(Hopcia *et al.*1994; Klassen *et al.*1993)
Gene expression	???	

Comparisons between chemical and biological endpoints when intact cells are irradiated may be more meaningful, and several observations can be made from the literature cited in Table I. Frequently, oxygen enhancement ratios (OERs) for single strand breaks (SSBs) in intact cells are higher than the OERs for survival (e.g., Meyn *et al.*1987; Prise *et al.*1992a), and when investigators have plotted SSBs versus survival for cells irradiated under the same conditions, the data do not give the same line for aerobic and hypoxic irradiations (e.g., Radford, 1986b; Sapora *et al.*1991). On the other hand, OERs for double strand breaks (DSBs) are usually close to those for survival, and several investigators, using such diverse endpoints for DSBs as sedimentation, neutral elution and pulsed field gel electrophoresis (PFGE), have shown the same line when DSBs were plotted versus survival for cells irradiated in air and in hypoxia (Sapora *et al.*1991; Radford, 1986a; Whitaker and McMillan, 1992). These studies suggest that, in the least, the precursor free radical species that cause DSBs are chemically modified by oxygen in the same fashion as the precursors to cell killing, and the studies often are interpreted as evidence for causality between DSBs and loss of survival. It would be quite interesting to have information on the types of end groups generated at DNA breaks in cellular DNA and whether these change with gassing conditions during irradiation.

Several studies have shown that production of DNA-protein crosslinks (DPCs) in irradiated mamalian cells is decreased in the presence of oxygen (Fornace and Little, 1977; Meyn *et al.*1987; Xue *et al.*1988), the opposite from the effect of oxygen on radiation-induced cell killing and DSBs. Does this mean DPCs are not a lethal lesion? Or does it mean the increased level of DPCs in the absence of oxygen is a protective mechanism because the DPCs hold the DNA together, thus facilitating repair of DSBs?

Some interesting new data have been published recently on the effects of oxygen on the frequency and spectrum of radiation-induced mutations. The OER for the frequency of radiation-induced mutagenesis in mammalian cells appears to be similar to that for cell killing and DSBs (Denault and Liber, 1993), and plots of mutation frequency versus survival fall on the same line for cells irradiated in air and hypoxia (Sapora *et al.*1991). However, the spectrum of mutations produced in mammalian cells irradiated in air appears to be different from that in cells irradiated in hypoxia. This difference did not achieve statistical significance, perhaps because of the fairly small number of mutants studied, but suggests that irradiation in hypoxia protects preferentially against deletions over point mutations (Denault and Liber, 1993). Such a difference in the mutational spectrum after irradiation in air versus hypoxia is consistent with results obtained in studies where purified DNA was irradiated: irradiation of plasmid DNA in solution in oxygen increased mutation frequency about 6-fold compared to irradiation in nitrogen, and the mutational spectrum showed a decrease in the number of frameshifts and larger rearrangements with irradiation in hypoxia (Retel *et al.*1993). One must make these comparisons between effects in plasmids in solution and effects in mammalian cells with care. The predominant type of mutation when DNA is irradiated in solution and assayed in a bacterial system is base pair substitutions, probably because most of the systems used are not suitable for detection of large deletions (Retel *et al.*1993). In contrast, when radiation-induced mutations are assayed at a heterozygous locus in intact mammalian cells, the predomiant type of mutation is deletions, often relatively large ones (Denault and Liber, 1993). If deletions are largely a result of strand breaks, and point mutations result largely from base damage, the observations by Denault and Liber are consistent with the observations that breaks are protected by hypoxia and base damage (as measured using an assay for enzyme sensitive sites (ESS)) is not (Hentosh, 1988). However, the ESS assay is rather insensitive and non-specific. It would be helpful to have data on the changes in specific types of radiation-induced base alterations in cells irradiated in air and hypoxia, similar to the data that exists for irradiation of purified DNA and DNA in chromatin (Fuciarelli *et al.*1990; Gajewski *et al.*1990). Such data might give some indication of the specific base damages that are involved in increasing mutations in cells irradiated in oxygen.

An important question about which there does not appear to be any published data yet is whether radiation-induced changes in gene expression are modified by the presence of oxygen. If the OER for radiation-induced changes in gene expression is similar to that for DNA damages, it would be support for the idea that DNA lesions themselves trigger the gene activity changes.

Is triggering of apoptosis changed by changing radiation chemistry?

In recent years there has been increasing interest in apoptosis in cancer research in general and in radiation studies in particular. Apoptosis is a mode of cell death that differs from necrosis both morphologically and biochemically (Kerr *et al.*1972; Walker *et al.*1988; Kerr and Harmon, 1991). The earliest morphological change in cells undergoing death by apoptosis is condensation and margination of the chromatin. This is followed by condensation of the cytoplasm, with membrane-bound organelles remaining intact, and this differs from necrosis where there is swelling of the cell and its cytoplasmic compartments and disintegation of cellular organelles. Subsequent to the nuclear compaction in apoptosis, there is nuclear fragmentation and development of protuberances, or blebs, on the cell surface, which then break down to produce apoptotic bodies. The apoptotic bodies may vary in size and shape and some contain nuclear fragments. In tissues, apoptotic bodies are usually rapidly phagocytosed by neighboring cells and degraded, or in epithelial tissues, may be sloughed off into the lumen. An important biochemical event frequently, although not always, seen in apoptosis is double stranded cleavage of DNA at the linker regions between nucleosomes (Wyllie, 1980; Arends *et al.*1990). This cleavage produces a characteristic "ladder" pattern of DNA fragments on agarose gel electrophoresis, where the fragments represent multiples of the 180-200 base pair fragments of DNA from nucleosomes. In cells that undergo this DNA cleavage, the fragmentation provides a method for quantifying apoptosis by separating low molecular weight, fragmented DNA, from the high molecular weight DNA present in non-apoptotic cells (Sellins and Cohen, 1987).

Apoptosis can be induced by a number of diverse agents including ionizing radiation, UV, hyperthermia, and some cancer chemotherapeutic agents (e.g., Sellins and Cohen, 1987; Stephens *et al.*1991; Martin and Cotter, 1991; Sellins and Cohen, 1991; Eastman, 1990). However, the nature of the cellular lesion or lesions that trigger cells to undergo apoptosis is still unknown (Wyllie *et al.*1980; Kaufmann, 1989). There is some evidence that damage to DNA may be a trigger of apoptosis since many of the agents that cause it are known to damage DNA, e.g., cis-platin crosslinks DNA, adriamycin is a topoisomerase II inhibitor, H_2O_2 caused DNA strand breaks, UV causes pyrimidine dimers, etc., and because an inhibitor of poly(ADP-ribosyl)ation can alter apoptosis induction (Jones *et al.*1989). We have begun to test whether ionizing radiation-induced apoptosis can be modified by chemical agents, e.g., oxygen and thiols, in a fashion that is similar to their modification of other radiation-induced cell and DNA damages. Figure 1 shows data on radiation-induced DNA fragmentation, characteristic of apoptosis, in HL-60 human leukemia cells irradiated *in vitro* in the presence and absence of oxygen. Clearly, the cells are protected to a significant degree by hypoxia. The OER for radiation-induced apoptosis in this data is about 2.7, nearly identical to the OER of 2.8 obtained in the same cell line using a colony formation assay to assess radiation damage. The similarity of the OER values suggests that DNA damage is the trigger for ionizing radiation-induced apoptosis, although much more investigation is needed to confirm this preliminary conclusion.

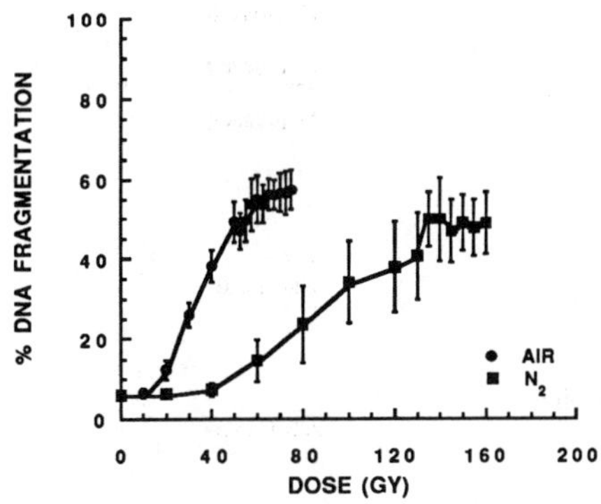

Fig. 1 Dose response curves for the induction of apoptosis (measured as % DNA fragmentation) in HL-60 cells exposed to X-rays in air or hypoxia. Each data point is the mean ± standard deviation of four separate experiments.

Thiols as modifying agents

Table II presents an overview of the types of studies of radiation-induced DNA and cellular damages that can be modified by the presence of thiols during the irradiation. This compilation is limited to studies where the irradiation occurred in air because the amount of data for irradiation in hypoxia is much more limited. However, when considering mechanisms involved in thiol radioprotection in air, at least two things must be borne in mind. First, both •OH scavenging and H atom donation could be occurring as chemical protective reactions. Under the conditions used in most studies of isolated DNA, the chemical reactions would be dominated by •OH scavenging (Zheng *et al.*1992), but in intact cells both reactions could occur. Second, when considering data on irradiations in air in the presence of thiols, one must always be cautious that the experimental conditions did not allow for significant oxidation of the thiol spontaneously or radiation-induced since this would not only create hypoxia, but also would generate H_2O_2 which could cause complicating effects.

Table II: Thiols (excluding WR-1065/2721) as Modifiers of Radiation Damage[a]

Endpoint	Damage in presence of thiol compared to in its absence?	References
Purified DNA in solution		
Single strand breaks	Decreased	(Spotheim-Maurizot *et al.* 1991; Zheng *et al.*1992)
Types of end groups	???	
Double strand breaks	Decreased	(Spotheim-Maurizot *et al.*1991)
Base release	Decreased	(Zheng *et al.*1988)
Base damage (GC-MS)	???	
Biological activity	Decreased	(Held *et al.*1981; Held *et al.*1984)
Mutations	???	
Chromatin/isolated nuclei		
Single strand breaks	???	
Double strand breaks	Deceased	(Radford, 1987)
Base damage (GC-MS)	???	
DNA-protein crosslinks???		
Mammalian cells - chemical endpoints		
Single strand breaks	Decreased	(Roots and Okada, 1972; Held *et al.*1986; Radford, 1986a; Murray *et al.*1990)
Double strand breaks	Decreased	(Radford, 1986b; Murray *et al.*1990; Murray *et al.*1994; Radford, 1985)
Base damage (ESS)	Decreased slightly	(Radford, 1986a)
Base damage (GC-MS)	???	
DNA-protein crosslinks	Decreased	(Radford, 1986a)
Mammalian cells - biological endpoints		
Survival	Decreased	(Chapman *et al.*173; Radford, 1986b; Denault and Liber, 1993; Aguilera *et al.*1992; Murray *et al.*1990; Murray *et al.*1994; Radford, 1985)
Chromosome aberrations	Decreased	(Sasaki and Matsubara, 1977; Virsik and Harder, 1982)
Mutagenesis	Decreased	(Denault and Liber, 1993)
	Spectrum changed	(Denault and Liber, 1993)
Neoplastic transformation	???	
Apoptosis	???	
Gene expression	???	

[a] Data for irradiation in air only.

Obviously, the amount of protection by thiols depends on the particular drug under consideration, its concentration, its charge, etc., so only qualitative comments could be made in Table II. As with the discussion of modification of radiation response by oxygen, consideration of the chemical and biological endpoints when intact cells are irradiated is more helpful at this time, although such studies can be complicated by cellular factors such as intracellular drug distribution and possible biotransformations of the drugs. Base damage, as measured using the ESS assay, is generally poorly protected against by thiols when intact cells are irradiated (Radford, 1986b; Murray *et al.*1990), but no more specfic information exists as to the actions of thiols on specific base alterations. Such information might be helpful in suggesting which base lesions would be candidates for the irreparable damage that is observed in survival curves.

Although our initial data showed similar amounts of protection by the thiol dithiothreitol (DTT) for SSBs and cell killing following irradiation in air (Held *et al.*1986), subsequent data of others generally showed less protection by several thiols against SSBs than against cell killing (e.g., Murray *et al.*1990; Radford, 1986b), consistent with suggestions mentioned above that SSBs are not causally related to cell killing. On the other hand, thiol protection factors for DSBs are very similar to those for survival, and a linear relationship between induction of DSBs (measured by neutral elution or PFGE) and cell killing has been demonstrated (Radford, 1985; Murray *et al.*1990; Murray *et al.*1994). There does not appear to be any literature data on whether thiols alter the nature of the end groups on breaks.

Again, as with oxygen, some interesting new data have recently been published on the the effects of thiols, cysteamine in particular, on radiation-induced mutagenesis. The freqency of radiation-induced mutations in mammalian cells is reduced about 3-fold by 25 mM cysteamine, and the mutation spectrum is altered significantly (Denault and Liber, 1993). The thiol appears to protect more against partial gene deletions/rearrangements than against either total gene deletions or point mutations. Until additional information on the effects of thiols on base damage becomes available, it is difficult to suggest an explanation for the effects of cysteamine on the mutational spectrum. Data on the mutational spectrum for cells irradiated in the presence of additional types of thiols and for DNA irradiated in solution would be informative.

There does not appear to be any literature data on the effects of thiols on radiation-induced oncogenic transformation or on gene expression.

A note of caution should be raised when applying this radiation chemical approach to study of the cellular effects of certain aminothiol radioprotectors, such as WR-1065 and related compounds. Some research, particularly from Grdina's group, shows that WR-1065 and related compounds have interesting biological effects in addition to their ability to modify radiation chemistry. For example, WR-1065 significantly decreases the frequency of radiation-induced mutations when added to mammalian cells as long as 3 h after irradiation (Grdina *et al.*1985), it decreases the rate of repair of DNA SSBs while protecting against the induction of those breaks (Grdina and Nagy, 1986), it slows cell cycle progression (Grdina and Nagy, 1986), it inhibits endogenous endonuclease activity (Grdina *et al.*1989), it can inhibit topoisomerase II activity (Grdina *et al.*1994a), and it can inhibit (Ramakrishnan and Catravas, 1992) or cause apoptosis (Sylvester et al., unpublished) depending on cell type and experimental conditions. These effects clearly are not related to the radiation chemical modes of protection which require that the thiol be present at the time of irradiation in order to be effective. Grdina et al. (Grdina *et al.*1994b) have pointed out the similarity of WR-1065 and its disulfide to polyamines and suggest that this may account for some of its actions. Although, clearly these biological effects of WR-1065 are interesting and may be important, they make it difficult to extrapolate with any confidence from pure radiation chemical studies using WR-1065 in solution to the cellular situation. Thus, it would be preferable to use thiols which do not have cellular effects when performing studies of radiation chemical mechanisms.

Copper as a radiation modifying agent

There is substantial evidence that mammalian cells contain copper (e.g., Margalioth *et al.*1983), much of it in the nucleus (e.g., Leighton *et al.*1990), and it has been known for may years that metal ions, such as copper(II), silver(I) and platinum(II) complexes, can increase the radiation sensitivity of bacterial and mammalian cells (Cramp, 1967; Hesslewood *et al.*1978; Richmond and Powers, 1974; Richmond and Powers, 1976; Douple and Richmond, 1979). In some instances the sensitization in hypoxia was almost as great as the oxygen effect. Of particular interest are reports that copper enhances the damaging effects of radiation in bacteriophage irradiated in the presence and absence of oxygen (Samuni *et al.*1984), and in bacteria irradiated in anoxia (Cramp, 1965; Cramp, 1967), and increases sensitivity of mammalian cells as measured in a clonogenic assay (Hesslewood *et al.*1978), using chromosome aberrations (Sideris *et al.*1990), or DNA-protein crosslinks (Chiu *et al.*1992; Chiu *et al.*1993) (Table III).

Table III: Copper as a Modifier of Radiation Damage

Endpoint	Damage in presence of Cu^{2+} compared to in its absence?	References
Purified DNA in solution		
Biological activity	Increased	this work
Chromatin/isolated nuclei		
DNA-protein crosslinks	Increased	(Chiu *et al.*1992; Chiu *et al.*1993)
Mammalian cells - chemical endpoints		
Double strand breaks	Increased	this work
Mammalian cells - biological endpoints		
Survival	Increased	(Hesslewood *et al.*1978); this work
Chromosome aberrations	Increased	(Sideris *et al.*1990)

The data shown in Fig. 2 demonstrate that Cu^{2+} significantly increases the radiation-induced biological inactivation of bacterial transforming DNA in a copper concentration-dependent manner. The sensitization is slightly greater when the DNA is irradiated in oxygen than when in hypoxia (Table IV). Our recent studies show that incubation of V79 cells in the presence of copper for 2 h prior to irradiation in hypoxia, increases the radiation sensitivity (colony forming assay) of the cells in a copper concentration-dependent manner: enhancement radios of 1.2 and 1.8 at 0.1 and 1.0 mM Cu^{2+}, respectively, were obtained (Fig. 3). Under the same conditions, there is no effect of the addition of 0.1 mM zinc, cobalt or nickel on radiation sensitivity of cell survival. Preliminary data using the neutral elution assay to measure radiation-induced DNA damage show that when V79 cells are treated with 1 mM Cu^{2+} for 2 h prior to irradiation in hypoxia, the DNA damage is increased over 2-fold compared to untreated cells, i.e., an enhancement similar to that for cell killing.

TABLE IV: Effect of Copper on Radiation Sensitivity of Transforming DNA

	Enhancement Ratio[a]	
	N_2	O_2
No Additive	1.0	1.0
1.0 μM Cu^{2+}	1.72[b]	N.D.[d]
10 μM Cu^{2+}	2.26 ± 0.03[c]	2.42 ± 0.03
100 μM Cu^{2+}	2.31 ± 0.19	2.98 ± 0.09

a Ratio of D_0 in the absence and presence of added copper
b Single experiment
c Mean ± SD of 2-3 experiments
d Not determined

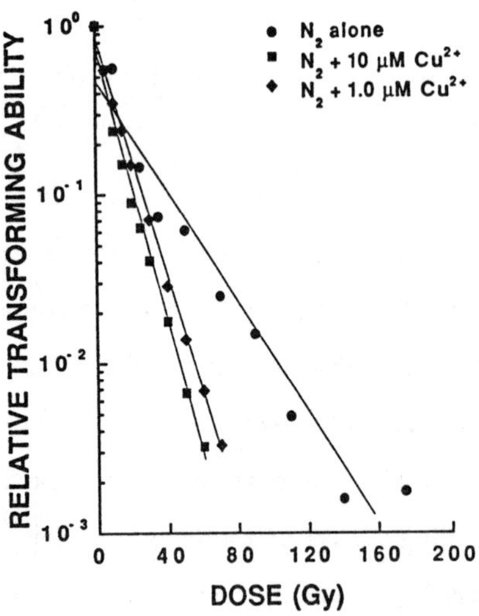

Fig. 2 Loss of biological activity of bacterial transforming DNA irradiated under hypoxic conditions is increased by the presence of low concentrations of copper.

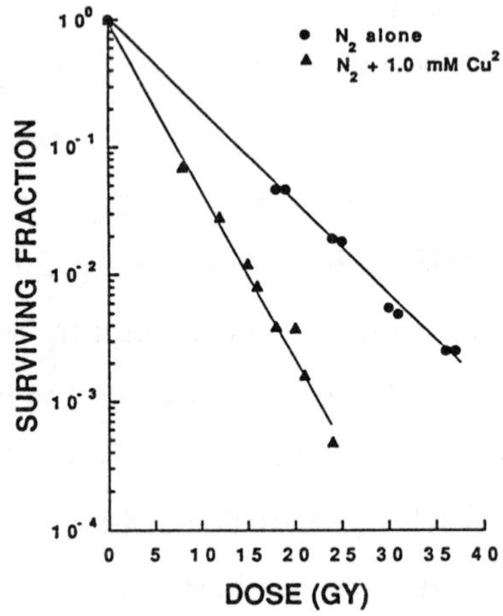

Fig. 3 Sensitivity of V79 cells (colony formation assay) irradiated in hypoxia is increased substantially when cells are exposed to Cu^{2+} for 2 h prior to irradiation.

The mechanism(s) responsible for sensitization by copper is(are) not understood, and at least two can be proposed. First, copper could enhance radiation chemical reactions that increase DNA damage in a "site-specific" manner. The involvement of •OH in radiation sensitization by metal ions has been indicated by results of studies with radical scavengers (Richmond and Powers, 1974; Richmond *et al.*1975). Two hypotheses, not mutually exclusive have been proposed to explain the findings. The first, called the electron sequestration theory (Powers, 1972), suggests the following. Normally, in areas of radiation energy depositions, high local concentrations of water radiolysis products will be formed, and within those spurs, blobs, etc. there will be some recombination of radiolysis products, e.g.:

$$e^-_{aq} + {}^\bullet OH \rightarrow OH^- \qquad (1)$$

However, metal ions (M^n) react rapidly with e^-_{aq}:

$$M^n + e^-_{aq} \rightarrow M^{n-1} \qquad (2)$$

and this later reaction can compete with reaction (1) to cause an increase in •OH concentration in a localized region. If the metal ion is bound to DNA, this could result in "site-specific" (Samuni *et al.*1981; Samuni *et al.*1983) generation of •OH in DNA in the vicinity of the bound metal.

An alternative theory suggests that Cu^+ is produced by radiation

$$Cu^{2+} + e^-_{aq} \rightarrow Cu^+ \qquad (3)$$

or by reaction of Cu^{2+} with radiation-induced organic radicals (von Sonntag, 1987)

$$Cu^{2+} + {}^\bullet C(R)(R){-}OH \rightarrow Cu^+ + C(R)(R){=}O + H^+. \qquad (4)$$

Cu^+ then reacts with H_2O_2 which has been produced either by radiation or during normal cell metabolism:

$$Cu^+ + H_2O_2 \rightarrow Cu^{2+} + OH^- + {}^\bullet OH \qquad (5)$$

The damaging •OH produced in this Fenton reaction would also be produced in a "site-specific" manner e.g., at DNA, since copper binds efficiently to DNA (e.g., Stoewe and Prutz, 1987).

Ward (1985; 1988) has postulated that DNA lesions in which damage is present on both DNA strands in a local region (locally multiply damaged sites - LMDS) may be produced by ionizing radiation when volumes of high radical density, e.g., spurs, overlap DNA. These LMDS or clustered lesions may be particularly difficult for cells to repair accurately, and, hence, may be more lethal than singly damaged sites. We suggest that the occurrence of either reactions 1 and 2 or 3 to 5 in copper bound to DNA will, in essence, increase the likelihood of LMDS-type damage.

The second possible mechanism for sensitization by copper is that radiation-induced changes in copper at the DNA-nuclear matrix attachment site might alter that attachment, thus altering the amount and type of DNA damage. Several recent papers have suggested that the association between DNA and the nuclear matrix relates to radiation sensitivity: more sensitive cell lines have a looser association, based on studies of DNA accessibility to DNAse I and the ability of nucleoids to maintain positive DNA supercoils (Vaughan *et al.*1991; Milner *et al.*1993). It has been postulated that copper plays a role in the attachment of DNA to the nuclear matrix (Lebkowski and Laemmli, 1982; Dijkwel and Wenink, 1986). This suggestion is based on studies (Lebkowski and Laemmli, 1982; Lewis and Laemmli, 1982; Dijkwel and Wenink, 1986) showing that addition of copper stabilizes DNA, causes contraction of the DNA halo in the fluorescent halo assay, and increases DNA sedimentation in sucrose gradients; removal of copper reverses those effects. Since radiation will change the valence state of copper (reaction 3), it is possible that this change will alter the nature of the DNA-nuclear matrix attachment, thus altering production of DNA damage at that site. There is some literature to support this proposal. George *et al.* (1987) showed that histone-free DNA-protein complexes isolated from lymphocytes in sucrose gradients were less dense when isolated from irradiated cells than from unirradiated cells. Addition of Cu^{2+} before isolation increased the density of the complex, while dithiothreitol or diethyldithiocarbamate, both metal chelating thiols, decreased the density. Those authors (George *et al.*1987; Cramp *et al.*1989) postulated that, when irradiated, chromatin becomes less dense because the endogenous copper, originally present at the DNA-nuclear attachment site as Cu^{2+} which forms a square planar 3D conformation, is converted by the radiation to Cu^{+} (reaction 3), which assumes a tetrahedral conformation. This conformational change of the copper could only occur with rearrangement of the topography of the DNA or the protein in the matrix. Thus, it is possible that copper can modulate radiation-induced DNA damage induction and cell killing due to its critical localization at the DNA-nuclear matrix attachment site.

Comparison of Tables I and II with III show that much less research has been conducted on the effects of copper on radiation-induced DNA and cellular damage than has been done using oxygen or thiols. Studies on the effects of copper on such DNA endpoints as base damage and strand breaks and on cellular endpoints such as mutagenesis, oncogenic transformation or gene expression might shed light on its importance. It has recently been suggested that the demonstration of mutagenic hot spots when genes are irradiated in solution followed by transfection into mammalian or bacterial cells but the lack of evidence for hot spots when genes are irradiated in cells may be due to induction of base change mutations in DNA in solution by reactions catalyzed by adventitious transition metal ions, e.g., copper, bound to the DNA (Hutchinson, 1992).

Conclusions

With the advent of new, increasingly sensitive and specific analytical methods to identify and quantify particular types of radiation-induced products in DNA, opportunities now exist to obtain new information on the radiation chemical and molecular mechanisms involved in the biological appearance of DNA damage. Chemical modifiers will be expected to continue to play a role in these studies by allowing a means for changing a number of different endpoints and correlating the outcomes.

Acknowledgements

Research in the author's laboratory has been supported by PHS grant numbers CA42167 and CA18614 awarded by the NCI, DHHS. The contents of this paper are solely the responsibility of the author and do not necessarily represent the official views of the National Institutes of Health. The author thanks Ms. K.L. Hopcia for assistance.

References

Aguilera JA, Newton GL, Fahey RC, Ward JF (1992) Thiol uptake by Chinese hamster V79 cells and aerobic radioprotection as a function of the net charge on the thiol. Radiat Res 130:194-204.

Arends MJ, Morris RG, Wyllie AH (1990) Apoptosis. the role of the endonuclease. Am J Path 136:593-608.

Blok J, Loman H (1973) The effects of γ-radiation in DNA. Curr Top Radiat Res Q 9:165-245.

Borsa J, Sargent MD, Einspenner M, Azzam EI, Raaphorst GP (1984) Effects of oxygen and misonidazole on cell transformation and cell killing in C3H 10T1/2 cells by X rays *in vitro*. Radiat Res 100:96-103.

Buchko GW, Weinfeld M (1993) Influence of nitrogen, oxygen, and nitroimidazole radiosensitizers on DNA damage induced by ionizing radiation. Biochem 32:2186-2193.

Chapman JD, Reuvers AP, Borsa J, Greenstock CL (1973) Chemical radioprotection and radiosensitization of mammalian cells growing in vitro. Radiat Res 56:291-306.

Chiu S-M, Xue L-Y, Friedman LR, Oleinick NL (1992) Chromatin compaction and the efficiency of formation of DNA-protein crosslinks in γ-irradiated mammalian cells. Radiat Res 129:184-191.

Chiu S-M, Xue L-Y, Friedman LR, Oleinick NL (1993) Copper Ion-mediated sensitization of nuclear matrix attachment sites to ionizing radiation. Biochem 32: 6214-6219.

Cramp WA (1965) Radiosensitization by copper ions, and consequent reversal of the oxygen effect. Nature 206:636-637.

Cramp WA (1967) The toxic action on bacteria of irradiated solutions of copper compounds. Radiat Res 30: 221-236.

Cramp WA, George AM, Khan H, Yatvin MB (1989) The role of copper in radiation and drug induced changes in the quaternary structure of DNA. Free Radicals, Metal Ions and Biopolymers, Beaumont PC, Deeble DJ, Parsons BJ, Rice-Evans C (eds) Richelieu Press, London, pages 127-141.

Denault CM, Liber HL (1993) The effects of hypoxia and cysteamine on X-ray mutagenesis in human cells. I. Dose response and Southern blot analysis of induced mutants. Radiat Res 135:98-107.

Deschner EE, Gray LH (1959) Influence of oxygen tension on X-ray-induced chromosomal damage in Ehrlich ascites tumor cells irradiated *in vitro* and *in vivo*. Radiat Res 11:115-146.

Dijkwel PA, Wenink PW (1986) Structural integrity of the nuclear matrix: Differential effects of thiol agents and metal chelators. J Cell Sci 84:53-67.

Douple EB, Richmond RC (1979) Radiosensitization of hypoxic tumor cells by cis- and trans-dichlorodiammineplatinum(II). Int J Radiat Oncol Biol Phys 5:1369-1372.

Eastman A (1990) Activation of programmed cell death by anticancer agents: Cisplatin as a model system. Cancer Cells 2:275-280.

Edgren MR (1991) Oxygen enhancement of radiosensitivity and nuclear GSH content. The Early Effects of Radiation on DNA, Fielden EM, O'Neill P (eds) Springer-Verlag, Berlin, pages 363-364.

Fornace AJ, Little JB (1977) DNA crosslinking induced by X-rays and chemical agents. Biochim Biophys Acta 477:343-355.

Fuciarelli AF, Wegher BJ, Blakely WF, Dizdaroglu M (1990) Yields of radiation-induced base products in DNA: effects of DNA conformation and gassing conditions. Int J Radiat Biol 58:397-415.

Gajewski E, Rao G, Nackerdien Z, Dizdaroglu M (1990) Modification of DNA bases in mammalian chromatin by radiation-generated free radicals. Biochem 29:7876-7882.

George AM, Sabovljev SA, Hart LE, Cramp WA, Harris G, Hornsey S (1987) DNA quaternary structure in the radiation sensitivity of human lymphocytes - a proposed role of copper. Br J Cancer 55 (Suppl. VIII):141-144.

Grdina DJ, Nagy B, Hill CK, Wells RL, Peraino C (1985) The radioprotector WR-1065 reduces radiation-induced mutations at the hypoxanthine-guanine phospho-ribosyl transferase locus of V79 cells. Carcinogenesis 6:929-939.

Grdina DJ, Nagy B (1986) The effect of 2-[(aminopropyl)amino]ethanethiol (WR1065) on radiation-induced DNA damage and repair and cell progression in V79 cells. Br J Cancer **54**:933-941.

Grdina DJ, Sigdestad CP, Dale PJ, Perrin J M (1989) The effect of 2[(aminopropyl)-amino]ethanethiol on fission-neutron-induced DNA damage and repair. Br J Cancer 59:17-21.

Grdina DJ, Constantinou A, Shigematsu N, Murley JS (1994a) Inhibition of topoisomerase IIα activity in CHO K1 cells by 2-[(aminopropyl)amino]ethanethiol (WR-1065). Radiat Res 138:44-52.

Grdina DJ, Shigematsu N, Dale P, Newton GL, Aguilera JA, Fahey RC (1994b) Thiol uptake and protection agains radiation-induced cell killing and mutagenesis at the *hprt* locus of CHO-AA8 cells. Cancer Res (submitted)

Hawes C, Howard A, Gray LH (1966) Induction of chromosome structural damage in Ehrlich ascites tumour cells. Mutat Res 3:79-89.

Held KD, Synek RW, Powers EL (1978) Radiation sensitivity of transforming DNA. Int J Radiat Biol 33:317-324.

Held KD, Harrop HA, Michael BD (1981) Effects of oxygen and sulfhydryl-containing compounds on irradiated transforming DNA. I. Actions of dithiothreitol. Int J Radiat Biol 40:613-622.

Held KD, Harrop HA, Michael BD (1984) Effects of oxygen and sulfhydryl-containing compounds on irradiated transforming DNA. II. Glutathione, cysteine and cysteamine. Int J Radiat Biol 45:615-626.

Held KD (1986) Role of free radical scavengers in radioprotection of DNA: Chemical and biological aspects. Free Radicals, Aging and Degenerative Diseases, Johnson JE, Walford R, Harman D, Miquel J (eds) Alan R. Liss, New York, pages 527-568.

Held KD, Bren GD, Melder DC (1986) Interactions of radioprotectors and oxygen in cultured mammalian cells. II. Effects of dithiothreitol on radiation-induced DNA damage and comparison with cell survival. Radiat Res 108:296-306.

Held KD, Hopcia KL (1993) Role of protein thiols in intrinsic radiation protection of DNA and cells. Mutat Res 299:261-269.

Hentosh P (1988) Induction and repair of DNA damage in γ-irradiated human lymphoblasts: irradiation in the presence and absence of misonidazole. Radiat Res 115:436-447.

Hesslewood IP, Cramp WA, McBrien DCH, Williamson P, Lott KAK (1978) Copper as a hypoxic cell sensitizer of mammalian cells. Br J Cancer 37 (Suppl. III):95-97.

Hopcia KL, Sylvester FC, Held KD (1994) Radiation-induced apoptosis in HL-60 cells: Oxygen effect and relationship between apoptosis and loss of clonogenicity. Radiat Res (submitted)

Hutchinson F (1992) Published data on mutagenesis by ionizing radiation of plasmids in solution probably reflects in part the specificity of adventitious transition metal ions complexed to the DNA. Mutat Res 281:261-266.

Jones DP, McConkey DJ, Nicotera P, Orrenius S (1989) Calcium-activated DNA fragmentation in rat liver nuclei. J Biol Chem 264:6398-6403.

Kaufmann SH (1989) Induction of endonucleolytic DNA cleavage in human acute myelogenous leukemia cells by etoposide, camptothecin, and other cytotoxic anticancer drugs: A cautionary note. Cancer Res 49:5870-5878.

Kerr JFR, Wyllie AH, Currie AR (1972) Apoptosis: A basic biological phenomenon with wide-ranging implications in tissue kinetics. Br J Cancer 26:239-257.

Kerr JFR, Harmon BV (1991) Definition and incidence of apoptosis: An historical perspective. Apoptosis: The Molecular Basis of Cell Death, Tomei LD, Cope FO (eds) Cold Spring Harbor Laboratory Press, pages 5-29.

Klassen NV, Walker PR, Ross CK, Cygler J, Lach B (1993) Two-stage cell shrinkage and the OER for radiation-induced apoptosis of rat thymocytes. Int J Radiat Biol 64:571-581.

Lebkowski JS, Laemmli UK (1982) Evidence for two levels of DNA folding in histone-depleted HeLa interphase nuclei. J Molec Biol 156:309-324.

Leighton MJ, Peters TJ, Hill R, Smith GD, Holt W, Jones DM (1990) Subcellular distribution of copper in the liver of fetal deer in the last month of gestation. J Vet Sci 49:298-305.

Lewis CD, Laemmli UK (1982) Higher order metaphase chromosome structure: Evidence for metalloprotein interactions. Cell 29:171-181.

Margalioth EJ, Schenker JG, Chevion M (1983) Copper and zinc levels in normal and malignant tissues. Cancer 52:868-872.

Martin SJ, Cotter TG (1991) Ultraviolet B irradiation of human leukaemia HL-60 cells *in vitro* induces apoptosis. Int J Radiat Biol 59:1001-1016.

Meyn RE, VanAnkeren SC, Jenkins WT (1987) The induction of DNA-protein crosslinks in hypoxic cells and their possible contribution to cell lethality. Radiat Res 109:419-429.

Michael BD, Prise KM, Fahey RC, Folkard M (1991) Radical multiplicity in radiation-induced DNA strand breaks: Implications for their chemical modification. The Early Effects of Radiation on DNA, Fielden EM, O'Neill P (eds.) Springer-Verlag, Berlin, pages 333-346.

Milner AE, Gordon DJ, Turner BM, Vaughan ATM (1993) A correlation between DNA-nuclear matrix binding and relative radiosensitivity in two human squamous cell carcinoma cell lines. Int J Radiat Biol 63:13-20.

Murray D, Prager A, Van Ankeren SC, Altschuler EM, Kerr MS, Terry NHA, Milas L (1990) Comparative effect of the thiols dithiothreitol, cysteamine and WR-151326 on survival and on the induction of DNA damage in cultured Chinese hamster ovary cells exposed to g-radiation. Int J Radiat Biol 58:71-91.

Murray D, Simpson R, Rosenberg E, Carraway A, Britten R (1994) Correlation between γ-ray-induced DNA double-strand breakage and cell killing after biologically relevant doses: analysis by pulsed-field gel electrophoresis. Int J Radiat Biol 65:419-426.

Nackerdien Z, Rao G, Cacciuttolo MA, Gajewski E, Dizdaroglu M (1991) Chemical nature of DNA-protein cross-links produced in mammalian chromatin by hydrogen peroxide in the presence of iron or copper ions. Biochem 30:4873-4879.

Powers EL (1972) The hydrated electron, the hydroxyl radical, and hydrogen peroxide in radiation damage in cells. Isr J Chem 10:1199-1211.

Prise KM, Davies S, Michael BD (1992a) A comparison of the chemical repair rates of free radical precursors of DNA damage and cell killing in Chinese hamster V79 cells. Int J Radiat Biol 61:721-728.

Prise KM, Davies S, Stratford MRL, Michael BD (1992b) The role of non-protein sulphydryls in determining the chemical repair rates of free radical precursors of DNA damage and cell killing in Chinese hamster V79 cells. Int J Radiat Biol 62: 297-306.

Prise KM, Davies S, Michael BD (1993) Evidence for induction of DNA double-strand breaks at paired radical sites. Radiat Res 134:102-106.

Radford IR (1985) The level of induced DNA double-strand breakage correlates with cell killing after X-irradiation. Int J Radiat Biol 48:45-54.

Radford IR (1986a) Evidence for a general relationship between the induced level of DNA double-strand breakage and cell killing after X-irradiation of mammalian cells. Int J Radiat Biol 49:611-620.

Radford IR (1986b) Effect of radiomodifying agents on the ratios of X-ray-induced lesions in cellular DNA: use of lethal lesion determination. Int J Radiat Biol 49:621-637.

Radford IR (1987) Use of 'nuclear monolayers' to identify factors influencing DNA double-strnd breakage by X-rays. Int J Radiat Biol 52:853-858.

Ramakrishnan N, Catravas G. N (1992) N-(2-mercaptoethyl)-1,3-propanediamine (WR-1065) protects thymocytes from programed cell death. J Immunol 148:1817-1821.

Retel J, Hoebee B, Braun JEF, Lutgerink JT, van den Akker E, Wanamarta AH, Joenje H, Lafleur MVM (1993) Mutational specificity of oxidative DNA damage. Mutat Res 299:165-182.

Richmond RC, Powers EL (1974) Modification of radiation sensitivity of bacterial spores by silver salts. Radiat Res 58:470-480.

Richmond RC, Simic MG, Powers EL (1975) Radiation sensitivity of *Bacillus megaterium* spores in the presence of Co(III) complexes. Radiat Res 63:140-148.

Richmond RC, Powers EL (1976) Radiation sensitization of bacterial spores by cis-dichlorodiammineplatinum(II). Radiat Res 68:251-257.

Roots R, Okada S (1972) Protection of DNA molecules of cultured mammalian cells growing *in vitro*. Int J Radiat Biol 21:329-342.

Samuni A, Chevion M, Czapski G (1981) Unusual copper-induced sensitization of the biological damage due to superoxide radicals. J Biol Chem 256:12632-12635.

Samuni A, Aronovitch J, Godinger D, Chevion M, Czapski G (1983) On the cytotoxicity of vitamin C and metal ions. a site-specific Fenton mechanism. Eur J Biochem 137:119-124.

Samuni A, Chevion M, Czapski G (1984) Roles of copper and $O_2^{-\cdot}$ in the radiation-induced inactivation of T7 bacteriophage. Radiat Res 99:562-572.

Sapora O, Barone F, Belli M, Maggi A, Quintiliani M, Tabocchini MA (1991) Relationships between cell killing, mutation induction and DNA damage in X-irradiated V79 cells: the influence of oxygen and DMSO. Int J Radiat Biol 60:467-482.

Sasaki MS, Matsubara S (1977) Free radical scavenging in protection of human lymphocytes against chromosome aberration formation by gamma-ray irradiation. Int J Radiat Biol 32:439-445.

Sellins KS, Cohen JJ (1987) Gene induction by γ-irradiation leads to DNA fragmentation in lymphocytes. J Immunol 139:3199-3206.

Sellins KS, Cohen JJ (1991) Hyperthermia induces apoptosis in thymocytes. Radiat Res 126:88-95.

Siddiqi MA, Bothe E (1987) Single- and double-strand break formation in DNA irradiated in aqueous solution: Dependence on dose and OH radical scavenger conditions. Radiat Res 112:449-463.

Sideris EG, Katsaros N, Tsolomyty A (1990) Binding position of certain metal ions on the DNA and its possible relationship to radiation-induced genetic lesions in the V-79 cells in culture. Frontiers in Radiation Biology, Riklis E (ed), pages 451-457.

Spotheim-Maurizot M, Franchet J, Sabattier R, Charlier M (1991) DNA radiolysis by fast neutrons. II. Oxygen, thiols and ionic strength effects. Int J Radiat Biol 59:1313-1324.

Stephens LC, Ang KK, Schultheiss TE, Milas L, Meyn RE (1991) Apoptosis in irradiated murine tumors. Radiat Res 127:308-316.

Stoewe R, Prutz WA (1987) Copper-catalyzed DNA damage by ascorbate and hydrogen peroxide: Kinetics and yield. Free Radicals Biol Med 3:97-105.

Vaughan ATM, Milner AM, Gordon DG, Schwartz JL (1991) Interaction between ionizing radiation and supercoiled DNA within human tumor cells. Cancer Res 51:3857-3861.

Virsik RP, Harder D (1982) Effect of L-cysteine on the dose effect relationship for chromosome aberrations in irradiated human lymphocytes. Int J Radiat Biol 42:211-214.

von Sonntag C (1987) The Chemical Basis of Radiation Biology, Taylor & Francis, London.
Walker NI, Harmon BV, Gobe GC, Kerr JFR (1988) Patterns of cell death. Methods Achiev Exper Path 13:18-54.
Ward JF (1985) Biochemistry of DNA lesions. Radiat Res 104:S103-S111.
Ward JF (1988) DNA damage produced by ionizing radiation in mammalian cells: Identities, mechanisms of formation, and reparability. Prog Nucleic Acid Res Molec Biol 35:95-125.
Whitaker SJ, McMillan TJ (1992) Oxygen effect for DNA double-strand break induction determined by pulsed-field gel electrophoresis. Int J Radiat Biol 61:29-41.
Wyllie AH (1980) Glucocorticoid-induced thymocyte apoptosis is associated with endogenous endonuclease activation. Nature 284:555-556.
Wyllie AH, Kerr JFR, Currie AR (1980) Cell death: The significance of apoptosis. Int Rev Cytol 68:251-306.
Xue L-Y, Friedman LR, Oleinick NL (1988) Repair of chromatin damage in glutathione-depleted V-79 cells: Comparison of oxic and hypoxic conditions. Radiat Res 116:89-99.
Zheng S, Newton GL, Gonick G, Fahey RC, Ward JF (1988) Radioprotection of DNA by thiols: relationship between the net charge on a thiol and its ability to protect DNA. Radiat Res 114:11-27.
Zheng S, Newton GL, Ward JF, Fahey R C (1992) Aerobic radioprotection of pBR322 by thiols: Effect of thiol net charge upon scavenging of hydroxyl radicals and repair of DNA radicals. Radiat Res 130:183-193.

LETHAL OR PROTECTIVE EFFECTS OF PROLONGED TREATMENT WITH HYPOXIC CELL SENSITIZERS

M. R. Edgren
Department of Medical Radiation Physics
Karolinska Institute
171 76 Stockholm
Sweden

Introduction

AK-2123 [N-(2-methoxyethyl)-2-(3-nitro-1-triazolyl)acetamide] is a hypoxic cell radiosensitizer which is currently being tested in several oncology clinics and which has a lower toxicity than misonidazole (MISO) *in vivo*. The positive experiences reported recently (Garcia-Angulo 1991, Huan and Hua 1994) certainly warrant further clinical evaluations. The experimental observations reported so far (Shibamoto et al. 1986 and 1987) need further experimental studies to clarify the sensitization mechanism, especially as recent intratumoral strategies used in the clinical administration of the sensitizers can result in a large local concentration of the drug that may persist for a prolonged period of time between and after radiation exposures. Model experiments *in vitro* using V79 cells were performed with AK-2123 under these conditions. Misonidazole (MISO) and metronidazole (METRO), well known hypoxic cell radiosensitzers (Adams and Stratford 1986), were used for comparison of the effects. Clonogenic survival and induction and repair of DNA damage were used as end-points.

Methods

V79-379A Chinese hamster cells were maintained as exponentalliy growing monolayer cultures in Eagle's MEM with Earle's salts supplemented with 10% foetal calf serum and antibiotics. The hypoxic cell sensitizers MISO [1-(2-nitro-1-imidazolyl)-3-methoxy-2-propanol], METRO [1-(2-hydroxyethyl-2-methyl-5-nitro-imidazole] and AK-2123 were freshly dissolved in growth medium in a concentration of 5 mmol dm^{-3}. The cells were irradiated with a Siemens Stabilipan unit working at 220 kV and 15 mA. With 4 mm Al filtration, 36.5 cm focus-target distance the dose-rate was 3.21 Gy/min (HVL 0.57 mm Cu). The cell cultures were aerobically incubated with the drugs 30 min or 20 h before, or 24 h after, irradiation.

Irradiations were performed under aerobic and hypoxic conditions as previously described (Sölen et al. 1987).

The clonogenic survival data were analysed by fitting the linear-quadratic model and the sensitizer enhancement ratios calculated as the mean value from the survival at 10^{-1} and 10^{-2}. Cellular glutathione (GSH) was measured using the Tietze assay (1969). Determination of DNA breaks was performed mostly in parallel with the alkaline unwinding technique (Ahnström and Erixon 1973) and the precipitation technique (Olive 1988). Details of the measurements have been published recently (Imamura and Edgren 1994). Flow cytometry (Castro et al. 1993) was performed with normal human fibroblasts in order to evaluate any cell phase disturbances after the prolonged treatment periods.

Results and Discussion

When unirradiated cells were aerobically incubated with the three drugs, only AK-2123 was found to have a time dependent cytotoxic effect (Table 1). In order to obtain aerobic cytotoxicity with MISO and METRO, much higher concentrations are needed (data not shown)(Balmukanov et al. 1992). The cellular GSH content changed in a time and concentration dependent manner with all three drugs (Table 1 and to be published elsewhere).

Table 1. Effect of aerobic incubation with 5 mmol dm^{-3} of radiosensitizer on cytotoxicity and cellular glutathione level.

	Incubation time (hours)	Cytotoxicity (surviving fraction)	Relative amount of GSH
AK-2123	20	0.62 ± 0.02[a] (19)[b]	1.52 ± 0.04 (25)
-"-	24	0.39 ± 0.02 (25)	1.50 ± 0.03 (18)
MISO	20	0.96 ± 0.01 (14)	1.20 ± 0.05 (11)
-"-	24	0.95 ± 0.02 (20)	1.56 ± 0.04 (18)
METRO	20	1.02 ± 0.01 (8)	1.24 ± 0.07 (12)
-"-	24	1.07 ± 0.03 (20)	1.27 ± 0.05 (18)

[a] Mean ± SEM [b] number of repeat experiments

A comparison of the enhancement of radiosensitivity by AK-2123, MISO and METRO (each 5 mmol dm^{-3}) is shown in Table 2. Treatment with the drugs before or after irradiation was made under aerobic conditions. The plating efficiencies of drug-treated, unirradiated cells were used as controls in order to exclude the effect of cytotoxicity on the enhancement ratios. There was no significant sensitization by either of these substances when the cells were irradiated in air without a prolonged preincubation period.

When the radiation exposures were made under hypoxic conditions using the same drug concentration a sensitization by a factor around 1.9 for AK-2123 and MISO and 1.6 for METRO was found. When the cells were incubated with the drugs for 20 h prior to irradiation, AK-2123 sensitized the cells to aerobic irradiation, but had no effect on hypoxically irradiated cells. Both MISO and METRO show a protective effect regardless of irradiation condition. However, METRO was somewhat more protective than MISO. All three drugs had a sensitizing effect when either added immediately before or just after exposure regardless of condition of radiation. AK-2123 being the most efficient one of the three drugs. A similar preincubation sensitizing effect has been seen earlier with MISO (Hall and Biaglow 1977) and a post-incubation sensitizing effect has been noted with a 2-nitroimidazole derivative (Hodgkiss et al. 1992).

Table 2. Sensitizer enhancement ratios for survival.

Treatment	Sensitizer enhancement ratio AK-2123	MISO	METRO
<u>aerobic irradiation</u>			
30 min	1.00 ± 0.01[a] (14)[b]	1.01 ± 0.01 (10)	0.97 ± 0.01 (8)
20 h before	1.29 ± 0.04 (7)	0.92 ± 0.03 (6)	0.86 ±0.01 (4)
24 h after	1.30 ± 0.03 (6)	1.13 ± 0.02 (4)	1.14 ± 0.01 (4)
only 24 h after	1.27 ± 0.03 (4)	1.08 ± 0.01 (4)	1.01 ± 0.01 (4)
<u>hypoxic irradiation</u>			
30 min	1.85 ± 0.06 (13)	1.94 ± 0.04 (8)	1.61 ± 0.02 (12)
20 h.before	1.88 ± 0.06 (4)	1.58 ± 0.05 (4)	1.28 ± 0.02 (4)
24 h after	2.40 ± 0.09 (5)	2.12 ± 0.06 (4)	1.75 ± 0.03 (8)
only 24 h after	1.41 ± 0.03 (4)	1.17 ± 0.01 (4)	1.07 ± 0.03 (4)

[a] Mean ± SEM [b] number of repeat experiments

The elevated GSH levels, obtained by a 20 h incubation period with AK-2123 and MISO, did not modify the enhancement ratios for DNA damage as measured by two techniques. The data are summarized in Table 3. The comparison of the sensitizing effect was made after just one irradiation dose (30 Gy in air and 90 Gy in hypoxia), since we have earlier shown that there is a dose dependent enhancement of AK-2123, especially when using the precipitation technique (Imamura and Edgren 1994). No breaks were induced by these drugs under the long incubation period using either of the assays.

Table 3. Sensitizer enhancement ratios for the induction of single-strand breaks in cells treated with sensitizers for different times before and during radiation exposure.

	Sensitizer enhancement ratio			
	Aerobic irradiation		Hypoxic irradiation	
	AK-2123	MISO	AK-2123	MISO
Unwinding				
30 min	0.98 (3)	1.04 (3)	2.49 ± 0.36[a] (4)[b]	2.59 ± 0.36 (4)
20 h	1.00 ± 0.09 (6)	1.02 ± 0.15 (4)	2.50 ± 0.29 (6)	2.65 ± 0.26 (5)
Precipitation				
30 min	1.07 ± 0.05 (6)	1.04 ± 0.03 (6)	2.55 ± 0.35 (6)	2.29 ± 0.24 (6)
20 h	1.01 ± 0.05 (10)	1.08 ± 0.04 (8)	2.56 ± 0.34 (8)	2.30 ± 0.21 (6)

[a] Mean ± SEM [b] number of repeat experiments

The elevated GSH levels have also no effect on the rejoining of DNA breaks as seen in Table 4.

Table 4. Percent rejoined DNA breaks within 60 min after a 20 h pre-irradiation incubation period.

Assay	Aerobic irradiation			Hypoxic irradiation		
	Control	AK-2123	MISO	Control	AK-2123	MISO
Alk. unwinding	97	94	92	93	95	98
Precipitation	96	95	100	97	92	92

Mean of 2-4 repeat experiments.

As both assays used for DNA damage measure the average effect on the cell population after nonphysiologically high irradiation doses, we will in the future use the very sensitive comet assay (Östling and Johanson 1984) in which the sensitivity of each can be measured. The sensitizing effect of AK-2123 will also be investigated on other cell types.

Both ionizing radiation and anti-cancer drugs can induce a death similar to apoptosis (Kerr and Harmon 1991, Smets 1994). We therefore performed some preliminary experiments with normal human skin fibroblasts which show a similar cytotoxicity and elevated GSH content after prolonged incubation with 5 mmol dm^{-3} AK-2123 (to be published elsewhere) as the hamster cells. Flow cytometry performed on attached cells 24 h after the end of the treatment period showed an increased number of cells in the G_2/M phase of the cell cycle (data not shown). However, no apoptotic fragments of the nuclei were noted. Most of the apoptotic cells have been lost by detachment as visualized in the fluorescent microscope and also probably due to the formaldehyde fixation of cells followed by a protease treatment and centrifugation (Castro et al. 1993). A time dependent rounding up of the normally outstretched fibroblasts were seen after incubation with AK-2123 and their nuclei looked like a flower. All these have to be further investigated and quantitated in order to explain the cytotoxic effect of the prolonged incubation with especially AK-2123.

Acknowledgements

The authorwish to thank Dr. T. Kagiya for the generous gift of AK-2123 and La Roche Co. for misonidazole. This work was supported by the Swedish Cancer Society and the Swedish Radiation Protection Institute.

References

Adams GE, Stratford IJ (1986) Hypoxia-mediated nitroheterocyclic drugs in the radio- and chemotherapy of cancer. An overview. Biochem Pharmacol 35:71-76.

Ahnström G, Erixon K (1973) Radiation induced strand breakage in DNA from mammalian cells: Strand separation in alkaline solution. Int J Radiat Biol 23:285-289.

Balmukanov SB, Edgren M, Scott OCA, Révész L (1992) Intratumoral stratgies for

clinical radiosensitization. Radiation Research II Dewey WC, Edington M, Fry RJM, Hall EJ, Whitmore GF (eds) Academic Press New York pp 578-583.
Castro J, Heiden T, Wang N, Tribukait B (1993) Preparation of cell nuclei from fresh tissues for high-quality DNA flow cytometry. Cytometry 14:793-804.
Garcia-Angulo AH (1991) Sensitized radiotherapy with both moderate high dose of AK-2123 and external beam radiotherapy combined with hydralazine induced hypoxia. Radiosensit Newsletters 10:5-6.
Hall EJ, Biaglow J (1977) Ro-07-0582 as a radiosensitizer and cytotoxic agent. Int J Radiat Oncol Biol Phys 2:521-530.
Hodgkiss RJ, Kelleher E, Parrick J (1992) Hypoxia-specific inhibition of recovery from radiation damage by a novel 2-nitroimidazole with a theophylline side chain. Int J Radiat Biol 61:797-804.
Huan LC, Hua BY (1994) Clinical pharmacokinetic study and sensitive effect of AK-2123. Int J Radiat Oncol Biol Phys 29:607-610.
Imamura M, Edgren MR (1994) Radiation dose dependence of the enhancement of DNA damage in cells treated with the hypoxic radiosensitzer AK-2123, a 3-nitrotriazole derivative. Int J Oncol 4:97-100.
Kerr JFR and Harmon BV (1991) Definition and incidence of apoptosis: An historical perspective. In Apoptosis: The molecular basis of cell death. Tomei LD, Cope FO (eds) Cold Spring Harbor Laboratory Press pp 5-29.
Olive PL (1988) DNA precipitation assay: A rapid and simple method for detecting DNA damage in mammalian cells. Envir Mol Mut 11:487-495.
Östling O, Johanson K-J (1984) Microelectrophoretic study of radiation-induced DNA damages in individual mammalian cells. Biochem Biophys Res Commun 123:291-298.
ShibamotoY, Sakano K, Kimura R, Nishidai T, Nishimoto SI, Ono K, Kagiya T, Abe M (1986) Radiosensitization *in vitro* and *in vivo* by 3-nitrotriazoles. Int J Radiat Oncol Biol Phys 12:1063-1066.
Shibamoto Y, Nishimoto S, Mi F, Sasai K, Kagiya T and Abe M (1987) Evaluation of various types of new hypoxic cell sensitizers using the EMT6 single cell-spheroid-solid tumour system. Int J Radiat Biol 52: 347-357.
Smets LA (1994) Programmed cell death (apoptosis) and response to anti-cancer drugs. Anti-Cancer Drugs 5:3-9.
Tietze F (1969) Enzymic method for quantitative determination of nanogram amounts of total and oxidized glutathione: Applications to mammalian blood and other tissues. Anal Biochem 27:502-522.

INTERACTION OF THIOLS AND NON-THIOL •OH SCAVENGERS IN THE MODIFICATION OF RADIATION-INDUCED DNA DAMAGE

R. E. Krisch, I. S. Ayene, & C. J. Koch
Department of Radiation Oncology
University of Pennsylvania School of Medicine
Philadelphia, PA 19104
USA

Introduction

Oxygen has been long known to sensitize cells to the lethal effects of ionizing radiation (Gray et al., 1953); and is widely believed to do so by the fixation of potentially reversible radical damage to cellular DNA (Howard-Flanders, 1960). A number of studies have suggested that this widely observed oxygen enhancement of cell killing requires the presence of reduced thiols (Hutchinson, 1961; Bump & Brown, 1990). Published *in vitro* studies of the modification of DNA damage by glutathione or other thiols have generally shown peak oxygen enhancement ratios (OERs) much higher than those observed for cell killing (e.g. Held et al., 1981, 1984a, 1984b). However, these studies measured loss of DNA transforming activity or induction of single-strand DNA breaks (SSBs), related endpoints which are not thought to represent lethal lesions, rather than double-strand breaks (DSBs), which are generally believed to be the dominant lethal lesions from ionizing radiation (Iliakis, 1991). In addition, non-thiol scavengers of OH radicals were not generally present. Typical cells contain concentrations of reduced glutathione, the major soluble thiol in cells, of only 0.5-10 mM (Meister & Anderson, 1983), but also contain near molar combined concentrations of non-thiol organic compounds, including proteins, polyamines, and carbohydrates, which are potent radioprotectors. There is also evidence that, in addition to their protective effects, some non-thiol •OH scavengers can produce radicals which are damaging to DNA under anoxic conditions (Nabben, 1983; Herskind & Westergaard, 1988; Held & Awad, 1991; Milligan & Ward, 1994; Ayene et al., 1994a). In the present investigation, we have adapted a previously used *in vitro* model system to simultaneously investigate the effects on radiation-induced single- and double-strand DNA breaks of various combinations of glutathione and glycerol, a widely used non-thiol scavenger, in the presence and absence of oxygen.

Methods

SV40 DNA (25 μg/ml) was irradiated in air or nitrogen, in the presence of various combinations of 0-750 mM glycerol and/or 0-20 mM GSH. The irradiation technique (Koch et al., 1984) was specifically designed to minimize oxygen depletion of aerated samples during high dose irradiation. After irradiation, we quantitated the three different forms of SV40 DNA (closed circular superhelices, nicked circles, and linear molecules) for each data point by a combination of neutral gel electrophoresis, fluorescence labeling of DNA with ethidium bromide, acquisition of high resolution video images by a Xillix Microimager CCD video camera (MI1400), digitization by a Matrox 1280 Frame-grabber Board, and computer analysis of these images by NIH Image 1.52 image analysis software. Induction efficiencies for SSBs were determined from loss of superhelices and those for DSBs from the appearance of linear forms. Induction efficiencies for both one hit SSBs and DSBs were then calculated from the linear terms of best fit quadratic polynomials to dose response curves for these data. This method weighted the lower dose data more heavily. Every induction efficiency calculated below represents the mean from a minimum of two independent experiments. Additional details of all aspects of our experimental approach are described in Ayene et al. (1994a, b).

Results and Discussion

Irradiation in glycerol

Tables 1 and 2 summarize our data for induction of one hit SSBs and DSBs respectively in SV40 DNA irradiated in the presence of glycerol and the absence of GSH. Under all of our experimental conditions, nearly all SSBs were one hit, as were nearly all DSBs except in the absence of glycerol and the presence of the lowest concentrations of GSH. One hit DSBs are believed to dominate in radiation induced damage to DNA in the presence of substantial concentrations of •OH scavengers because of the clustering of OH radicals at the site of a single ionization cluster or LMDS (Ward, 1981; Krisch et al., 1991). The major trend in the present experiments for both SSBs and DSBs is for decreasing induction efficiency with increasing glycerol concentration in either air or nitrogen, as expected from the increasingly efficient scavenging of OH radicals. The results in air are essentially superimposable on our prior published data (Krisch et al., 1991). These results, as well as comparable data from elsewhere (e.g. Milligan et al., 1994; Held et al., 1981) have generally been interpreted in terms of competition for •OH between scavenger molecules, leading to protection, and DNA molecules, leading to damage. However, this simple competition kinetics model leads to a prediction that breakage induction efficiencies for oxygenated DNA should decrease by the same ratio that the scavenger concentration increases. Our data in Tables 1 & 2, as well as most other comparable published data demonstrate instead a significantly smaller rate of decrease, consistent with a current model for non-homogeneous reaction kinetics (Udovicic et al., 1994).

The results in nitrogen for SSBs and DSBs demonstrate decreased breakage efficiencies compared to those in air at the lowest glycerol concentrations (≤ 0.15 mM), followed by a shift towards more damage in nitrogen at all higher glycerol concentrations. This shift is clearly demonstrated in the last two columns of each Table, which demonstrate a change from a negative to a positive value for the differences in induction efficiencies in nitrogen and air and a corresponding change from >1 to <1 for the OERs. These results confirm and extend those of Milligan and Ward (1994) who reported increased induction of SSBs under anoxia for a variety of non-thiol OH scavengers and interpreted their results in terms of the formation of damaging secondary scavenger radicals under anoxia. A more detailed comparison and analysis

Table 1. Induction efficiencies and oxygen enhancement ratios (OERs) for one hit single strand break (SSB) formation in SV40 DNA irradiated in aqueous solution in the presence of increasing concentrations of glycerol and in the presence or absence of oxygen.

[Glycerol] (mM)	G (SSB) (SSB Gy^{-1} Da^{-1})			OER (SSB)
	Air	N_2	N_2-O_2	O_2/N_2
0	5.44×10^{-08}	4.12×10^{-08}	-1.32×10^{-08}	1.32
0.15	3.18×10^{-08}	2.44×10^{-08}	-7.35×10^{-09}	1.3
1.5	1.47×10^{-08}	1.62×10^{-08}	1.47×10^{-09}	0.91
75	1.22×10^{-09}	1.66×10^{-09}	4.39×10^{-10}	0.73
750	1.94×10^{-10}	4.85×10^{-10}	2.92×10^{-10}	0.41
2500	9.08×10^{-11}	1.77×10^{-10}	8.66×10^{-11}	0.51

Table 2. Induction efficiencies and OERs for one hit double strand break (DSB) formation in the same experiments as in Table 1.

[Glycerol] (mM)	G (DSB) (DSB Gy^{-1} Da^{-1})			OER (DSB)
	Air	N_2	N_2-O_2	O_2/N_2
0	1.03×10^{-09}	7.67×10^{-10}	-2.60×10^{-10}	1.34
0.15	3.88×10^{-10}	2.78×10^{-10}	-1.11×10^{-10}	1.39
1.5	1.71×10^{-10}	1.70×10^{-10}	-9.32×10^{-13}	1.01
75	2.36×10^{-11}	2.94×10^{-11}	5.79×10^{-12}	0.81
750	5.43×10^{-12}	9.37×10^{-12}	3.94×10^{-12}	0.57
2500	3.32×10^{-12}	5.20×10^{-12}	1.88×10^{-12}	0.63

of our data for both SSBs and DSBs (Ayene et al., 1994a) suggests that, following anoxic irradiation in glycerol, DSBs are induced in DNA, presumably via damaging secondary radicals formed when glycerol molecules react with primary OH radicals. In addition, we find that at high glycerol concentrations these DSBs are induced via a one hit mechanism. This analysis also suggests that at glycerol concentrations ≤ 750 mM the dominant mechanism for the formation of glycerol radical induced DSBs involves a SSB induced by a glycerol radical on one DNA strand together with a SSB induced by an OH radical on the opposite strand, both arising from the same cluster (LMDS); while at the 2,500 mM glycerol concentration the dominant mechanism appears to involve two glycerol radicals on opposite strands, again both from the same cluster.

Irradiation in glutathione (GSH)

Glutathione (GSH) is both an efficient scavenger of OH radicals and, under anoxic conditions, an efficient agent of chemical repair of radiation damage. It is the dominant intracellular soluble reduced thiol compound and is generally regarded as the major intracellular radioprotector (see, e.g., Bump and Brown, 1990). Table 3 summarizes our data for induction of SSBs and DSBs respectively in SV40 DNA irradiated in the presence of GSH and the absence of glycerol. These data will be presented in more detail elsewhere (Ayene et al., 1994b). Briefly, the Table shows, for oxic irradiation, similarly increasing protective effects of increasing concentrations of GSH against either SSBs or DSBs. For anoxic irradiation, the Table shows even more rapidly increasing protection against either SSBs or DSBs with increasing GSH

Table 3. Induction efficiencies for one hit SSB and DSB formation in SV40 DNA irradiated in the presence of increasing concentration of glutathione (GSH) and in the presence or absence of oxygen.

[GSH] (mM)	G (SSB) (SSB Gy^{-1} Da^{-1})		G (DSB) (DSB Gy^{-1} Da^{-1})	
	Air	N_2	Air	N_2
0	5.44×10^{-08}	4.12×10^{-08}	1.03×10^{-09}	7.67×10^{-10}
0.5	2.52×10^{-08}	5.59×10^{-09}	2.85×10^{-10}	6.89×10^{-11}
5	3.42×10^{-09}	4.29×10^{-10}	1.17×10^{-10}	1.37×10^{-11}
20	1.21×10^{-09}	5.05×10^{-10}	4.30×10^{-11}	2.71×10^{-11}

concentration, up to 5 mM. This difference between oxic and anoxic irradiation is consistent with our present understanding of the radiation chemistry. Two mechanisms of protection apply: first, GSH can protect in both air and N_2 by scavenging •OH; second, GSH can protect by free-radical repair whereby the -SH donates a hydrogen atom to a carbon centered radical in DNA. This second mechanism is in competition with fixation of the same radicals by oxygen. Hence, more protection occurs in N_2. For the highest GSH concentration examined (20 mM), the induction efficiencies for either SSBs or DSBs actually increase compared to the corresponding values for 5 mM GSH. This suggests that in addition to its protective effects, high concentrations of GSH somehow also induce strand break damage into DNA. There is some evidence supporting this conclusion for SSBs in anoxically irradiated DNA (e.g. Rosenkrantz & Rosenkrantz, 1971; Prütz, & Mönig, 1987) but none that we know of for one hit DSBs. A plausible mechanism to account for our results, but one with only very limited supporting evidence (See discussion of work of Becker et al., next section), is that a GSH molecule reacts with •OH to form a potentially damaging carbon centered radical, but that this radical can be inactivated by oxygen, analogous to the proposed mechanism for anoxic damage from glycerol and DMSO radicals (Ayene et al., 1994a; Milligan & Ward, 1994).

Irradiation in glycerol and glutathione

Table 4 summarizes our data for induction of one hit SSBs and DSBs when the same concentrations of GSH as in Table 3 are combined with 750 mM glycerol. This concentration was chosen to give an •OH scavenging efficiency somewhat higher than that of the intracellular environment, but comparable overall radioprotection (Ayene et al 1994c). The addition of glycerol alters the results substantially from those in Table 3. Breakage induction efficiencies are now generally lower, as would be expected from the additional •OH scavenging efficiency of the glycerol. However, the relative increases in damage seen in going from 5 to 20 mM GSH, under anoxic irradiation are now much more striking, both for SSBs and DSBs. In addition, the induction efficiencies for SSBs and DSBs for oxic irradiation now show parallel increases with increasing GSH concentration, rather than the expected decreases. This latter effect suggests that GSH induces SSBs and DSBs in irradiated oxygenated DNA via an independent second mechanism that requires the presence of a non thiol •OH scavenger. Analogous experiments with GSH and 75 mM glycerol give consistent results, intermediate between those for 0 and 750 mM glycerol. All of these results will be presented in more detail elsewhere (Ayene et al., 1994c). Becker et al. (1988, 1994) have recently proposed that irradiated GSH can produce different damaging radicals in the presence or absence of oxygen, based on EPR studies of frozen DNA.

Table 4. Induction efficiencies for all one hit SSB and DSB formation in SV40 DNA irradiated in the presence of the same concentrations of GSH as in Table 3 plus 750 mM glycerol.

[GSH] (mM)	G (SSB) (SSB Gy^{-1} Da^{-1})		G (DSB) (DSB Gy^{-1} Da^{-1})	
	Air	N_2	Air	N_2
0	1.86×10^{-10}	5.60×10^{-10}	5.43×10^{-12}	9.37×10^{-12}
0.5	1.74×10^{-10}	2.60×10^{-10}	7.03×10^{-12}	3.95×10^{-12}
5	2.42×10^{-10}	8.15×10^{-11}	1.25×10^{-11}	3.72×10^{-12}
20	4.62×10^{-10}	3.74×10^{-10}	2.24×10^{-11}	2.24×10^{-11}

Table 5 presents mean OERs calculated from induction efficiencies for both SSBs and DSBs for all measured combinations of glycerol and GSH. Our observations for glutathione alone confirmed those from other laboratories for SSBs using thiol alone: The OER for either SSB or DSB peaked sharply, with maximum values of 8.0 and 8.6 respectively at 5 mM GSH. At the highest glycerol concentration, (750 mM) which roughly corresponds to total effective non-enzymatic intracellular radioprotection,

OERs increased more slowly with increasing GSH concentration to plateau values of 3.0 and 3.4 respectively at 5 mM GSH, closely mimicking intracellular observations.

Table 5. OERs for induction of one hit SSBs and DSBs in SV40 irradiated in various combined concentrations of glycerol and GSH.

[Glycerol] (mM)	[GSH] (mM)	OER (SSB)	OER (DSB)
0	0	1.32	1.34
	0.5	4.50	4.13
	5	7.97	8.57
	20	2.39	1.58
75	0	0.57	0.80
	0.5	1.67	2.23
	5	5.69	7.09
	20	2.80	1.73
750	0	0.33	0.58
	0.5	0.67	1.78
	5	2.97	3.37
	20	1.24	1.00

Conclusions

Based on the above findings for radiation induced DSBs (and SSBS), we conclude that a soluble intracellular thiol compound, glutathione, and a non-thiol •OH scavenger, glycerol, each exhibit well known protective effects and little investigated sensitizing effects towards DNA in aqueous solution. Our observations suggest that to fully account for the observed non-enzymatic modification of radiation induced cell killing and its dependence on oxygen, we need to invoke the sum total of all of these individual protective and sensitizing effects from soluble intracellular thiol compounds and non-thiol •OH scavengers, as well as poorly understood joint interactions of these compounds.

Acknowledgments

The authors gratefully acknowledge the technical assistance of Aaron A. Ambrad and Alma Gilo. This research was supported in part by NCI Grant No. CA49498 to C.J. Koch.

References

Ayene, IS, Koch, CJ, & Krisch, RE: Role of scavenger-derived radicals in the induction of double strand breaks in irradiated DNA. *Radiat. Res.* (submitted), 1994a.

Ayene, IS, Koch, CJ, & Krisch, RE: Modification by glutathione of radiation induced single and double strand breaks in SV40 DNA. *Radiat. Res.* (submitted), 1994b.

Ayene, IS, Koch, CJ, & Krisch, RE: Modification of radiation induced DNA strand breaks by the combination of glutathione and glycerol. *Radiat. Res.* (submitted), 1994c.

Becker, D, Swarts, S, Champagne, M, & Sevilla, MD: An ESR investigation of the reactions of glutathione, cysteine and penicillamine thiyl radicals: competitive formation of RSO•, R•, $RSSR^{-}$•, and RSS•. *Int. J. Radiat. Biol.* 53, 767-786, 1988.

Becker, D, Summerfield, S, Gillich, S, Wang, W, & Sevilla, MD: The influence of oxygen on the repair of direct damage to DNA by thiols in model systems. *Abstr. 42nd Annual Meeting, Radiation Research Society*, p. 157, 1994.

Bump, EA, & Brown, MB: Role of glutathione in the radiation response of mammalian cells *in vitro* and *in vivo*. *Pharmac. Ther.* **47**, 117-136, 1990.

Gray, LH, Conger, AD, Ebert, M, Hornsey, S, & Scott, OCA: The concentration of oxygen dissolved in tissues at the time of irradiation as a factor in radiotherapy. *Br. J. Radiol.* **26**, 638-648, 1953.

Held, KD, Harrop, HA, & Michael, BD: Effects of oxygen and sulphydryl-containing compounds on irradiated transforming DNA. *Int. J. Radiat. Biol.* **40**, 613-622, 1981.

Held, KD, Harrop, HA, & Michael, BD: Effects of oxygen and sulphydryl-containing compounds on irradiated transforming DNA. II. Glutathione, cysteine and cysteamine. *Int. J. Radiat. Biol.* **45**, 615-626, 1984.

Held, KD, Harrop, HA, & Michael, BD: Effects of oxygen and sulphydryl-containing compounds on irradiated transforming DNA. III. Reaction Rates. *Int. J. Radiat. Biol.* **45**, 627-636, 1984a.

Held, KD, Awad, S: Effects of polyamines and thiols on the radiation sensitivity of bacterial transforming DNA. *Int. J. Radiat. Biol.* **59**, 699-710, 1991.

Herskind, C, & Westergaard, O: Variable Protection by OH scavengers and radiation-induced inactivation of isolated transcriptionally active chromatin: the influence of secondary radicals. *Radiat. Res.* **114**, 28-41, 1988.

Howard-Flanders, P: Effect of oxygen on the radiosensitivity of bacteriophage in the presence of sulfhydryl compounds. *Nature* **186**, 485-487, 1960.

Hutchinson, F: Sulfhydryl groups and the oxygen effect on irradiated dilute solutions of enzymes and nucleic acids. *Radiat. Res.* **14**, 721-731, 1961.

Iliakis, G: The role of DNA double-strand breaks in ionizing radiation-induced killing of eukaryotic cells. *BioEssays* **13**, 641-648, 1991.

Koch, CJ. Competition between radiation protectors and radiation sensitizers in mammalian cells. In: *Radioprotectors and Anticarcinogens*. (OF Nygaard & MG Simic, eds) Academic Press, pp. 275-296, 1983.

Krisch, RE, Flick, MB, & Trumbore, CN: Radiation chemical mechanisms of single- and double-strand DNA break formation in irradiated SV40 DNA. *Radiat. Res.* **126,** 251-259, 1991.

Milligan, JR, & Ward, JF: Yield of single-strand breaks due to attack on DNA by scavenger-derived radicacals. *Radiat. Res.* **137,** 295-299, 1994.

Nabben, FJ, van der Stroom, HA, & Loman, H: Inactivation of biologically active DNA by isopropanol and formate radicals. *Int. J. Radiat. Biol.* **43,** 495-504, 1983.

Prütz, WA, & Mönig, H: On the efect of oxygen or copper (II) in radiation-induced degaradation of DNA in the presence of thiols. *Int. J. Radiat. Biol.* **52,** 677-682, 1987.

Rosenkranz, HS, & Rosenkranz, S: Degradation of DNA by cysteine. *Arch. Biochem. Biophys.* **146,** 483-487, 1971.

Udovicic, L, Mark, F, & Bothe, E: Yields of single-strand breaks in double-stranded calf thymus DNA irradiated in aqueous solution in the presence of oxygen and scavengers. *Radiat. Res.* 140, 166-171, 1994.

Ward, JF: Some biochemical consequences of the spatial distribution of ionizing radiation-produced free radicals. *Radiat. Res.* **86,** 185-195, 1981.

Chapter 12

Structure/Function Relationships at Higher Levels of Organization

HIGHER ORDER DNA STRUCTURE AND RADIATION DAMAGE

Nancy L. Oleinick, Song-mao Chiu, Liang-yan Xue,
Libby R. Friedman, and Usha Balasubramaniam
Division of Radiation Biology, Department of Radiology
Case Western Reserve University School of Medicine
10900 Euclid Avenue
Cleveland, OH 44106-4942

Introduction

In the nucleus of eukaryotic cells, DNA is complexed with histones and a large variety of non-histone proteins to form chromatin (Pienta and Coffey, 1984). The first level of chromatin structure is the nucleosomes, the basic repeat units of chromatin, formed by the wrapping of the double-helical DNA strand around octamers of core histones (two dimers of histones H2A and H2B and a tetramer of histones H3 and H4) to form nucleosomal core particles. The fifth histone, H1, helps to anchor DNA at the entry and exit to the core, interacts with DNA in the linker region, and is believed to be essential to form higher order chromatin structure (Felsenfeld, 1992). The nucleosomes are further coiled into a 30-nm fiber, which is composed of solenoids containing six nucleosomes per turn, and other higher order structures. Thus, the DNA of highly condensed metaphase chromosomes is about 10,000-fold compacted with respect to protein-free DNA.

Within the interphase nucleus, the chromatin is not evenly packed. The transcriptionally active or potentially active domains, for example, have more relaxed chromatin structures (Felsenfeld, 1992). Eukaryotic chromatin is also attached at intervals of 30-100 kbp to a salt-resistant fibrous protein network called the nuclear matrix (Berezney, 1991). The periodic anchorage of chromosomes to the nuclear matrix creates series of supercoiled chromatin loops which function as independent topological domains; this structure has been suggested to be important for the maintaince of torsional stress as well as independent transcriptional activation (Bonifer *et al.*, 1991; Roti Roti *et al.*, 1993). In addition, many important cellular functions, such as DNA replication and RNA transcription, processing, and splicing, have been demonstrated to take place at the nuclear matrix (reviewed by Berezney, 1991).

The heterogenous chromatin environment has a marked influence on the distribution of DNA damage within the genome of cells exposed to ionizing radiation. For example, DNA sequences that are transcriptionally active sustain more DNA single-strand breaks (SSB) (Chiu *et al.*, 1982, Bunch *et al.*, 1992) and more DNA-protein crosslinks (DPC) (Chiu *et al.*, 1986) than transcriptionally inactive regions. Newly replicated DNA prior to maturation of the chromatin structure is also more susceptible to radiation-induced base damage (Warters and Childers, 1982).

Copper ion has been shown to be important for stabilization of the nuclear matrix structure and folding of the attached DNA (Lebkowski and Laemmli, 1982; Dijkwell and Wennink, 1986), presumbly by stabilizing the attachment of DNA loops to topoisomerase II, a major component of the nuclear matrix thought to be located at the base of DNA loops (Earnshaw and Heck, 1985; Earnshaw *et al.*, 1985; Berrios, 1987). We have observed that treatment of nuclei with a low level of copper ion results in an increase in the amount of matrix-associated DNA tightly bound to matrix proteins and also enhances the production of DPC upon subsequent γ-irradiation (Chiu *et al.*, 1992, 1993). Since the DPC production is sensitive to OH radical scavengers, it was proposed that DPC induction in copper ion-treated nuclei occurs preferentially at the sites of Cu^{++} binding to chromatin wherein OH radicals are produced repeatedly through a copper-catalyzed Fenton reaction. A role for bound copper ion in the determination of cellular radiosensitivity through the production of damage to DNA quaternary structure has been proposed (George *et al.*, 1987) The production of radiation-induced DNA damage, therefore, is influenced by the chromatin environment, features of which include association of DNA with chromosomal proteins, attachment to the nuclear matrix, the site-specific binding of metal ions, and the compaction of chromatin structure.

To elucidate the contribution of the chromatin environment to radiation-induced DNA damage production, several laboratories have artificially manipulated the chromatin to mimic open *vs.* condensed regions of native chromatin and then subsequently determined their susceptibility to DNA damage (reviewed in Oleinick and Chiu, 1992, 1994; Oleinick *et al.*, 1994). For example, when isolated nuclei were treated with buffers of decreasing ionic strength, the higher order (solenoid) structure was expanded, but the nucleosomes remained intact with their full complement of histones, and the chromatin remained attached to the nuclear matrix (Chiu *et al.*, 1992). Irradiation of the expanded chromatin resulted in 4-13 fold increases in the yield of SSB, double-strand breaks (DSB), or DPC as compared to condensed chromatin or whole cells (see references quoted in Oleinick *et al.*, 1994). In the presence of 0.1-5.0 mM $MgCl_2$ recondensation of expanded chromatin occurs and is accompanied by the restoration of radioresistance (Chiu *et al.*, 1992; Warters and Lyons, 1992). A similar radiosensitization was observed in butyrate-treated cells whose chromatin was expanded by hyperacetylation of histones (Nackerdien *et al.*, 1989). Interestingly, the butyrate-treated cells were also more sensitive to the lethal effects of ionizing radiation.

In contrast to the response of chromatin to hypotonic conditions, hypertonic buffers strip different classes of chromatin proteins from the DNA; *i.e.*, 0.35 M NaCl extracts some non-histone proteins, 0.6 M NaCl extracts histone H1 as well, 1.2 M NaCl removes core histones H2A and H2B, and 2.0 M NaCl strips all remaining histones (H3 and H4) from the DNA. Nuclear matrix proteins are not lost by these procedures, and DNA loops remain anchored to the nuclear matrix. For study of the radiation response of such protein-depleted nuclei, Elia and Bradley (1992) introduced the technique of embedding cells in agarose plugs for nuclear preparation, hypertonic extraction, and irradiation. This procedure has permitted manipulation of the protein-stripped chromatin without clumping of the DNA followed by direct analysis of radiation-induced DSB by pulsed field gel electrophoresis (PFGE). Protein-depleted chromatin was markedly radiosensitized relative to native chromatin with respect to DSB induction (Elia and Bradley, 1992; Xue *et al.*, 1994), and the radiosensitization, although reduced in magnitude, was still apparent even in the presence of 0.1 M DMSO as OH radical scavenger (Oleinick *et al.*, 1994). The results suggested that protein-deficient chromatin is radiosensitive under cellular scavenging conditions.

Work until now has implicated chromatin structure and/or chromatin proteins as both radioprotectors of cellular DNA for DSB induction and substrates for DPC production. In the present study, we have attempted further to differentiate between the possible roles of chromatin proteins by reconstitution of chromatin. One or more of the above-mentioned hypertonic salt extracts or commercial histones were added back to dehistonized nuclei, following which the preparations were irradiated and the yields of DPC and DSB were measured.

Materials and Methods

All of the procedures for growth of V79 cells, labeling of the cells with ^{3}H-thymidine, embedding the cells in agarose plugs, preparation of nuclei, extraction of proteins by treatment of the plugs with hypertonic buffers, irradiation of the plugs, and assay for DSB and DPC have been described in detail (Xue *et al.*, 1994). Nuclear extracts were prepared from plugs containing 1-2 x 10^6 cells each by extraction with buffer A (10 mM Tris-HCl, 1 mM $MgCl_2$, pH 7.4, 0.5 mM phenylmethylsulfonyl fluoride, 2 mM mercaptoethanol) containing various concentrations of NaCl. When the complete range of extractable nuclear proteins was desired, the plugs were extracted with 2.0 M NaCl; when primarily core histones were desired, the plugs were first extracted with 1.2 M NaCl, the extract was discarded, and the plugs were then re-extracted with 2.0 M NaCl. For reconstitution of chromatin, plugs containing 5-10 x 10^5 cells each were used, and dehistonized chromatin was prepared by extraction of agarose-embedded nuclei with buffer A containing 2.0 M NaCl. To the plugs containing dehistonized chromatin were added 5 volumes of one or more protein preparations in dialysis casing. The plugs were subjected to stepwise dialysis at 4°C against Buffer A containing decreasing NaCl from 2.0 to 0.14 M over a period of 30 hours. Following dialysis, the plugs were washed three times with phosphate-buffered saline (PBS), the agarose was melted, and the contents were assayed as previously described (Xue *et al.*, 1994).

Results

Although many methods have been used to remove different classes of proteins from chromatin, the method reported by Elia and Bradley (1992) was adopted by our laboratory because of the ease in manipulation of chromatin and the reproducibility of the results. Cells are embedded in agarose plugs, and all of the procedures for lysis and extraction are carried out in the plugs. By monitoring the residual proteins remaining in the chromatin after extraction, using polyacrylamide gel electrophoresis, we have confirmed that many non-histone proteins (NHP), histone H1, histones H2A and H2B, and histones H3 and H4 can be progessively removed from chromatin by increasing NaCl concentrations (Elia and Bradley, 1992; Xue *et al.*, 1994).

Extraction with salt concentrations ≤ 0.6 M to remove NHP and H1 resulted in only a small increase in the yield of radiation-induced DSB, whereas removal of core histones from chromatin with higher salt concentrations enhanced DSB production by radiation by about 40-fold. Deproteinization with a protease further increased the yield of DSB to about 50-70-fold greater than condensed chromatin. Data on SSB show a similar magnititude of enhancement in the yield after removal of these classes of chromatin proteins and have been summarized in our previous reviews (Oleinick and Chiu, 1994; Oleinick *et al.*, 1994). However, it is striking that the yield of DPC was not much affected by salt extraction below 1.2 M NaCl, although NHP, H1, and the majority of core histones were removed. Further increases in NaCl concentration resulted in a drastic reduction in the formation of DPC, and at 2 M NaCl, when the chromatin was fully depleted of histones, the induction of DPC was blocked (Xue *et al.*, 1994).

Dehistonized chromatin was used as the starting point for chromatin reconstitution, because the DNA is minimally protected by chromatin proteins; *i.e.*, its radiosensitivity is only slightly less than that of fully deproteinized DNA with respect to DSB production. Furthermore, the maximum amount of DPC can be induced in chromatin extracted with 1.2 M NaCl, but radiation-induced DPC formation is inhibited after 2.0 M NaCl extraction. Therefore, it was reasoned that the deficiency in DPC induction in the latter case is due to the extraction of DPC substrates, and that addition of a salt extract to the dehistonized chromatin would restore the capacity for DPC formation, if reconstitution of chromatin can be achieved. Likewise, reconstitution might restore the relative radioresistance of DNA from DSB induction. Reconstitution of chromatin was carried out by stepwise dialysis of a mixture of dehistonized DNA (embedded in agarose plugs) and histones against salt solutions of decreasing concentration from 2.0 M to isotonic condition. Following dialysis, the chromatin within the plugs was analyzed for the content of histones and the presence of nucleosomes. Finally, the plugs were irradiated and the DNA assayed for damaged products.

Electrophoretic analysis of reconstituted chromatins reveals that they consist of core histones of similar quantity and composition as that from native chromatin, whether they were reconstituted from a salt extract or from commercial histones (Fig. 1). As expected, the chromatin reconstituted from the salt extract was deficient in histone H1. However, when commercial histone H1 was added to the reconstitution mixture, elevated levels of this histone were recovered from the reconstituted chromatin (not shown). Therefore, it was possible to restore the overall content of histones to the DNA by reconstitution in the agarose plugs. We next asked whether the nucleosomal structure of the reconstituted chromatin was normal.

To probe the structural organization of the reconstituted chromatin, the chromatin was digested with micrococcal nuclease, which cleaves preferentially in the linker regions between nucleosomal cores. As shown in Fig. 2, digestion of native chromatin produces the characteristic "ladder" of DNA fragments which results from the positioning of nucleosomes at constant intervals of ~190 bp. Digestion of reconstituted chromatin proceeds to the same size limit (mononucleosome size) as does native chromatin, and the reconstituted chromatin appears resistant to further degradation, suggesting that DNA has been reconstituted to form nucleosomes. In contrast, the higher molecular weight fragments produced by digestion of the reconstituted chromatin migrated in the agarose gel as a smear of DNA, and no clear bands could be discerned. Such a result is indicative of irregular spacing of nucleosomal cores along the DNA chain.

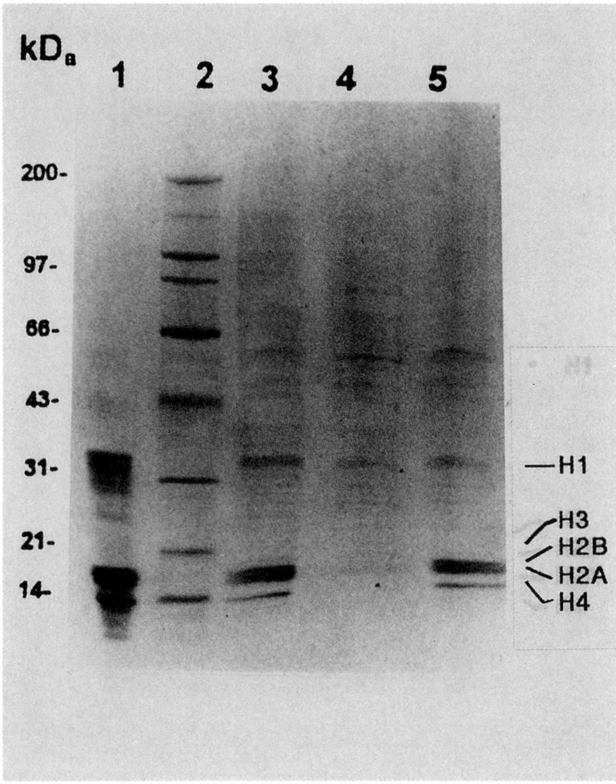

Figure 1. Polyacrylamide gel electrophoretic analysis of proteins from native, dehistonized, and reconstituted chromatin. V79 cells embedded in agarose plugs were lysed and extracted with Buffer A containing 0.14 M (lane 3) or 2.0 M (lane 4) NaCl, as described (Elia and Bradley, 1992; Xue *et al.*, 1994), producing native (lane 3) or dehistonized (lane 4) chromatin, respectively. For reconstitution of chromatin (lane 5), plugs containing dehistonized chromatin were placed in dialysis casing with 2.0 M salt extract (60 μg protein/ml) and dialyzed against Buffer A containing decreasing NaCl concentrations, as described in Methods. At the end of the dialysis, the plugs were washed at least three times in PBS to remove unbound proteins. The plugs were melted, and the proteins were solubilized and loaded onto the gel. Lane 1 contains histone standards (mixture of V-S and II-S, Sigma), and lane 2 contains protein molecular weight markers (Bio-Rad). Electrophoresis was performed on an 11.5% acrylamide minigel (Bio-Rad), and proteins were stained with Coomassie blue.

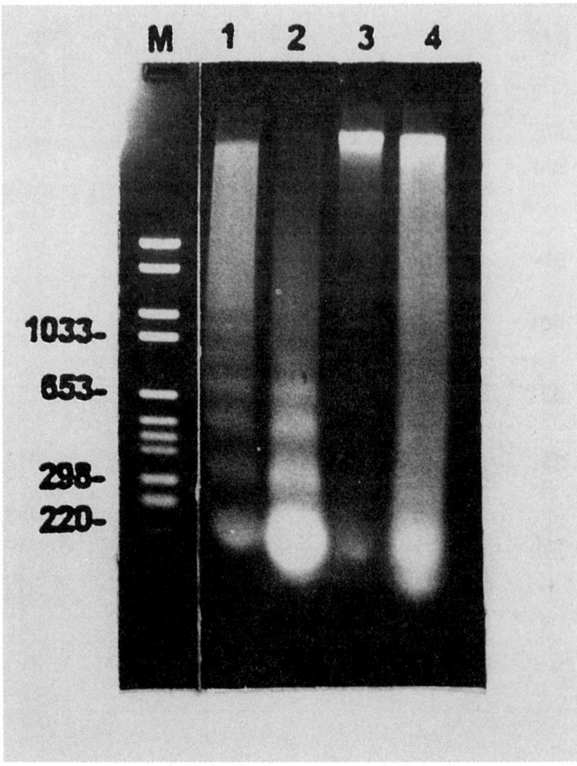

Figure 2. Gel electrophoretic analysis of DNA fragments generated by micrococcal nuclease digestion of native and reconstituted chromatin. Native (lanes 1 and 2) or reconstituted (lanes 3 and 4) chromatins were prepared as described in Fig. 1 and digested with micrococcal nuclease (1 unit/ml) in Buffer A containing 0.14 M NaCl and 2 mM $CaCl_2$ for 2 to 40 min at 34°C. The reaction was stopped by transferring the plugs to lysis buffer (1 X SSC, 20 mM EDTA, 0.5 mg/ml proteinase K) and incubating at 50°C overnight. The digested DNA was concentrated in a Centricon microconcentrator (Amicon) and analyzed on a 1.5 % agarose gel. The gel was stained with ethidium bromide and photographed. Lanes 1 and 3, samples from a 2- and 5-min digestion, respectively; lanes 2 and 4, 10- and 40-min digestions, respectively; lane M, molecular weight marker VI (Boehringer Mannheim).

As is apparent from the results in Fig. 3, full restoration of DPC formation was obtained in the reconsitituted chromatin. Although the salt extract (proteins extracted between 1.2 and 2.0 M NaCl) contains non-histone proteins as well as core histones, an important role for histones in the DPC response was confirmed by the experiment in which commercial histones were used for reconstitution. DPC induction in the reconstituted chromatin was as efficient as in native chromatin or in chromatin reconstituted using the salt extract. As a control for possible effects of non-DNA-associated protein, a reconstitution experiment was also carried with bovine serum albumin (BSA) instead of histones. In this case, no induction of DPC was found when the dialyzed material was irradiated. This observation indicates that DPC induction requires a close association between DNA and protein.

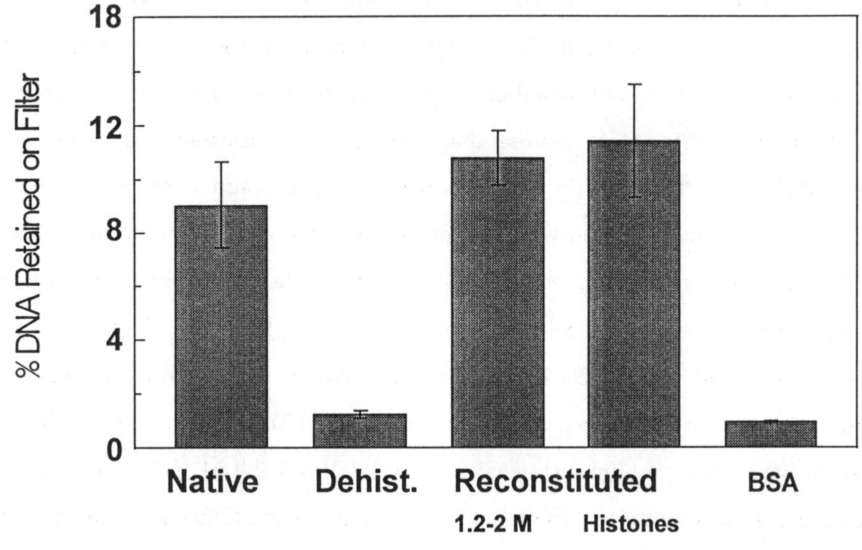

Figure 3. DPC induction in V79 native, dehistonized, and reconstituted chromatin. Chromatin was prepared as described in Fig. 1. Reconstituted chromatin was obtained using salt extract (1.2-2.0 M), histones (100 μg/ml), or BSA (100 μg/ml) as protein sources. The plugs were irradiated with 100 Gy, and DPC were assayed by a modified filter binding technique (Xue *et al.,* 1994). DPC from unirradiated samples were less than 1.5%.

The ability of histones to restore the capacity for DPC induction may appear to contradict our previous observations that DPC induction occurs preferentially at the sites of attachment of DNA loops to the nuclear matrix (Chiu *et al.*, 1986). However, it should be noted that the protein substrates for radiation-induced DPC may differ for native and expanded chromatin. It has been suggested that histones become crosslinked to DNA when expanded chromatin is irradiated (Mee and Adelstein, 1981; Chiu *et al.*, 1992), whereas in cells, the highly condensed state of the majority of the chromatin can inhibit DPC formation except at matrix attachment sites, which exist in a more open or accessible conformation. Alternatively, reconstitution of chromatin may generate a structure favoring DPC formation with histone or non-histone proteins.

The yield of DNA DSB in dehistonized chromatin is about 40-fold higher than in condensed chromatin (Xue *et al.*, 1994), and core histones appear to be responsible for the majority of the radioprotection of native condensed chromatin. Therefore, we next asked whether or not the restoration of the nucleosomal structure was sufficient to restore the relatively radioresistant status of native chromatin. The radiosensitivity of reconstituted chromatin with regard to DSB induction has been compared with that of dehistonized and native chromatins (Fig. 4). DNA DSB were determined by PFGE, and the data are expressed as FAR, the fraction of radioactive DNA released into the gel from the plug. Native chromatin was highly resistant to 5 Gy of γ-radiation (FAR < 0.05), while dehistonized chromatin was much more sensitive (FAR $= 0.65$). When chromatin was reconstituted with proteins extracted between 1.2 and 2.0 M NaCl, only a small decrease in radiosensitivity was observed over that of dehistonized chromatin. The addition of excess histone H1 or the use of the whole range of extracted nuclear proteins for reconstitution provided some additional protection, but these resultant chromatins were still considerably more radiosensitive than native chromatin, despite their similar composition of bound histones as native chromatin, indicating that simple association of histones with DNA is not sufficient to protect DNA to the same extent as native chromatin structure. When a five-fold excess of commercial histones was employed during the reconstitution, it was possible almost completely to restore the radioresistance of the DNA (compare Reconstituted Chromatins 4 and 5 of Fig. 4). However, the structure of Reconstituted Chromatin 5 was also abnormal, because in spite of the supranormal level of histones bound to DNA, it lacked ordered nucleosomes and was highly resistant to nuclease digestion (data not shown).

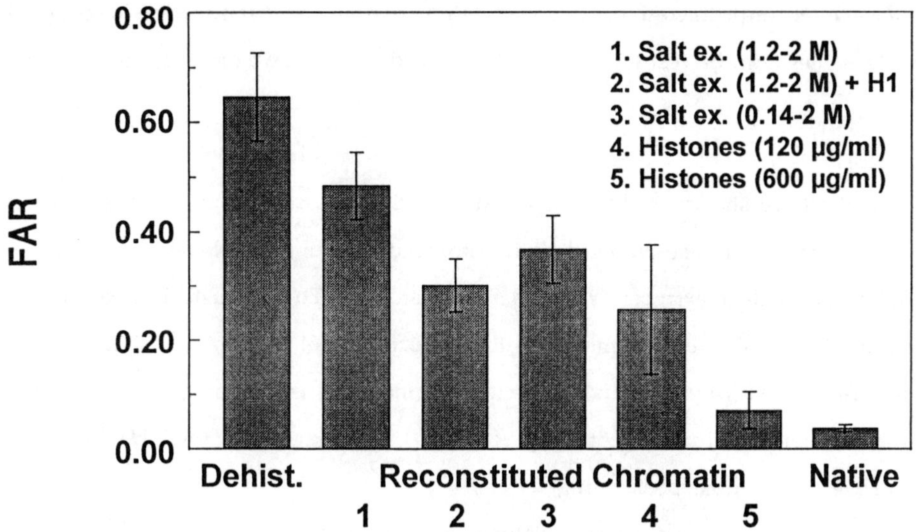

Figure 4. The yield of DNA DSB in irradiated V79 native, dehistonized, and reconstituted chromatin. Reconstituted chromatins were prepared from either 1.2-2.0 M salt extract (#1), the same extract supplemented with histone H1 (20 µg/ml, #2), whole range salt extract (0.14-2.0 M, #3), or histones (120 µg/ml, #4; 600 µg/ml, #5). The plugs were irradiated with 5 Gy, except for native chromatin, for which FAR was estimated from samples receiving 10-40 Gy, and DNA was analyzed by PFGE and FAR determined as described (Xue *et al.*, 1994).

An important role of chromatin compaction in radioprotection against DSB formation has been demonstrated previously, since relaxation of condensed chromatin in low ionic strength medium results in a ~5-fold increase in the yield of DSB (Warters and Lyons, 1992). However, the expanded DNA is still about 8-fold more resistant than fully deproteinized DNA, confirmng the importance of the packing of DNA into nucleosomes in radioprotection. A full content of nucleosomes probably is necessary to minimize the presence of unprotected DNA regions in chromatin. This notion is supported by the observations that a DNA-binding protein protects DNA from radiation damage at the site to which the protein is bound (Franchet-Beuzit *et al.*, 1993; Isabelle *et al.*, 1993). However, our data also implicate the ordered spacing of nucleosomes along the DNA as an additional condition required for radioprotection. The higher radiosensitivity of reconstituted chromatin than native chromatin may thus be due either to the presence of unprotected or exposed DNA regions or to improperly ordered nucleosomes and the resultant lack of higher order chromatin organization or both.

Conclusions

We have shown in this study that the original core histone composition can be restored to agarose-embedded dehistonized nuclei by dialysis with histone-containing nuclear extracts or commercial histones. The reconstituted chromatin appears to contain nucleosomes with the correct amount of DNA (~180-190 bp per nucleosome). However, the positioning of nucleosomes is irregular, suggesting that higher order structure is not fully restored. Addition of histone H1 does not improve nucleosome positioning.

Restoration of the original histone composition restores radiosensitivity with respect to DPC formation. Either histones become crosslinked under these conditions or they reconstitute a structure suitable for DPC formation. In contrast, restoration of the original histone composition does not restore normal radioprotection by histones against DSB formation. The greater yield of DSB in reconstituted chromatin compared to native chromatin implies the existence of unprotected or more open regions in the former. Furthermore, radioprotection by histones may be due largely to their roles in chromatin structural organization rather than to simple radical scavenging.

Acknowledgements

Research in the authors' laboratory has been supported by Research Grant R37-CA15378 and by Cancer Center Support Grant P30-CA43703 from the U.S. National Cancer Institute, DHHS.

References

Berezney R (1991) The nuclear matrix: A heuristic model for investigating genomic organization and function in the cell nucleus. J Cell Biochem 47:109-123.

Berrios M, Oshuroff N, Fisher PA (1987) In situ localization of DNA topoisomerase II, a major polypeptide component of the Drosophila nuclear matrix fraction. Proc Natl Acad Sci USA 82:4142-4146.

Bonifer C, Hecht A, Sauerssig H, Winter DM, Sippel AE (1991) Dynamic chromatin: the regulatory domain organization of eukaryotic gene loci. J Cell Biochem 47:99-108.

Bunch RT, Gewirtz DA, Povirk LR (1992) A combined alkaline unwinding /Southern blotting assay for measuring low levels of cellular DNA breakage within specific genomic regions. Oncol Res 3:7-15.

Chiu SM, Oleinick NL, Friedman LR, Stambrook PJ (1982) Hypersensitivity of DNA in transcriptionally active chromatin to ionizing radiation. Biochim Biophys Acta 699:15-21.

Chiu SM, Friedman LR, Sokany NM, Xue LY, Oleinick NL (1986) Nuclear matrix proteins are crosslinked to transcriptionally active gene sequences by ionizing radiation. Radiat Res 107:24-38.

Chiu SM, Xue LY, Friedman LR, Oleinick NL (1992) Chromatin compaction and the efficiency of formation of DNA-protein crosslinks in γ-irradiated mammalian cells. Radiat Res 129:184-191.

Chiu SM, Xue LY, Friedman LR, Oleinick NL (1993) Copper ion-mediated sensitization of nuclear matrix attachment sites to ionizing radiation. Biochemistry 32:6214-6219.

Dijkwel PA, Wenink PW (1986) Structural integrity of the nuclear matrix: differential effects of thiol agents and metal chelators. J Cell Sci 84:53-67.

Earnshaw WC, Halligan G, Cooke CA, Heck MMS, Liu LF (1985) Topisomerase II is a structural component of mitotic chromosome scaffolds. J Cell Biol 100:1706-1715.

Earnshaw WC, Heck MMS (1985) Localization of topoisomerase II in mitotic chromosomes. J Cell Biol 100:1716-1725.

Elia MC, Bradley MO (1992) Influence of chromatin structure on the induction of DNA double strand breaks by ionizing radiation. Cancer Res 52:1580-1586.

Felsenfeld G (1992) Chromatin as an essential part of the transcriptional mechanism. Nature 355:219-224.

Franchet-Beuzit J, Spotheim-Maurizot M, Sabattier R, Blazy-Baudras B, Charlier M (1993) Radiolytic footprinting. β Rays, γ photons, and fast neutrons probe DNA-protein interactions. Biochemistry 32:2104-2110.

George AM, Sabovljiev SA, Hart LE, Cramp WA, Harris G, Hornsey S (1987) DNA quaternary structure in the radiation sensitivity of human lymphocytes - a proposed role of copper. Brit J Cancer 55 Suppl VIII:141-144.

Heussen C, Nackerdien Z, Smit BJ, Bohm L (1987) Irradiation damage in chromatin isolated from V-79 Chinese hamster lung fibroblasts. Radiat Res 110: 84-94.

Isabelle V, Franchet-Beuzit J, Sabattier R, Laine B, Spotheim-Maurizot M, Charlier M (1993) Radioprotection of DNA by a DNA-binding protein: MC1 chromosomal protein from the archaebacterium Methanosarcina sp. CHTI55. Int J Radiat Biol 63:749-758.

Lebkowski JS, Laemmli UK (1982) Evidence for two levels of DNA folding in histone-depleted HeLa interphase nuclei. J Mol Biol 156:309-324.

Mee LK, Adelstein SJ (1981) Predominance of core histones in formation of DNA-protein crosslinks in -irradiated chromatin. Proc Natl Acad Sci USA 78:2194-2198.

Nackerdien Z, Michie J, Bohm L (1989) Chromatin decondensed by acetylation shows an elevated radiation response. Radiat Res 117:234-244.

Oleinick NL, Chiu SM (1992) Chromatin and nuclear structural factors which modify the radiosensitivity of DNA. In Radiation Research: A Twentieth Century Perspective. Vol II (ed WC Dewey, M Edington, RJM Fry, EJ Hall, GF Whitmore) Academic Press San Diego pp 149-154.
Oleinick NL, Chiu SM (1994) Nuclear and chromatin structures and their influence on the radiosensitivity of DNA. Radiat Protect Dosimetry 52:353-358.
Oleinick NL, Balasubramaniam U, Xue LY, Chiu SM (1994) Nuclear structure and the microdistribution of radiation damage in DNA. Int J Radiat Biol (In Press).
Pienta KJ, Coffey DS (1984) A structural analysis of the role of the nuclear matrix and DNA loops in the organization of the nucleus and chromosome. J Cell Sci Suppl 1:123-135.
Roti Roti JL, Wright WD, Taylor YC (1993) DNA loop structure and radiation response. Adv Radiat Biol 17:227-259.
Warters RL, Childers TJ (1982) Radiation-induced thymine damage in replicating chromatin. Radiat Res 90:564-574.
Warters RL, Lyons BW (1992) Variation in radiation-induced formation of DNA double-strand breaks as a function of chromatin structure. Radiat Res 130:309-318.
Xue LY, Friedman LR, Oleinick NL, Chiu SM (1994) The induction of DNA damage in γ-irradiated nuclei stripped of nuclear protein classes: Differential modulation of double-strand break and DNA-protein crosslink formation. Int J Radiat Biol 66:11-21.

POSSIBLE ROLE(S) OF NUCLEAR MATRIX AND DNA LOOP ORGANIZATION IN FIXATION OR REPAIR OF DNA DOUBLE-STRAND BREAKS

R.S. Malyapa, W.D. Wright and J.L. Roti Roti
Washington University School of Medicine
St. Louis, MO 63108 USA

Introduction

DNA double-strand breaks produced by ionizing radiation are considered to be a critical radiation-induced lesion responsible, in part, for cell killing (1). However, the manner in which structures within the nucleus involving DNA organization contribute to the balance between fixation or repair of these critical lesions remains largely obscure. The repair process requires both functional enzymes and substrate availability, i.e., access to and orientation of damage sites (2). Therefore, the ability to repair damaged DNA could be influenced not only by DNA integrity but also by the spatial organization of DNA. One aspect of spatial organization is the arrangement of eukaryotic nuclear DNA into loop domains (3) which are anchored to the nuclear matrix (4). There are periodic associations of the chromatin structure with the underlying matrix through nucleoprotein interactions called matrix associated regions (MARs). MARs allow the maintenance of contiguous DNA regions with different superhelical densities which, in turn, have been implicated in the spatial control of DNA replication, control of gene transcription and DNA damage repair (5). Changes in the DNA-matrix associations should alter the characteristic isolation between contiguous loop domains and could affect the stability of DNA in the presence of radiation-induced DNA damage, leading to inefficient repair.

Therefore, we investigated the possibility that radiation-induced DNA damage differentially affects DNA supercoiling ability in cells of differing radiosensitivities using radioresistant and radiosensitive mutants of different origins. This study was also designed to determine if differences in the composition of the nuclear matrix exist between cell lines of each origin. Results from these studies indicate that differences in the composition of the nuclear matrix proteins and DNA stability might be related to intrinsic radiation resistance.

Materials and Methods

Two groups of cell lines were used as model systems in this study: Chinese hamster fibroblast cell lines: CHO.K1, 4364, XR-122 (all 3 radioresistant), XR-1 and xrs-5 (both radiosensitive), and murine lymphoma cell lines: L5178Y/R, /SR, /S35 M10(neo 5)-1 (all 4 radioresistant); L5178Y/S, M10, and LX830 (all 3 radiosensitive). The origins and characteristics of these cell lines are summarized in Table 1. In addition, we have studied oncogene transformed rat embryonal cells (REC) in ascending order of radioresistance--primary < myc transformed < ras transformed < myc+ras transformed. However, space and time does not permit discussion of the results with REC lines. Radiosensitive hamster fibroblasts (6) and murine lymphoid cells (7, 8, 9) were studied because they are thought to be defective in DSB repair whereas the REC lines were studied as a model system for oncogene mediated radioresistance.

DSB REPAIR DEFECTIVE VARIANTS CONFERRED by VARIOUS PATHWAYS		
CELL TYPES (R vs. S)	ORIGIN	LESION
Human Fibroblasts:		
Normal vs. AT	Spontaneous in human population	Misrepair, Damage Fixation
Murine Lymphoma:		
L5178Y/R vs. L5178Y/S	Spontaneous in cell culture	Deficient in DSB repair
L5178Y/S35 /SR vs. /S	Selected after long-term low dose irradiation	DSB repair restored
L5178Y/R vs. M10	Mutagenized (MMS) and selected	Deficient in DSB repair
M10 (neo 5)-1 vs. M10	Human/mouse hybrid containing human chromosome #5	? DSB repair restored
L5178/R vs. LX830	Selected after x-irradiation	Deficient in DSB repair
CHO Fibroblasts:		
4364 vs. XR-1	Mutagenized (EMS) and selected	Deficient in DSB repair
XR-122 vs. XR-1	Human/Hamster hybrid containing human chromosome #5	DSB repair restored
CHO.K1 vs. xrs-5	Mutagenized (EMS) and selected	Deficient in DSB repair

Table I

The differential inhibition of DNA supercoiling between radioresistant and radiosensitive cell lines following radiation was measured in nucleoids using the fluorescent halo assay (FHA) (10). Nucleoid protein content and protein differences within each group of cell lines were analyzed by high resolution 2D-PAGE as described previously (11).

Results and Discussion

To determine whether radiation-induced DNA damage differentially affects DNA stability, we compared the ability of DNA to undergo supercoiling changes in the presence of varying concentrations of the DNA-intercalating dye propidium iodide (PI) in radiosensitive cells relative to radioresistant cells with and without radiation. The FHA was used to measure changes in DNA supercoiling ability (12). The relaxation of endogenous negatively coiled DNA loop domains was observed as an increase in the nucleoid diameter from 18 μm to 40 μm in the presence of 0.5 to 7.5 μg/ml of PI. The rewinding of these loops in the opposite sense to form positively supercoiled domains was indicated as a decrease in the nucleoid diameter from 40 μm to 24 μm at higher PI concentrations from 7.5 μg/ml to 50 μg/ml. Prior to irradiation, the unwinding and rewinding responses of DNA supercoils was the same for both radioresistant and radiosensitive cell lines. If DNA contains radiation induced single-strand DNA breaks, then rewinding is inhibited. Loops containing DNA strand breaks are impaired in their ability to rewind due to free rotation of the DNA double helix at the site of a single-strand break around the single sugar-phosphate backbone.

When cells were irradiated and assayed for their ability to undergo supercoiling changes in the presence of PI, the ability to rewind DNA was inhibited in a dose-dependent manner. To quantify the inhibition of DNA loop rewinding, an arbitrary parameter called "excess halo diameter" was computed for each of the radiation doses for each cell line by adding the halo diameters measured at 15, 20, 35 and 50 μg/ml PI for the nucleoids from the control and subtracting this value from the equivalent sum from the irradiated cells. The excess halo diameters as a function of dose is shown in Figs. 1 and 2 for the hamster fibroblasts and murine

CHO FIBROBLASTS

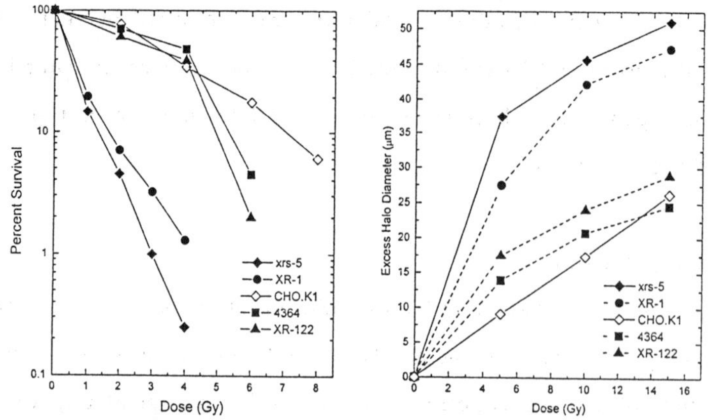

Fig. 1 (above) and Fig. 2 (below): The left side panels of both figures show clonogenic survival after γ-irradiation. The right side panels of both figures show the radiation-induced inhibition of DNA supercoil rewinding in cells of differing radiosensitivities. Radiation dose response curves were generated for each cell line and the excess halo diameter was calculated by summing the halo diameters at PI concentrations of 15 μg/ml and above for each radiation dose and subtracting the values for unirradiated cells. The resulting difference is then plotted against the radiation dose. The plotted points represent the mean of 3-5 individual experiments.

MURINE LYMPHOID CELLS

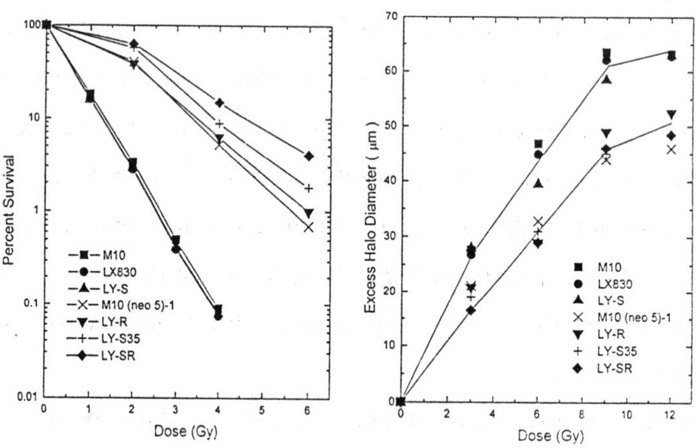

lymphoid cells, respectively. It can be seen that the inhibition of DNA supercoil rewinding was greater in the nucleoids from the radiosensitive cell lines as compared to their radioresistant counterparts, as seen by the greater excess halo diameters of the radiosensitive cell lines (Figs. 1 and 2). Since the amount of DNA damage is the same per unit DNA dose for the radiosensitive and radioresistant cell lines, the differences must be due to differences in DNA anchoring. Note that the cell lines which reacquired radioresistance due to the introduction of human chromosome, 5, XR-122, and M10(neo 5)-1 showed the same loop rewinding behavior as the radioresistant parental lines, 4364 and L5178Y/R.

DNA loops are anchored to the nuclear matrix by protein-DNA anchor points. It is reasonable, therefore, to expect that there would be differences in the protein composition of these anchoring regions. To confirm this expectation, the nuclear matrix proteins from radioresistant and radiosensitive cell lines in each group (hamster fibroblasts and murine lymphoid cells), were resolved by high-resolution 2D-PAGE. The protein gels were compared for consistent differences in polypeptide composition and abundance between the radioresistant and radiosensitive cell lines within each group.

In the hamster fibroblasts group, 12 proteins were absent and 1 in a lesser amount in the radiosensitive XR-1 in comparison to the radioresistant 4364. Of these 13, 7 proteins absent in XR-1 were also absent in xrs-5 cells in comparison to the radioresistant CHO.K1, in which all 7 were present similar to that of the 4364 cell line. A representative gel of CHO.K1 and xrs-5 is illustrated in Fig. 3. When the

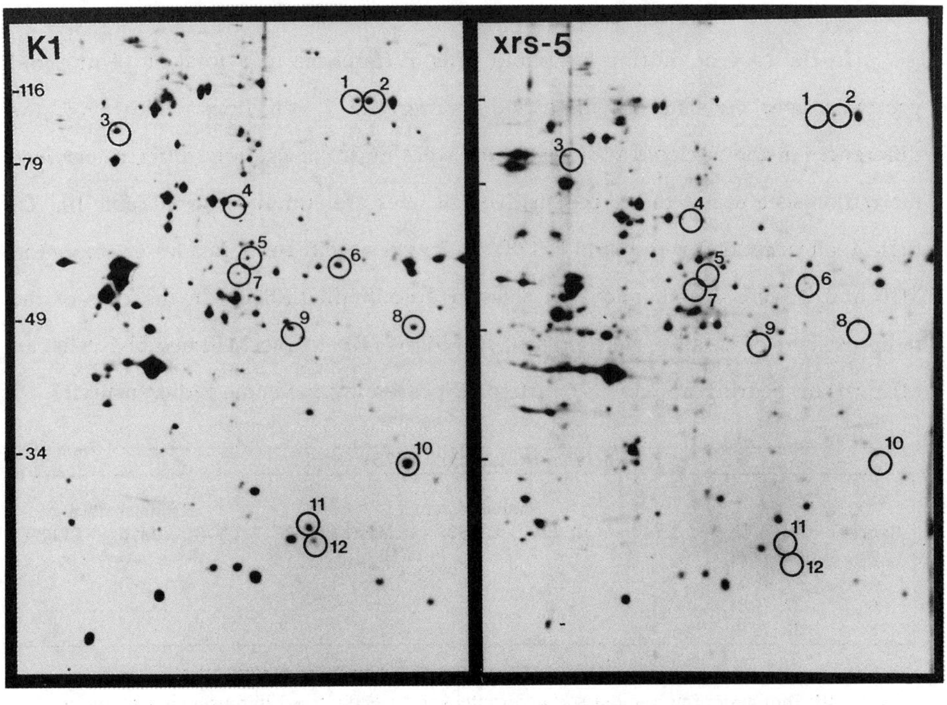

Fig. 3: Analysis of nucleoid proteins obtains from CHO.K1 (radioresistant) and xrs-5 (radiosensitive) cells by 2D-PAGE. Protein spots of differing abundance are identified by numbered circles.

nucleoid proteins of XR-122, a radioresistant variant of XR-1 bearing human chromosome 5 identical to that of 4364, were analyzed, 5 of the missing 12 proteins reappeared (7). The study on hamster fibroblasts was extended to the radioresistant CHO.K1 and the radiosensitive xrs-5. The proteins of interest are numbered and characterized by approximate isoelectric point and molecular weight (Table II).

Protein Variations in CHO Nucleoids					
		Radioresistant		Radiosensitive	
Spot#	Mr(pI)	4364	XR-122	XR-1	xrs-5
1	105(6.7)	+	+	-	-
2	105(6.8)	+	+	⇩	⇩
3	64(5.9)	+	+	-	-
6	56(6.8)	+	-	-	-
9	36(7.0)	+	-	-	-
11	35(5.8)	+	+	-	-
13	32(6.1)	+	+	-	-

Table II: Summary of nucleoid-associated protein changes observed in Chinese hamster cell lines. Proteins are characterized by relative mobility (Mr), isoelectric point (pI), and presence (+), absence (-), or reduced abundance (→) in the respective cell lines.

In the case of murine lymphoid cells, differences in a total of 12 nucleoid proteins were consistently identified among the 7 cell lines. However, the differences in the nucleoid proteins which were most consistently different between the radioresistant and the radiosensitive cell lines are summarized in Table III. Of these 4, all were totally absent in L5178Y/S, 3 were absent and 1 in a lesser amount in M10 and 2 were absent and 2 in a lesser amount in LX830. In the case of the radioresistant cells, all 4 were present in L5178Y/R. In the M10(neo 5)-1 cells, an M10 variant bearing human chromosome 5 possessing the same radiosensitivity

Nucleoid Proteins of L5178Y Origin								
		Radioresistant				Radiosensitive		
Spot #	Mr (pI)	LY-R	LY-SR	LY-S35	M10(neo5)-1	LY-S	M10	LX830
2	64 (7.6)	+	↓	-	+	-	-	-
4	47 (5.1)	+	↓	↓	-	-	-	↓
6	37 (5.8)	+	+(↓)	+	+	-	-	-
8	39 (6.8)	+	+	+	+	-	↓	↓

Table III: Summary of nucleoid-associated protein changes observed in murine lymphoma cell lines. Proteins are characterized by relative mobility (Mr), isoelectric point (pI), and presence (+), absence (-) or reduced abundance (→) in the respective cell lines.

and DNA loop rewinding characteristic as L5178Y/R, 3 of the 4 proteins missing in the radiosensitive M10 reappeared. The nucleoid protein composition in the other radioresistant revertants (L5178Y/SR and /S35) derived from the radiosensitive L5178Y/S had only 2 of the 4 proteins present and the other 2 were reduced in amount in L5178Y/SR and 1 absent and 1 reduced in L5178Y/S35.

The results presented here show that in the presence of DNA damage, differences in intrinsic radiosensitivity correlate with DNA stability in the presence of DNA damage as measured by DNA rewinding ability, and with changes in nucleoid protein composition (Table IV). Previous work has demonstrated that in both the hamster fibroblasts (13) and the murine lymphoid cell lines (7), initial DNA

PARAMETERS ASSOCIATED WITH DSB REPAIR VARIANTS			
CELL TYPES (R vs. S)	INHIBITION of DNA REWINDING	NUCLEOID PROTEIN CHANGES	SPECIFIC PROTEINS MISSING
CHO Fibroblasts:			
4364 vs. XR-1	YES	YES	YES
XR-122 vs. XR-1	YES	YES	YES
CHO.K1 vs. xrs-5	YES	YES	YES
Murine Lymphoma:			
L5178Y/R vs. L5178Y/S	YES	YES	YES
L5178Y/S35 & /SR vs. /S	YES	YES	YES
L5178Y/R vs. M10 & LX830	YES	YES	YES
M10 (neo 5)-1 vs. M10	YES	YES	YES
Human Fibroblasts:			
Normal vs. AT	YES	YES?	?

Table IV

damage is the same, and cannot account for the differences in DNA rewinding between the cell lines studied. Therefore, alternative explanations other than differences in the amount of DNA damage are needed. One possibility is the differences in the level of nuclear matrix-imposed DNA organization by the DNA-protein anchor points that allows adjacent DNA loop domains to maintain different supercoiling status. Recent results demonstrate that in nucleoids prepared by methods similar to those used in this study, most of the DNA involved in active transcription remains in a compact form associated with the nuclear matrix, as illustrated in Fig. 4 (14). Therefore, any ssb's in active genes would not contribute to the inhibition of DNA supercoil rewinding if the DNA-nuclear matrix anchor points effectively isolate supercoiling domains. In contrast, if DNA-nuclear matrix anchor points do not isolate supercoiling domains, as proposed for radiosensitive cells, then ssb's in active genes will contribute to the inhibition of DNA supercoiling. If the foregoing were true, then the difference between the excess halo

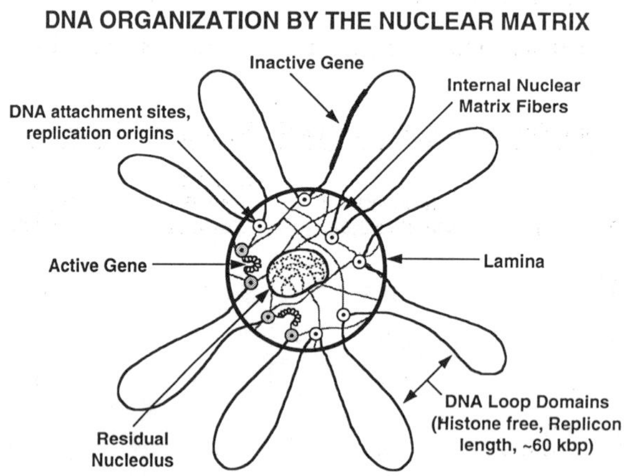

Fig. 4: An illustration of the current understanding of the structural organization of the nucleoid particles studied in this work.

diameter dose response functions would be due to dose modification; in other words, a fraction of the damage would not contribute to the measured effect. Fig. 5 shows that for the L5178Y cell lines, a dose modification calculation will produce a single dose response function for cell lines. (Please note that the dose modification factor is arbitrary.) However, the data from the CHO cell lines could not be transformed into a single dose response function by a dose modification calculation (not shown), because the excess halo diameter of the radiosensitive cell lines was too large. Therefore, these data are consistent with the idea that differences in DNA anchoring can account for the differences in DNA rewinding between irradiated radiosensitive and radioresistant cells.

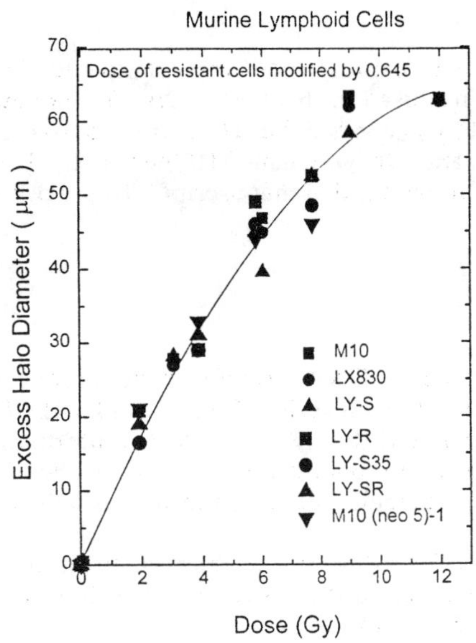

Fig. 5: Shows that for the murine lymphoma cell lines, a dose modification calculation will produce a single dose-response function. Please note that the dose modification factor is arbitrary.

It is known that potentially lethal radiation damage is subject to the competing processes of repair and fixation (15). Therefore, if one assumes that a double strand break is a potentially lethal lesion, it is possible to visualize that DNA supercoiling changes in a loop domain could re-orient the break and make it a poor substrate for repair. In the case of radioresistant cells, DNA-nuclear matrix anchor points effectively isolate the damaged loop domains from the adjacent regions, thereby preventing the propagation of supercoil changes into adjacent domains. However, in radiosensitive cells, alterations in the anchoring complex could allow transmission of supercoiling changes into adjacent loop domains affecting DSB reparability not only in the domain undergoing supercoiling changes but in adjacent domains as well. Assuming that changes in DNA supercoiling contribute to the fixation of potentially lethal damage, it is possible to explain how alterations in nuclear anchoring would contribute to radiosensitivity by allowing or not allowing supercoil changes in one loop domain to affect the reparability of double-strand breaks in other domains.

Acknowledgements

The authors wish to express their appreciation to the following individuals for providing the cell lines used in this study: Drs. T. Stamato and A. Giaccia--4364, XR-1, and XR-122; Dr. P. Jeggo--xrs-5; Dr. H. Evans--L5178Y/R, /S, /SR, /S35; Dr. K. Sato--M10 and LX830; and Dr. D. Chen--M10(neo 5)-1. The authors wish to also thank Ms. Kathy Bles for typing this manuscript. This work was supported by NCI grant CA51116.

References

1. G. Iliakis, The role of DNA double-strand breaks in ionizing radiation-induced killing of eukaryotic cells. *Bio Essays* **13,** 641-648 (1991).
2. P.L. Olive, J. Hilton and R.E. Durand, DNA conformation of Chinese hamster V79 cells and sensitivity to ionizing radiation. *Radiat. Res.* **107,** 115-124 (1986).
3. P.R. Cook and I.A. Brazell, Supercoils in human DNA. *J. Cell Sci.* **19,** 261-279 (1976).
4. K.J. Pienta and D.S. Coffey, A structural analysis of the role of the nuclear matrix and DNA loops in the organization of the nucleus and chromosome. *J. Ceil Sci.* Suppl. **1,** 123-135 (1984).
5. B. Nelkin, D. Pardol, S. Robinson, D. Small and B. Vogelstein, Nuclear structure and DNA organization. *In Tumor Ceil Heterogeneity and Applications* (A.L. Owens, D.S. Coffey and S.B. Baylin, eds.) pp. 441-457. Academic Press, New York, 1982.
6. T.D. Stamato, R. Weinstein, A.J. Giaccia and L. Mackenzie, Isolation of a cell cycle dependent gamma ray-sensitive Chinese hamster ovary cell. *Somat. Cell. Genet.* **9,** 165-173 (1983).
7. H.H. Evans, M. Ricanati and M.F. Horng, Deficiency in DNA repair in mouse lymphoma strain L5178Y-S. *Proc. Natl. Acad. Sci. USA* **84,** 7562-7566 (1987).
8. K. Sato and N. Hieda, Isolation and characterization of a mutant mouse lymphoma cell sensitive to methyl methanosulfonate and X-rays. *Radiat. Res.* 78,167-171 (1979).
9. K. Sato, N. Hieda-Shiomi and H. Hama-Inaba, X-ray sensitive mutant mouse cells with various sensitivities to chemical mutagens. *Mutat. Res.* **121,** 281-285 (1983).
10. J.L. Roti Roti and W.D. Wright, Visualization of DNA loops in nucleoids from HeLa cells: Assays for DNA damage and repair. *Cytometry* **8,** 461-467 (1987).
11. R.S. Malyapa, W.D. Wright and J.L. Roti Roti, Radiation sensitivity correlates with changes in DNA supercoiling and nucleoid protein content in cells of three Chinese hamster cell lines. *Radiat. Res.* (1 994), *In press.*
12. M. Kapiszewska, W.D. Wright, C.S. Lange and J.L. Roti Roti, DNA supercoiling changes in nucleoids from irradiated L5178Y-S and -R cells. *Radiat. Res.* **119,** 569-575 (1989).
13. A. Giaccia, R. Weinstein, J. Hu and T.D. Stamato, Cell cycle-dependent repair of double-strand DNA breaks in a X-ray sensitive Chinese hamster cell. *Somat. Cell Mol. Genet.* **5,** 485-491 (1985).
14. M.G. Gerdes, K.C. Carter, P.T. Moen and J.B. Lawrence, Dynamic changes in higher-level chromatin organization of specific sequences revealed by in situ hybridization to nuclear halos. *J. Cell Biol.* **126,** 289-304 (1994).
15. R.A. Phillips and L.J. Tolmach, Repair of potentially lethal damage in X-irradiated HeLa cells. *Radiat. Res.* **29,** 413-432 (1966).

THE ALTERATION OF CHROMATIN DOMAINS DURING DAMAGE REPAIR INDUCED BY IONIZING RADIATION

A.E. Cress[1], K.M. Olson[2] and G.B. Olson[2],
Depts. of Radiation Oncology[1] and Microbiology and Immunology[2],
The University of Arizona,
1501 N. Cambell Ave
Tucson, AZ 85724

Introduction

Several groups previously have reported the ability of chromatin structure to influence the production of damage induced by ionizing radiation. For example, Warters et al. have shown that the organization of the DNA into the basic nucleosome repeat structure and condensation of the chromatin fiber into a higher order structure protected DNA from double strand break induction as compared to the deproteinized DNA (Warters and Lyons 1992). Another group has found that relaxed chromatin was 5 to 10 fold more susceptable to the induction of DNA-protein crosslinks by gamma radiation than is chromatin residing in living cells (Chin et al. 1992). Additional investigators have found that the unfolding of the nucleosome core into an extended structure increases the distribution of cyclobutane-pyrimidine dimers after UV irradiation (Brown et al. 1993). These reports suggest that chromatin structure influences the induction of DNA damage by ionizing radiation.

It is also known that the ability of the chromatin structure to be repaired may require a more extensive process than is required for single and double strand DNA break repair. For example, Hittleman et al. have shown that whereas the single and double strand breaks in DNA can be repaired within hours of induction, the repair of the chromosome required considerably longer. (Hittleman and Pollard 1982). More recently, they have found that although AT and normal cells exhibited similar levels of initial DNA double strand breaks, AT cells showed nearly two-fold higher initial levels of chromosome damage than normal control cells. Their results suggested that one determining component of cellular radiosensitivity might include chromatin structure (Pandita and Hittleman 1992).

Our interest has been to determine whether chromatin structural alterations exist after ionizing radiation during a repair interval. Our earlier work investigated this question using biochemical techniques. The crosslinking of nuclear structural proteins to DNA after ionizing radiation was observed (Cress and Kurath 1988; Cress et al. 1990). In addition, we found that the chromatin structure in vitro as measured by sucrose density gradient sedimentation, was altered after ionizing radiation (Cress et al. 1989). These observations added to earlier studies in which digital imaging techniques showed an alteration in feulgen-positive DNA after irradiation (Olson and Bartels 1984; Olson et al. 1982) prompted the present study.

The object of this study was to detect whether the higher order structure of DNA into chromatin domains within interphase human cells was altered in interphase cells in response to a radiation induced damage. The present study takes advantage of the advances in the detection of chromatin domains in situ using DNA specific dyes and digital image processing of established human T and B cell lines.

Method

Exponentially growing human lymphocyte cell lines, Molt 4 and 8226, were grown in RPMI 1640 media containing 10% FBS, a reducing agent and antibiotics and were treated with ^{60}Co gamma rays. Immediately following radiation, the cells were returned to a 37°C incubator for 1.5 hr. The cells were harvested by centrifugation, washed with PBS and centrifuged onto microscope slides using a Shannon cytocentrifuge. The cells were fixed by standard methods including periodate-lysine-2% paraformaldehyde (PLP) or 50% methanol for 10 minutes at 4°C.

Slides stained with May Grunwald-Giemsa were used to evaluate general cell morphology and nuclear/cytoplasm ratios and were stained immediately by standard procedure. PLP fixed cells/slides for Feulgen staining were hydroyzed for 30 min in 5 N HCL, 1 min in 1N HCL, stained in pararosaniline (0.5g/100ml) pH 1.5 for 2 hr., bleached for 6 min in 5% sulfurous acid and cover-slipped with Pro-Texx pH 5. Methanol fixed cells/slides for fluorescent staining were placed in a mixture of 5 µM Hoechst 33258 and 5 µM 7-amino-actinomycin-D (AMD) (to stain AT-rich and GC-rich regions of chromatin) for 5 min, rinsed in 4 five min changes of PBS, dipped 3 times in distilled water and cover slipped with Pro-Texx pH 7. All stained slides were stored at 4°C until ready for analysis. All slides designated for a given stain and belonging to an experiment were stained in batch to minimize technical variability and artifacts.

Cells stained with Feulgen and May-Grunwald-Giemsa were analyzed using the CAS 100R work-station with a Riechert Diastar microscope at a magnification of 40x and a wavelength of 540 nm and 560 nm respectively. Fluorescent Hoechst 33258 and AMD images were acquired with a Nikon diaphot inverted epi-fluorescent phase microscope on-line to a Modified Quantex QX7 fluorescent image processor. Images were collected at excitation/dichroic/barrier wave lengths of 360/430/470 for Hoechst and 510/580/610 for AMD.

The 50 or 100 images were collected from each slide and placed into designated cell files to represent each experimental condition being tested. The digitized images were analyzed using the nuclear/cytoplasm routine of the CAS Cell Measurement Program and the Celedt and feature extraction programs which have been adopted from TICAS (Taxonomic Intracellular Analytical System).

Results and Discussion

The data in Figure 1 and Table 1 indicate a significant but modest radiation dependent change in the N/C ratios of human T cells. Non-stoichiometric stains such as May-Grunwald-Giemsa stains when used in controlled conditions and analyzed with

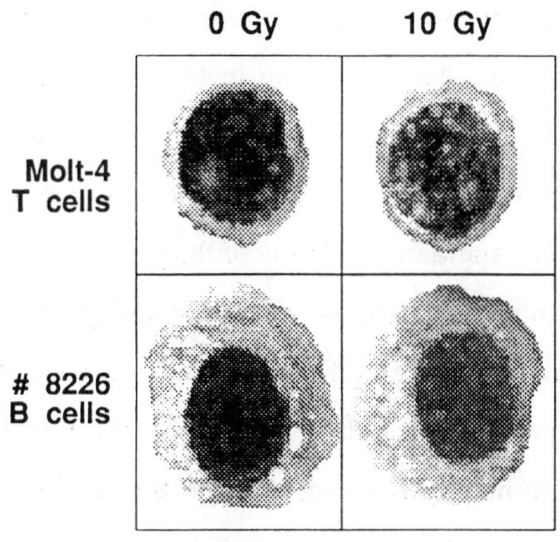

Figure 1. Digitized Images of May Grunwals-Giemsa stained irradiated and non-irradiated cultured human lymphocyte cell lines. The Molt 4 cells (T cells) and 8226 cells (B cells) were irradiated using ^{60}Co gamma rays and incubated for one hour at 37°C. The cells were cytospun, stained and the images were collected and analyzed using a CAS 100R image processor. Approximate magnification is 400X.

proper image conditions can measure differences not observable to the eye. The 13% ratio decrease over the 10 Gy range is due to an increase in the cytoplasm of the T cells. Such a change is not obvious to the eye, but can be detected by image techniques. May-Grunwald Giemsa in not a stochiometic stain and is not used for precise measurements but did indicate a radiation induced alteration in cellular features. Therefore, we continued the analysis using a stochiometric Feulgen stain to quantitate any irradiation induced alteration of the DNA within the nucleus.

Table 1. Alteration of nucleus/cytoplasm ratio in irradiated Human lymphocyte cell lines.

Cell Line	0 Gy	1 Gy	10 Gy
Molt 4 (T cell)	2.02	1.84	1.76
8226 (B cell)	0.52	0.52	0.53

The values are the mean from four experiments. Each data set represents 100 May-GrunwaldGiemsa stained cells/treatment/experiment (2400 cells total).

The data in Figure 2 indicate that T and B cells can not be differentiated by features such as the relative size of the nucleus and the heterogeneity of the chromatin structure as detected by the feature, stdev of the pixels. On the other hand, features which measure the degree of DNA clumping in the chromatin of the cells show a difference with B cells having a greater number and size of the clumps as measured by the grain size and number of grains than T cells. The heterogeneity of the chromatin structure as measured by the standard deviation of the pixels and the relative size of the nucleus was chosen as features to determine if a radiation dependent alteration occurs.

T and B-cells obtained from established cell lines can be stained with specific DNA stains and subsequently analyzed by a number of feature extraction programs to find certain features which may differentiate T from B cells (Fig 2) and find features which differentiate irradiated cells from non-irradiated cells (Fig 3). The data in Fig 3 indicate that alterations in the DNA staining pattern can be detected with 1 Gy of irradiation and the change in the heterogeneity increases with an increasing dose of radiation.

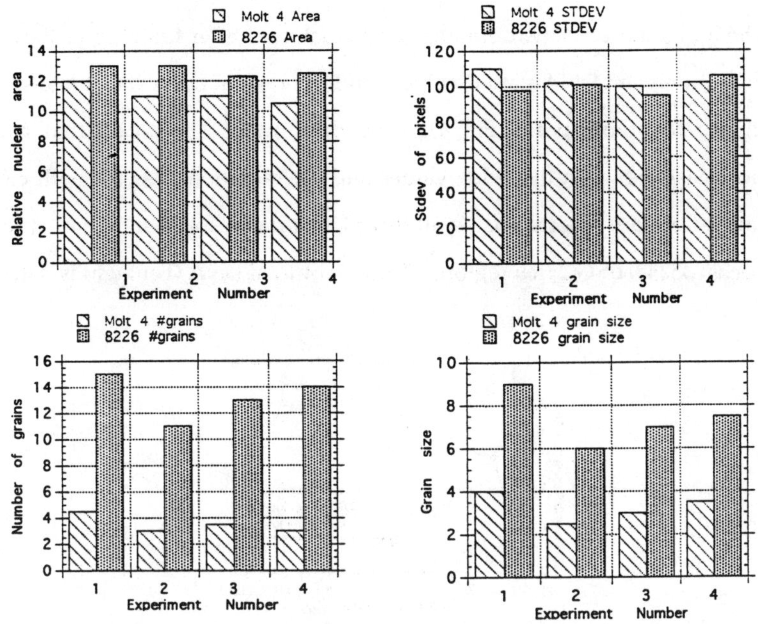

Figure 2. Comparison of four different image features of DNA staining in human lymphocyte cell lines. Molt 4 cells [▧] and 8226 cells [▩] were cultured, cytospun and stained for DNA using Feulgen. Feulgen stains the DNA stoichiometrically, independent of DNA packaging. The data represents determinations compiled from four different experiments. The following features were scored:

<u>Relative nuclear area:</u> total number of pixels with values greater than background
<u>Stdev of pixels:</u> standard deviation of the individual pixel values
<u># of grains:</u> number of non-contiguous pixel values
<u>Grain size:</u> total value of non-contiguous pixels divided by the number of grains

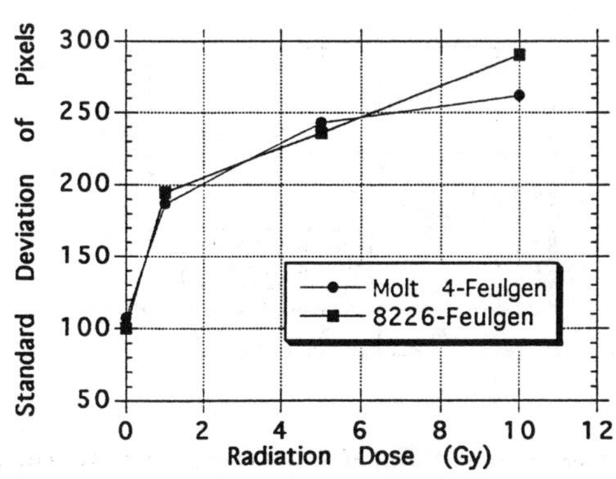

Figure 3. Alteration of one DNA staining feature (Stdev) in Human lymphocyte cell lines with radiation dose. Molt 4 cells [●] and 8226 cells [■] were damaged with ^{60}Co gamma rays and after a 1.5 hr incubation at 37°C, the cells were cytospun and processed for DNA staining using the Feulgen stain.

The alteration of DNA domains can be detected after ionizing radiation using both a non-fluorescent DNA stain (Feulgen) and fluorescent DNA stains (Hoechst 33258 and 7-amino-actinomycin D) (Fig 4). The data in Fig 4 indicate that a positive correlation exists between the stdev values and the radiation dose for both cell types. Similar alterations are detected using all three DNA stains, independent of whether AT rich (Hoechst 33258) or GC rich regions (7-AMD) or total DNA (Feulgen) is detected.

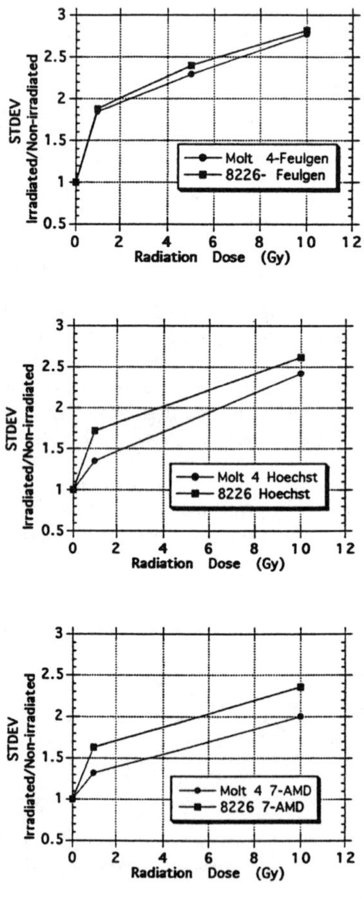

Figure 4. Allteration of DNA stainıng after ionizing radiation using three different stains in human lymphocyte cell lines. Molt 4 cells [●-●] and 8226 cells [■-■] were damaged with ^{60}Co gamma rays and after a 1.5 hr incubation at 37°C, the cells were cytospun and processed for DNA staining using Feulgen, Hoechst 33258 or 7-amino-acintomycin D. The stdev ratio of irradiated/nonirradiated cells was determined using the mean ratios from four experiments. The fluorescent images were obtained using a Quantex QX7 image processor. Approximate magnification is 400X.

The alterations in DNA domains can also be detected using a nuclear antigen called CINA (Fig 5). The data in Fig 5 indicate that the redistribution of an abundant nuclear antigen occurs coincident with the alterations in DNA distribution. The ability to use confocal microsopy coupled with digital image processing will allow the quantitation of these intranuclear alterations in live cells in the future.

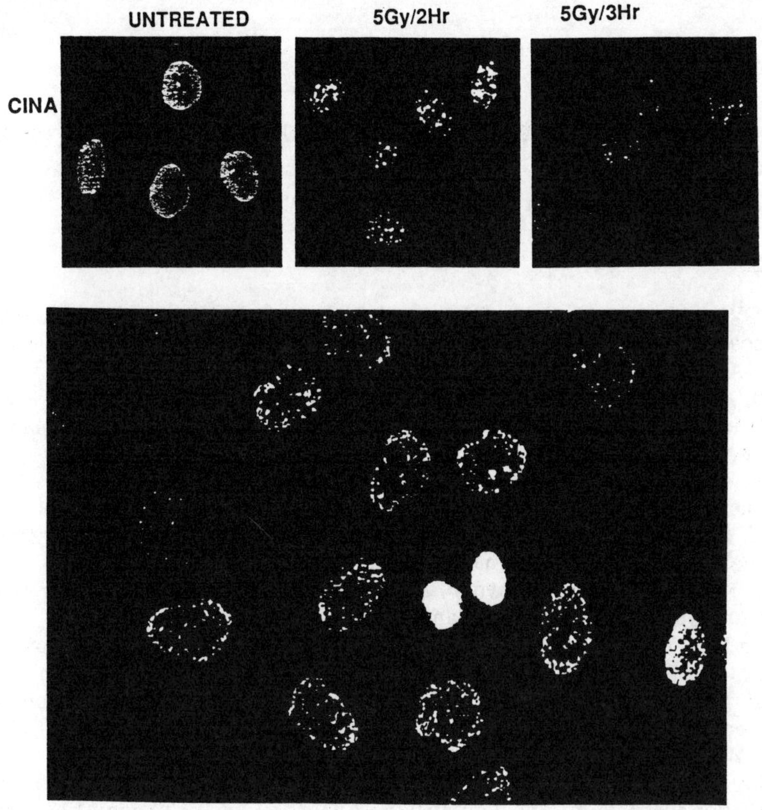

Figure 5. Altered distribution of a chromosome interphase nuclear antigen (CINA) after ionizing radiation damage. A549 cells were treated with ^{60}Co gamma rays, washed with PBS, fixed with 50% methanol at 4 degrees and stained for the distribution of the antigen. The top panel shows the distribution of the antigen which is altered after ionizing radiation damage as viewed by conventional microscopy. The bottom panel shows a confocal image of the nuclear antigen in unirradiated cells.

The slide based assay and use of target cell lines may provide a sensitive rapid biological dosimeter for DNA damaging agents. Our future work is directed at determining the mechanism responsible for the alteration in chromatin domains. Further, we are testing the hypothesis using confocal microscopy that the alteration of chromatin domains involves the pairing of specific chromosome regions in interphase chromatin.

Acknowledgement

Supported in part by DOE contract #307190, The Phoenix Friends of the Arizona Cancer Center and CA 23074. Financial support does not constitute an endorsement by DOE of the views expressed in the article.

References

Brown DW, Libertini LJ, Suquet C, Small EW, Smerdon MJ (1993) Unfolding of nucleosome cores dramatically changes the distribution of ultraviolet photoproducts in DNA. Biochem 32:10527-10531

Chiu SM, Xue LY, Friedman, LR, Oleinick NL (1992) Chromatin compaction and the efficiency of formation of DNA-protein crosslinks in gamma-irradiated mammalian cells. Rad Res 129:184-191

Cress AE, Kurath KM (1988) Identification of attachment proteins for DNA in Chinese Hamster ovary cells. J Biol Chem 263:19678-19683

Cress AE, Kurath KM, Hendrix MJ, Bowden GT (1989) Nuclear protein organization and the repair of radiation damage. Carcinogenesis 10:939-943

Cress AE, Kurath KM, Stea B, Bowden, GT (1990) The crosslinking of nuclear protein to DNA using ionizing radiation. J Can Res Clin Oncol 116:324-330

Hittleman WN, Pollard M (1982) A comparison of the DNA and chromosome repair kinetics after iradiation. Rad Res 92:497-509

Olson GB, Bartels PH (1984) Assessment of environmental insults on lymphoid cells as detected by computer assisted morphometric techniques. Biological Dosimetry Eisert WG, Mendelsohn ML (eds)

Olson GB, Bartels, PH, Anderson RE (1982) Computer assisted morphometric analysis of radiation injury in murine lymphocytes. Anal and Quant Cytology 4:181-187

Pandita TK, Hittleman WN (1992) Initial chromosome damage but not DNA damage is greater in ataxia telangiectasia cells. Rad Res 130:94-103

Warters RL, Lyons BW (1992) Variation in radiation-induced formation of DNA double-strand breaks as a function of chromatin structure. Rad Res 130: 309-318

REJOINING OF PREMATURELY CONDENSED CHROMOSOMES IN RADIOSENSITIVE XRS-5 CELLS

R. Okayasu*+ and G. Iliakis*

*Department of Radiation Oncology
Thomas Jefferson University
Philadelphia, PA 19107 USA

+Center for Radiological Research
College of Physicians and Surgeons of
Columbia University
New York, NY 10032 USA

Introduction

The premature chromosome condensation (PCC) technique allows the visualization of interphase chromosomes in G1 and G2 phases of the cell cycle, and has been used to detect chromosome damage induced by ionizing radiation (Johnson and Rao 1970, Waldren and Johnson 1974, Hittelman and Rao 1974, Cornforth and Bedford 1983, 1985, Pantelias and Maillie 1983, Pantelias 1986 and Iliakis et al 1988). The strength of the technique lies in its potential to bridge results obtained at the DNA level with those at the cell level. Thus the PCC method can provide useful information on the sequence of events that lead from DNA damage to chromatin fragmentation, to chromosome aberration and to ultimate reproductive cell death. Evidence suggests that the initial DNA damage responsible for interphase chromosome fragmentation is DNA double strand breaks (dsb) (Iliakis 1991). Several investigators have examined the correlation between the induction and repair of DNA dsb and the induction and repair of interphase chromosome breaks with radiation in order to elucidate the mechanism leading to cell death (Hittelman and Pollard 1982, Wlodek and Hittelman 1988 a,b, Iliakis et al 1991, Wang and Iliakis 1992, Pandita and Hittelman 1992 a,b).

One difficulty with comparisons between DNA and chromatin damage is that lengthy incubation periods at 37°C are necessary for chromosome condensation in the PCC method. Since DNA repair can take place during this incubation period to an unknown extent, it is not appropriate to directly compare the PCC breaks and DNA dsb. In order to conquer this problem, we have introduced two kinds of DNA repair inhibitors applied during the incubation period of PCC. By incubating fused cells in the presence of hypertonic medium, the number of interphase breaks in X-irradiated CHO cells was doubled in comparison to that without hypertonic treatment (Iliakis et al 1993). The hypertonic treatment has been known to inhibit

DNA dsb repair (Iliakis et al 1993). The same strategy was applied to another inhibitor 9-beta-D-arabinofuranosyladenine (ß-araA) which is known to inhibit DNA polymerases α and ß (Mueller et al 1975), affect repair of DNA dsb (Bryant and Bloecher 1982) and interphase chromosome breaks (Iliakis et al 1988), and fix potentially lethal damage (PLD) (Tolmach 1966, Iliakis 1980, 1988a). The number of PCC chromosome breaks were also approximately doubled in the presence of ß-araA during the incubation period. The combined experiments, in which both of these inhibitors were applied at the same time, show the additive effect indicating that these agents act independently in inhibiting DNA break repair (Okayasu and Iliakis 1993).

The kinetics of interphase chromosome break repair were also examined with X-irradiated plateau phase CHO cells by applying these inhibitors as a function of time, during the condensation period for PCC. As a result, two independent forms of interphase chromosome breaks were uncovered. One form, termed the α-form, consists of interphase chromosome breaks with slow rejoining kinetics ($t_{1/2}$ = 60-90 min) which are sensitive to post-irradiation treatment with ß-araA (Okayasu and Iliakis 1993). The α-form breaks were visualized by the standard PCC protocol (without any post irradiation treatment) in addition to being treated with ß-araA during the incubation period allowed for PCC (Iliakis et al 1988, Okayasu and Iliakis 1994).

Another form, termed the ß-form, consists of interphase chromosome breaks with fast rejoining kinetics ($t_{1/2}$ = 2-10 min) which are sensitive to post-irradiation treatment in hypertonic salt solution (Iliakis et al 1993, Okayasu and Iliakis 1993). The ß-form breaks are not visualized with the standard PCC protocol, but can be uncovered by hypertonic treatment during the period allowed for PCC. Independency for the rejoining of these two forms of breaks was also demonstrated by simultaneous application of these two inhibitors during the incubation period (Okayasu and Iliakis 1993).

Xrs-5 cells, a radiosensitive, DNA double strand break repair deficient mutant of CHO cells (Jeggo and Kemp 1983, Kemp et al 1984, Nagasawa et al 1989, Iliakis and Okayasu 1990, Iliakis et al 1992), have been studied with the PCC technique. This mutant displayed a higher number of initial chromosome breaks (approximately twice) with x-ray treatment as well as partial deficiency in the rejoining of interphase chromosome breaks with the standard PCC protocol (Iliakis and Pantelias 1990). Moreover, hypertonic treatment during the incubation period which allowed for PCC did not change the yield of PCC breaks in x-irradiated xrs-5 cells. Notably the number of PCC breaks after treatment with hypertonic media is similar in CHO and xrs-5 cells (Okayasu et al 1993).

Recently, a gene product responsible for the *xrs* phenotype was identified as a Ku-like DNA end binding protein (Getts and Stamato 1994). The present paper summarizes completed information regarding the induction and repair of the α- and ß- forms of PCC breaks in xrs-5 cells and demonstrates that this gene product predominantly affects the fast form (ß-form) of interphase chromosome breaks (Okayasu and Iliakis 1994).

Materials and Methods

Cell Culture and Irradiation

CHO and xrs-5 cells were grown as a monolayer in McCoy's 5A medium supplemented with 10% fetal calf serum and antibiotics. Plateau-phase cells were obtained by growing 1.0×10^5 cells for CHO and 1.5×10^5 cells for xrs-5 in 60 mm dishes for 4 days. Irradiations were carried out with a Siemens therapeutic x-ray machine operated at 250 kVp, 15 mA with a 2 mm Al filter (effective photon energy = 70 keV), at a dose rate of 2.0 Gy/min. Cells were irradiated on ice after trypsinization, in a suspension form.

Premature Chromosome Condensation (PCC)

Procedures developed for polyethylene glycol (PEG) mediated cell fusion and premature chromosome condensation (PCC) induction were employed (Pantelias and Maillie 1983). Briefly, 10^6 mitotic CHO-cells were mixed with an equal number of interphase cells in a tissue culture tube. After centrifugation at 200g for 5 min, the supernatant was discarded, and the pellet was resuspended in 0.1 ml of 50% ice cold PEG solution (MW = 1450, Sigma, prepared in PBS, 50% weight/volume) for 1 min. Subsequently, 2 ml of cold PBS was slowly added. The cell suspension was centrifuged (200g for 5 min) and the supernatant was discarded. The pellet was resuspended in 0.7 ml of growth medium with Colcemid (10^{-6} M). Finally the tube was placed in the incubator at 37°C for 60 min; in that time, cell fusion and induction of PCC was completed. The protocol described here is the standard procedure for PEG mediated induction of PCC.

Treatment with ß-araA: Cells were irradiated and treated for fusion as described above. Just before incubation at 37°C, ß-araA was added to the samples at the desired concentration from a 10 mM stock solution and was allowed to act during the entire period of chromosome condensation (60 min). To measure repair of ß-araA-sensitive interphase chromosome breaks, the drug was given at various time points after the beginning of incubation, and was allowed to act for the remainder of the condensation period (Okayasu and Iliakis 1993).

Treatment in hypertonic medium: Previously described protocols (Iliakis et al 1993) were used to treat cells in hypertonic medium. Briefly, cells were fused and exposed to hypertonic medium (McCoy's 5A with 10% serum in which the NaCl concentration was adjusted to 500 mM) starting at the beginning of a 20 min incubation period. After this time, hypertonic medium was diluted with warm, (37°C) fresh isotonic medium and condensation was allowed to proceed to completion for an additional 40 min.

Combined treatment with araA and hypertonic medium: The protocol outlined above for hypertonic treatment was employed for this set of experiments. The only modification was that the hypertonic medium was supplemented with ß-araA at the appropriate concentration. To measure repair of ß-araA-sensitive lesions (α-form) in cells exposed to combined treatment, a set of tubes containing test cells fused with mitotic cells using the standard procedure was prepared and exposed to hypertonic medium for 20 min starting at the beginning of incubation period. ß-araA was added at various time points after the beginning of incubation, and was allowed to act for the remaining time of incubation. This set of samples was centrifuged to wash out hypertonic medium after completion of hypertonic treatment, and cells were resuspended in fresh isotonic medium containing (when appropriate) ß-araA to complete chromosome condensation. To measure repair of hypertonic treatment-sensitive interphase chromosome breaks (ß-form) in the presence of ß-araA, a previously described protocol was used (Iliakis et al 1993). Briefly, a 20 min treatment in hypertonic medium was administered at various times after the beginning of incubation. All media used, including the hypertonic medium and the isotonic medium used to dilute the former, were supplemented with ß-araA at the appropriate concentration. Fresh growth medium was made hypertonic (500 mM) by adding the appropriate amount of NaCl from a 5.0 M stock solution.

After the incubation period for chromosome condensation, cells were treated in hypotonic (75 mM) KCl solution and fixed in 10 ml methanol:acetic acid (3:1). Cells were washed once more in fixative, dropped on precleaned wet slides, air dried, and stained with 2% Giemsa (Sigma). Analysis for PCC fragments was performed by means of light microscopy. Routinely, 25-30 cells were scored for each experiment. The results obtained are presented as mean ± standard error of three independent experiments.

Results

Yields of interphase chromosome breaks in x-irradiated CHO and xrs-5 cells treated under various post-irradiation conditions are summarized in Table 1. For these results, CHO or xrs-5 cells were irradiated, fused with CHO mitotic cells and incubated at 37°C for 1 hr for PCC induction. During the incubation period, when appropriate, cells were treated with hypertonic medium and/or ß-araA (see Materials and Methods). CHO cells treated with x-rays alone resulted in 2.0 breaks/cell/Gy, while xrs-5 cells with radiation alone resulted in 3.7 breaks/cell/Gy. These values confirm our previous observations (Iliakis and Pantelias 1990). When irradiated fused CHO cells were treated with hypertonic medium during the incubation period, the number of breaks increased to 3.6 breaks/cell/Gy. On the other hand, the same hypertonic treatment with xrs-5 cells did not result in any change in the number of breaks. This is consistent with our previous observation that xrs-5 cells are deficient in repairing one form of interphase chromosome breaks (defined as the ß-form of breaks) which is sensitive to hypertonic condition

(Okayasu et al 1993). The ß-form interphase chromosome breaks are expressed constitutively in xrs-5 cells.

TABLE 1: Yields of interphase chromosome breaks under various postirradiation conditions in CHO and *xrs-5* cells. (Reproduced from Okayasu and Iliakis 1994)

Postirradiation Conditions	Interphase Chromosome Breaks Per Cell Per Gy	
	xrs-5	CHO
Standard Conditions	3.7 ± 0.1	2.0 ± 0.1
1 mM ß-araA During PCC	5.4 ± 0.1	3.6 ± 0.1
500 mM NaCl During PCC	3.7 ± 0.1	3.6 ± 0.1
1 mM ß-araA + 500 mM NaCl During PCC	5.6 ± 0.2	5.3 ± 0.1

When irradiated fused CHO cells were treated with ß-araA for the entire condensation period, the number of interphase breaks increased from 2.0 to 3.6 breaks/cell/Gy. Xrs-5 cells under the same treatment condition showed an increase in the number of breaks from 3.7 to 5.4 breaks/cell/Gy. The increase in the number of breaks with ß-araA treatment for CHO and xrs-5 cells was practically identical (1.6 for CHO and 1.7 for xrs-5) suggesting that ß-araA acts similarly in these two cell lines. Since interphase chromosome breaks sensitive to ß-araA treatment have been defined as α-form breaks (Okayasu and Iliakis 1993), ß-araA uncovers the same amount of interphase chromosome breaks of the α-form in both cell lines.

Results are also given in Table 1 when irradiated fused CHO and xrs-5 cells were simultaneously treated with hypertonic medium and ß-araA during the condensation period for PCC (see Material and Methods). With combined treatments, CHO cells yield 5.3 breaks/cell/Gy which is increased from 3.6 breaks/cell/Gy with ß-araA treatment alone. Xrs-5 cells yield 5.6 breaks/cell/Gy for the combined treatment, and this value is almost identical to the number of breaks in xrs-5 cells with araA treatment alone. Hypertonic treatment did not affect the yield of interphase chromosome breaks even in the presence of ß-araA in xrs-5 cells. It is worth noting that the yield of interphase chromosome breaks with the combined treatment is similar in CHO and xrs-5 cells (5.3 vs 5.6 breaks/cell/Gy) despite the significant difference in non-treated cells (2.0 vs 3.7 breaks/cell/Gy). These results indicate that the difference in the yield of interphase chromosome breaks among these two cell lines can be explained by the deficiency in repairing the ß-form of interphase breaks in xrs-5 cells.

In order to further confirm the deficiency in a specific form of interphase chromosome repair associated with xrs-5 cells, we measured the kinetics of rejoining of interphase chromosome breaks after hypertonic medium and/or ß-araA treatment during the condensation period for PCC. For this purpose, irradiated fused cells were treated with the inhibitor(s) at various intervals starting at the beginning of incubation. Since the results of rejoining kinetics with one inhibitor treatment (either ß-araA or hypertonic medium) have been given elsewhere (Okayasu and Iliakis 1993, 1994), data with the combined treatment which are the mirror images with the singular treatment, are given here.

Closed circles in Fig. 1 show rejoining kinetics of interphase chromosome breaks of irradiated (10 Gy) fused CHO cells treated with hypertonic medium as a function of time during the incubation period in the presence of ß-araA (see Materials and Methods). ß-araA has been shown to prevent interphase chromosome breaks of the α-form (having slow rejoining kinetics). These results indicate that rejoining of hypertonic salt sensitive interphase chromosome breaks (ß-form) occurs normally even in the presence of ß-araA treatment. The rejoining kinetics given here is a mirror image of that with hypertonic treatment alone (Iliakis et al 1993). Closed triangles in the Fig. 1 give rejoining kinetics of interphase chromosome breaks of irradiated fused CHO cells treated with ß-araA as a function of time during the incubation period in the presence of hypertonic media (see Material and Methods). Hypertonic media has been shown to prevent the ß-form of interphase chromosome breaks. The rejoining of ß-araA sensitive interphase chromosome breaks (α-form) in the presence of hypertonic media occurs similarly to that without hypertonic treatment (Okayasu and Iliakis 1993). These results confirm that x-irradiation induces two independent forms of interphase chromosome breaks which are discriminated not only on the basis of their sensitivity to ß-araA or hypertonic medium, but also on the basis of their repair kinetics.

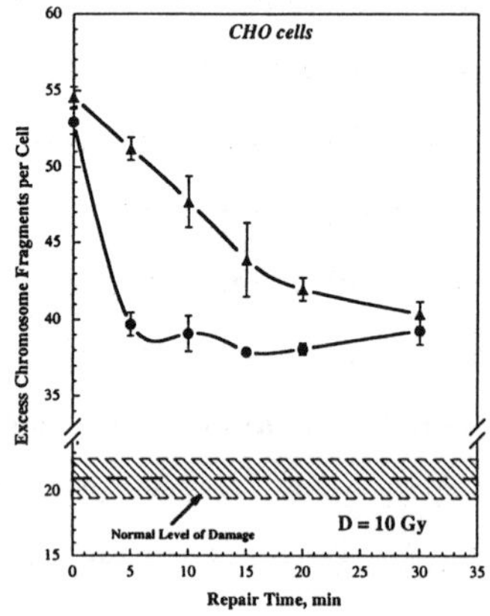

Figure 1: Rejoining kinetics of interphase chromosome breaks in irradiated CHO cells exposed to hypertonic treatment in the presence of 1 mM ß-araA (closed circles). Rejoining kinetics of interphase breaks in irradiated CHO cells exposed to ß-araA in the presence of hypertonic medium (closed triangle). The dashed area shows the range of expected values for interphase chromosome breaks of irradiated CHO cells without inhibitor treatment. (Reproduced from Okayssu and Iliakis 1993).

Corresponding results of rejoining kinetics of interphase chromosome breaks with xrs-5 cells are given in Fig. 2. Closed circles depict rejoining kinetics of interphase chromosome breaks of irradiated (10 Gy) fused xrs-5 cells treated with hypertonic media as a function of time during the incubation period in the presence of ß-araA. Except for some oscillating responses of unknown origin in the kinetics, no rejoining of this type of interphase chromosome breaks was observed. This is clearly different than results obtained with CHO cells where rapid rejoining of hypertonic sensitive interphase breaks occurred (Fig. 1). The rejoining kinetics of the ß-form breaks with hypertonic treatment alone in xrs-5 cells is similar to what observed in this figure (Okayasu et al 1993). Closed triangles depict the kinetics of interphase chromosome breaks of the α-form (sensitive to ß-araA treatment) in the presence of hypertonic medium. Rejoining of α-form of breaks occur similarly to that with CHO cells (Fig. 1). The kinetics (closed triangles) shown are also identical to that of xrs-5 cells with ß–araA treatment alone (Okayasu and Iliakis 1994).

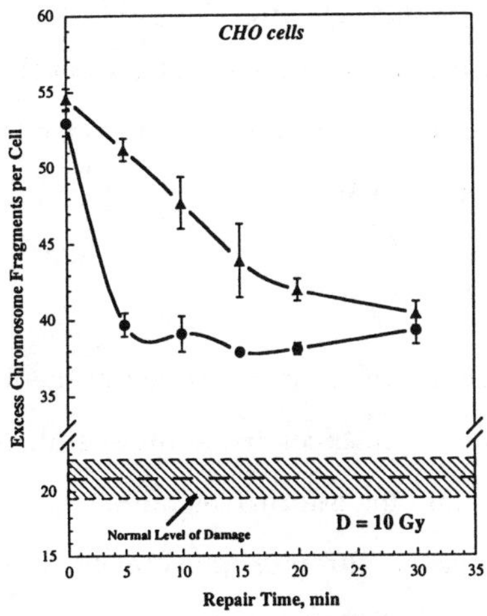

Figure 2. Rejoining kinetics of interphase chromosome breaks sensitive to hypertonic treatment in irradiated xrs-5 cells in the presence of ß-araA (closed circles). Rejoining kinetics of interphase chromosome breaks sensitive to ß-araA in irradiated xrs-5 cells in the presence of hypertonic medium (closed triangles). The dashed area shows the range of expected values for interphase chromosome breaks of irradiated xrs-5 cells without inhibitor treatment. (Reproduced from Okayasu and Iliakis 1994).

These results further confirm that xrs -5 cells are only deficient in repairing the ß-form of interphase chromosome breaks, but proficient in repairing the α-form of interphase chromosome breaks. The results also verify independence of the α-form interphase chromosome breaks from the ß-form breaks. The conditions uncovering the ß-form interphase chromosome breaks (treatment in hypertonic media) do not affect the rejoining kinetics of the α-form breaks (ß-araA sensitive breaks).

Since a portion of interphase chromosome breaks with irradiated xrs-5 cells rejoined in our previous studies using the standard PCC protocol (without any drug or hypertonic salt treatment) (Iliakis and Pantelias 1990), we also tested if these rejoined breaks belong to the α-form breaks (sensitive to ß-araA). For this purpose, xrs-5 cells were treated with ß-araA before (1 hr prior to x-irradiation) and various amount of times after irradiation (1 to 6 hr), and then cells were processed for PCC under the standard protocol. No rejoining of PCC breaks were observed with ß-araA treatment in irradiated xrs-5 cells in this protocol (Okayasu and Iliakis 1994). On the other hand, about 1/3 of the initial breaks were rejoined ($t_{1/2}$ = about 30 min) in the standard PCC protocol without any drug treatment. These results indicate that these rejoined breaks belong to the α-form, suggesting that xrs-5 cells sustain and repair the α-form of interphase chromosome breaks at the level similar to those of CHO cells. Detailed calculation on the number of rejoined breaks have been given elsewhere (Okayasu and Iliakis 1994).

Discussion

The results presented above are summarized in the following points:

1. Ionizing radiation induces two independent forms of interphase chromosome breaks (the α- and the ß-forms) as measured by the PCC technique.

2. The α-form interphase chromosome breaks consists of breaks obtained with the standard PCC protocol without any treatment during the incubation period as well as breaks uncovered by the post-irradiation ß-araA treatment during the incubation period for PCC. The α-form interphase chromosome breaks have slow rejoining kinetics ($t_{1/2}$ > 30 min), and are sensitive to post-irradiation ß-araA treatment but not sensitive to hypertonic treatment.

3. The ß-form interphase chromosome breaks can be uncovered by post-irradiation incubation with hypertonic medium during the incubation period for PCC. The ß-form breaks have fast rejoining kinetics ($t_{1/2}$ = 2 to 10 min), and are sensitive to post-irradiation hypertonic treatment but not sensitive to ß-araA treatment.

4. Xrs-5 cells constitutively express the ß-form interphase chromosome breaks. The repair kinetics of irradiated xrs-5 cells are not affected by post-irradiation hypertonic treatment, i.e., xrs-5 cells are deficient in repairing the ß-form interphase chromosome breaks.

5. Xrs-5 cells are proficient in the α-form interphase chromosome breaks. Xrs-5 cells can repair the α-form of breaks at to a similar extent as wild type CHO cells.

Since we have shown a good agreement between residual interphase chromosome breaks and cell survival in xrs-5 cells (Okayasu et al 1993), we can infer that the radiosensitive phenotype of xrs-5 cells are directly related to the deficiency in repairing the ß-form interphase chromosome breaks. An important implication of these results is that the product of the *xrs* gene contributes specifically to the rejoining of the ß-form interphase chromosome breaks and has little effect on the α-form of interphase chromosome breaks. As for the α-form interphase chromosome breaks, our recent experiments have shown that all the α-form breaks can be visualized if mitotic cells with high mitosis-promoting activity (e.g., HeLa cells) are used as inducers instead of CHO cells (Cheng et al 1993).

Our observations described above are also consistent with the formation of two forms of potential lethal damage (PLD) termed α and ß in irradiated cells (Iliakis 1985a,b). The α and ß-forms of interphase chromosome breaks are related to the α and ß-forms of PLD. In fact it was found that ß-araA enhances killing in irradiated xrs-5 cells but, hypertonic treatment has a limited effect on cell lethality (Iliakis and Okayasu 1990, Okayasu et al 1993). These observations are consistent with the deficiency in the ß-form interphase chromosome breaks in xrs-5 cells but not in the α-form. Similar correlations can also be drawn between fast and slowly repairing DNA dsb and the ß- and α-forms of interphase chromosome breaks, respectively (Metzger and Iliakis 1991).

Recently, Getts and Stamato identified a protein or protein complex associated with the *xrs* phenotype (Getts and Stamato 1994). According to their report, radiosensitive xrs-5 cells lack a Ku-like DNA end binding protein or complex of proteins which seems to facilitate the repair process of DNA dsb caused by radiation or other agents. It appears from our studies that this protein (or complex of proteins) can only be responsible for the rejoining of the one form (ß-form) of DNA dsb which has a fast rejoining kinetics. More information on this protein and the gene (XRCC5) coding the protein can be found in another recent article (Taccioli et al

1994). One possible explanation would be that binding of this protein to DNA dsb (the fast rejoining DNA dsb) can stabilize the DNA structure such that the repair machinery can act appropriately.

Acknowledgments

This work was supported by NCI grants, 2RO1 CA 42026, and 5T32 CA 09137 awarded by NIH, DHHS. The authors wish to thank Ms. Nancy Mott for secretarial assistance.

References

Bryant PE, Blöcher D (1982) The effects of 9-beta-D-arabinofuranosyladenine on the repair of DNA strand breaks in x-irradiated Ehrlich ascites tumor cells. *Int. J. Radiat. Biol.* **42**: 385-394.

Cheng X, Pantelias GE, Okayasu R, Cheong N, Iliakis G (1993) Mitosis-promoting factor activity of inducer mitotic cells may affect the radiation yield of interphase chromosome breaks in the premature chromosome condensation assay. *Cancer Res.* **53:** 1-5.

Cornforth MN, Bedford JS (1983) X-ray-induced breakage and rejoining of human interphase chromosomes. *Science* **222**: 1141-1143.

Cornforth MN, Bedford JS (1985) On the nature of a defect in cells from individuals with ataxia telangiectasia. *Science* **227**: 1589-1591.

Getts RC, Stamato TD (1994) Absence of a Ku-like DNA end binding activity in the xrs double strand DNA repair-deficient mutant. *J. Biol. Chem.* **269**:15981-15984.

Hittelman WN, Rao PN (1974) Premature chromosome condensation. I. Visualization of X-ray-induced chromosome damage in interphase cells. *Mutat. Res.* **23**: 251-258.

Hittelman WN, Pollard M (1982) A comparison of DNA and chromosome repair kinetics after gamma irradiation. *Radiat. Res.* **92**: 497-509.

Iliakis G (1985) Evidence for the induction of two types of potentially lethal damage after exposure of plateau phase Chinese hamster V79 cells to gamma-rays. *Radiat. Environ. Biophys.* **24**: 185-202.

Iliakis G, Bryant PE, Ngo FHQ (1985) Independent forms of potentially lethal damage fixed in plateau-phase Chinese hamster cells by post irradiation treatment in hypertonic salt solution or araA. *Radiat. Res.* **104**: 329-345.

Iliakis G (1988) Radiation-induced potentially lethal damage: DNA lesions susceptible to fixation. *Int. J. Radiat. Biol.* **53**: 541-584.

Iliakis G, Pantelias G, Seaner R (1988) Effect of arabinofuranosyladenine on radiation-induced chromosome damage in plateau-phase CHO cells measured by premature chromosome condensation: Implications for repair and fixation of alpha PLD. *Radiat. Res.* **114**: 361

Iliakis GE, Okayasu R (1990) Radiosensitivity throughout the cell cycle and repair of potentially lethal damage and DNA double-strand breaks in an X-ray sensitive CHO mutant. *Int. J. Radiat. Biol.* **57**: 1175-1211.

Iliakis G (1991) The role of DNA double strand breaks in ionizing radiation-induced killing of eukaryotic cells. *BioEssays*, **13**: 1-8.

Iliakis G, Bloecher D, Metzger L, Pantelias G (1991) Comparison of DNA double strand break rejoining as measured by pulsed field gel electrophoresis, neutral sucrose gradient centrifugation and non-unwinding filter elution in irradiated plateau phase CHO cells. *Int. J. Radiat. Biol.* **59**: 927-939.

Iliakis G, Mehta R, Jackson M (1992) Level of DNA double-strand break rejoining in Chinese hamster *xrs-5* cells is dose-dependent: Implications for the mechanism of radiosensitivity. *Int. J. Radiat. Biol.* **61**: 315-321.

Iliakis GE, Pantelias GE (1990) Production and repair of chromosome damage in an X-ray sensitive CHO mutant visualized and analyzed in interphase using the technique of premature chromosome condensation. *Int. J. Radiat. Biol.* **57**: 1213-1223.

Iliakis G, Okayasu, R, Varlotto J, Shernoff C, Wang Y (1993) Hypertonic treatment during premature chromosome condensation allows visualization of a class of interphase chromosome breaks repaired with fast kinetics in irradiated CHO cells. *Radiat. Res.* **135**: 160-170.

Jeggo PA, Kemp LM (1983) X-ray sensitive mutants of Chinese hamster ovary cell line. Isolation and cross-sensitivity to other DNA-damaging agents. *Mutation Res.* **112**: 313-327.

Johnson RT, Rao RN (1970) Mammalian cell fusion: Induction of premature chromosome condensation in interphase nuclei. *Nature,* **226**: 717-722.

Kemp LM, Sedgwick SG, Jeggo PA (1984) X-ray sensitive mutants of Chinese hamster ovary cells defective in double-strand break rejoining. *Mutation Res.* **132**: 189-196.

Metzger L, Iliakis G (1991) Kinetics of DNA double-strand break repair throughout the cell cycle as assayed by pulsed field gel electrophoresis in CHO cells. *Int. J. Radiat. Biol.* **59**: 1325-1339.

Mueller WEG, Rohde HJ, Beyer R, Maidhof A, Lachman M, Taschner H, Zahn RK (1975) Mode of action of 9-β-D-arabinofuranosyladenine on the synthesis of DNA, RNA, and protein in-vivo and in-vitro. *Cancer Res.* **35**: 2160-2168.

Nagasawa H, Chen D, Strniste GF (1989) Response of x-ray-sensitive CHO mutant cells to gamma radiation. 1. Effects of low dose rates and the process of repair of potentially lethal damage in G1 phase. *Radiat. Res.* **118**: 559-567.

Okayasu R, Varlotto J, Iliakis G (1993) Hypertonic treatment does not affect the radiation yield of interphase chromosome breaks in DNA double-strand break repair deficient *xrs-5* Cells. *Radiat. Res.* **135**: 171-177.

Okayasu R, Iliakis G (1993) Ionizing radiation induces two forms of interphase chromosome breaks in CHO cells that rejoin with different kinetics and show different sensitivity to treatment in hypertonic medium or β-araA. *Radiat. Res.* (1993) **136**: 262-270.

Okayasu R, Iliakis, G (1994) Evidence that the product of the xrs gene is predominantly involved in the repair of a subset of radiation-induced interphase chromosome breaks rejoining with fast kinetics. *Radiat. Res.* **138**: 34-43.

Pandita T, Hittelman WN (1992) Initial chromosome damage but not DNA damage is greater in ataxia telangiectacia cells. *Radat. Res.* **130**:94-103.

Pandita T, Hittelman WN (1992) The contribution of DNA and chromosome repair deficiencies to the radiosensitivity of ataxia-telangiectasia. *Radat. Res.* **130**:214-223.

Pantelias GE, Maillie HD (1983) A simple method for premature chromosome condensation induction in primary human and rodent cells using polyethylene glycol. *Somatic Cell. Genet.* **9**: 533-547.

Pantelias GE (1986) Radiation induced cytogenetic damage in relation to change in interphase chromosome conformation. *Radiat. Res.* **105**: 341-350.

Phillips RA, Tolmach LJ (1966) Repair of potential lethal damage in x-irradiated HeLa cells. *Radiat. Res.* **29**: 413-432.

Taccioli GE, Gottlieb TM, Blunt T, Priestley A, Demengeot J, Mizuta R, Lehmann AR, Alt FW, Jackson SP, Jeggo PA (1994) Ku 80: Product of the XRCC5 gene and its role in DNA repair and V(D)J recombination. *Science* **265**: 1442-1445.

Waldren CA, Johnson RT (1974) Analysis of interphase chromosome damage by means of premature chromosome condensation after x- and ultra-violet irradiation. *Proc. Natl. Acad. Sci. USA,* **71**:1137-1141.

Wang Y, Iliakis G (1992) Effect of 5'-iododeoxyuridine on the repair of radiation-induced potential lethal damage interphase chromatin breaks, and DNA double strand breaks in Chinese hamster ovary cells. *Int. J. Radiat. Oncol. Biol. Phys.* **23**: 353-360.

Wlodek D, Hittelman WN (1988) The relationship of DNA and chromosome damage to survival of synchronized X-irradiated L5178Y cells. I. Initial damage. *Radiat. Res.* **115**:550-565.

Wlodek D, Hittelman WN (1988) The relationship of DNA and chromosome damage to survival of synchronized X-irradiated L5178Y cells. II. Repair. *Radiat. Res.* **115**:565-575.

FURTHER INVESTIGATIONS INTO THE NATURE OF THE TRIGGER OF INCREASED RADIORESISTANCE: THE EFFECT OF ^{14}C AND ^{3}H THYMIDINE ON LOW DOSE HYPERSENSITIVITY

Kirsten Skov[*], Cameron Koch[†] and Brian Marples[*]
[*]Medical Biophysics, B.C. Cancer Research Centre
601 West 10th Avenue
Vancouver, Canada V5Z 1L3

[†]Radiation Oncology
University of Pennsylvania
Philadelphia, PA 19104-6072

Introduction

Increased radioresistance (IRR) At very low doses, survival is characterised by a hyper-radiosensitive response (HRS), followed by increased radioresistance, the extent of which appears to vary with cell line (1-4), LET (5,6) and prior exposure to chemicals or radiation (1,3,6,7). Proposed explanations for the phenomena, as recently summarized (7,4), include the induction of repair; an induced cell cycle delay; increased fidelity of repair; down-regulation of some factor or the inhibition of apoptosis. However, unlike the substructure described by Skarsgard'sgroup (8) it is not explained by mixed cell populations. Therefore, our Studies (4) are directed towards determining the trigger for the response, and its nature.

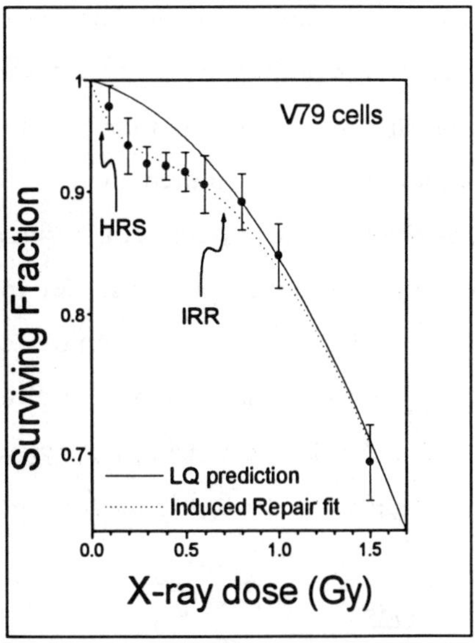

The trigger: Considerable evidence suggests that IRR is a response to DNA damage, possibly single-strand breaks (SSBs) (or OH damage), since H_2O_2 pretreatment modifies these phenomena, as do priming doses of x rays (9). These agents have also been shown to induce the adaptive response (10) and it is known that H_2O_2 pretreatment radioprotects (11,12). Further indirect support for the involvement of DNA in this effect is the variable IRR response in repair deficient mutant cell lines (2); the inhibition of IRR by 3-aminobenzamide (3); the effect of the topoisomerase I inhibitor camptothecin on IRR (13) and the LET dependence of the phenomenon (5,6,14,). In most respects, the HRS/IRR phenomenon is reminiscent of the adaptive response (10). Arguments have also been presented against the notion that SSBs are the trigger for IRR (see discussion). If SSB induction is non-linear at very low doses, or if SSB repair rate is increased by a priming dose treatment (either reflecting the IRR/HRS structure), then measurement of SSBs directly (15) using methods sensitive enough for this dose range (16) might show relevant anomalies. However, a preliminary study did not detect any substructure in the initial damage, nor the rate of repair after priming (15), although it was proposed that the radioactive label required in the assay could itself serve as the priming agent, thereby masking any structure. Priming effects from incorporated radioactive labels have been reported previously for the adaptive response using various endpoints (17,18). Therefore it was important to ask whether cells with radioactivity incorporated in the DNA exhibit a response in the HRS/IRR region which differs from unlabeled cells using a survival endpoint. Such a difference might further implicate DNA damage as the trigger for the increased radioresistant response.

Methods

Survival: V79-379A cells were used in this study. Survival was determined using the microscopic assay employing a 50-cell colony survival criterion (3,5,13).

Labeling: Cells were labeled with thymidine (Amersham) using the methodology employed to measure DNA damage in the elution studies (14), at our standard level for ^{14}C (0.05 mCi mL^{-1}, 56 mCi $mmole^{-1}$), and at 1/5 of the standard ^{3}H level (0.05 mCi mL^{-1}, 44 Ci $mmole^{-1}$). After incubation at 37°C overnight, cells were chased with fresh unlabeled medium for 1 hour.

Irradiation: Cells were X-irradiated as a suspension culture under aerobic conditions.

Breaks: SSBs in DNA are measured using alkaline elution, with modifications (16) permitting data to be obtained below 1 Gy.

Results

Survival Figure 1 shows data for labeled and unlabeled cells. Plating efficiencies (PE) are included, as survival data alone may be misleading or result in loss of important information (19). As might be expected, the ^{14}C label at this concentration was relatively non-toxic (PE 0.77 ± 0.05 *vs* control 0.85 ± 0.05), while tritium, used at 1/5 the level employed in many DNA studies, may be slightly more toxic than ^{14}C, with plating efficiencies of 0.72 ± 0.03 compared with 0.86 ± 0.02 for unlabeled cells.

Breaks The size of the priming doses of x rays and pi-mesons (pions) which have been shown to eliminate hypersensitivity are sufficient to produce strand breaks in DNA. Moreover, strand breaks produced after single doses of x-rays and pions were found to correlate with the extent of the increased radioresistant response. We have also measured the strand breaks caused by 10^{-4} M peroxide since this overcomes the HRS response (9), however it appears that possibly as many as 60 times the number of SSBs are produced by the H_2O_2 treatment than that caused by a comparable x ray priming dose of 0.5 Gy (20).

Discussion

Survival and DNA damage are being measured at very low doses. Results from survival studies in independent laboratories support the existence of IRR (increased radioresistance), which may be explained as induced repair of radiation damage, although no direct evidence implicating DNA damage and repair has yet been found for the IRR phenomenon. Here, we report results which lend support for the involvement of DNA damage. The survival response of unlabeled x-irradiated-cells show the HRS and IRR structures in the very low dose region, whereas cells labeled with DNA precursors show less structure, suggesting that the damage caused by ^{14}C or ^{3}H-thymidine elicits a protective effect. While these results do not prove that DNA damage is the trigger for the increased resistance, they support such a hypothesis.

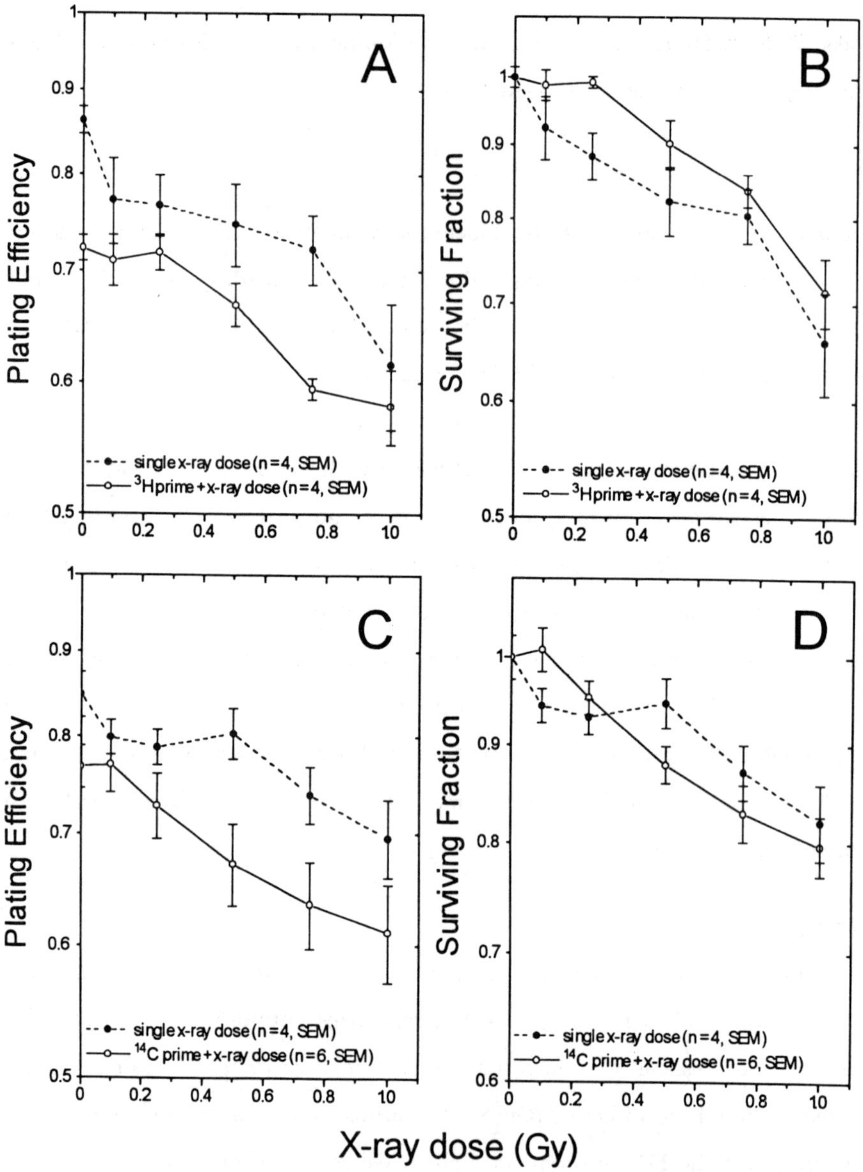

Figure 1 The effect of incorporation of labeled thymidine on the clonogenic activity of V79 cells: (a), (b) ^{3}H; (c) (d) ^{14}C on plating efficiency (a,c) and survival (b,d).

The dose delivered by the ^{14}C label is estimated to be < 0.25 Gy. These data also show that hypersensitivity can be eliminated by very low dose rate treatments, in addition to acute doses of x-rays and pions reported previously. Likewise, adaptive responses have been shown in human lymphocytes in response to acute and chronic treatments of ionising radiations (17,18).

On the nature of the trigger of IRR SSBs are obvious candidates as triggers for the IRR response, but in a recent presentation (4,20) we summarized contradictory evidence. For example, the low dose oxygen enhancement ratio (OER) for SSB is ~4-6, but the OER of the IRR "trigger dose" is much lower (~ 1.4) (15,16,21). Furthermore, there are as much as 60 times more breaks produced by H_2O_2 at the concentration (10^{-4} M) required to induce IRR than by x rays (20). The IRR response is diminished in cell lines defective in either excision repair or double-strand break repair, while single-strand break repair deficient EM9 cells exhibited a more pronounced hypersensitive response and a degree of IRR (2). Furthermore, pretreatment with cisplatin, which produces DNA adducts and crosslinks, also appeared to induce radioresistance (22).[1]

Since the correlation appears not to be *quantitative* (more breaks do not necessarily trigger an earlier response) we may need to invoke a *qualitative* component, such as chemical differences in the types of single-strand breaks or the location of the damaging lesion (4,12). Alternatively, breaks produced during excision repair (such as with cisplatin) or other lesion(s) such as DSB, base damage, adducts, crosslinks, or special sub-classes thereof which act independently, or possibly as precursors for SSBs, must be considered as potential trigger(s) for IRR.

It has been known for some time that ^{3}H thymidine inhibits growth of cultured cells (23), and causes lower survival, breaks in DNA, cell cycle perturbations, and chromosome aberrations (24), as summarized by Ribeiro *et al.* (25), who also stated "the apparent accommodative response by the cells to such DNA damage may involve an inducible repair mechanism which was maintained at a much higher level in the presence of [3H]TdR" "This accommodative mechanism must be considered when one uses radiolabeled cells to assess DNA damage of the repair of such damage by other genotoxic agents" (25). Our results confirm these statements.

---

[1]although treatment in hypoxia led to sensitization by apparently preventing IRR (20).

Summary

In this study, we have shown that the hypersensitive region of the low dose survival curve is diminished in ^{14}C and ^{3}H thymidine-labeled cells. This implies that such cells are relatively radioresistant at very low doses of x-rays, presumably due to induced radioprotection, which has been proposed to be via induced repair, or increased time for repair. If this is true, then caution must be exercised when comparing results from labeled cells (e.g. studies of DNA damage) and unlabeled cells (e.g. survival studies), as the labeled cells may be already "turned on" by the radioactivity. This result, while possibly the first using survival in mammalian cells as an endpoint, is perhaps not surprising in that it agrees with earlier studies on the adaptive response using other endpoints. However, given the clinical relevance of the doses in these studies, it is important to note that various insults may lead to radioresistance.

Acknowledgements

The authors thank NCI and MRC of Canada for financial support and H. Zhou and H. Adomat for technical assistance.

References

1. Molecular, cellular, and genetic basis of radiosensitivity at low doses: A case of induced repair? *Radiat. Res.* 138 Supplement, April, 1994.
2. Skov, K., Marples, B., Matthews, J.B., Joiner, M.C. and Zhou, H.: A preliminary investigation into the extent of increased radioresistance or hyper-radiosensitivity in cells of hamster cell lines known to be deficient in DNA repair. *Radiat. Res.* 138: S126-9, 1994.
3. Marples B.: The response of mammalian cells to low doses of ionising radiation. Ph.D. Thesis, University of London, 1991.
4. Skov, K.A. and Marples, B.: Increased radioresistance at clinical doses: Investigations from the survival perspective. *Radiation Oncology Investigations* (accepted).
5. Marples, B. and Joiner, M.C.: The response of Chinese hamster cells to low radiation doses: Evidence of enhanced sensitivity of cells to low radiation doses. *Radiat. Res.* 133: 41-51, 1993.
6. Marples, B., Lam, G.K.Y., Zhou, H. and Skov, K.A.: The response of Chinese hamster V79-379A cells exposed to negative Pi-mesons: Evidence that increased radioresistance is dependent on linear energy transfer. *Radiat. Res.* 138: S81-4, 1994.
7. Joiner, M.C., Evidence for induced radioresistance from survival and other end points: An introduction. *Radiat. Res.* 138: S5-S8, 1994.

8. Skarsgard, L.D., Harrison, I. and Durand, R.E.: The radiation response of asynchronous cells at low dose: evidence of substructure. *Radiat. Res.* 127: 248-256, 1991.
9. Marples, B., and Joiner, M.C.: The elimination of low-dose hypersensitivity in V79-379A cells by pre-treatment with x-rays or hydrogen peroxide. *Radiat. Res.* (Submitted, 1994)
10. Wolff. S.: Failla Memorial Lecture: Is radiation all bad? *Radiat. Res.* 131: 117-123, 1992.
11. Gupta, S.S. and Bhattacharjee, S.B.: Induction of repair function by hydrogen peroxide in Chinese hamster cells. *Int. J. Radiat. Biol.* 53: 935-942, 1988.
12. Boreham, D.R. and Mitchell, R.E.J.: DNA lesions that signal the induction of radioresistance and DNA repair in yeast. *Radiat. Res.* 128: 19-29, 1991.
13. Skov, K.A., Zhou, H. and Marples, B.: The effect of two topoisomerase inhibitors on low-dose hypersensitivity and increased radioresistance in Chinese hamster V79 cells. *Radiat. Res.* 138: S117-120, 1994.
14. Marples, B., Lam, G., Adomat, H., Primmett, D. and Skov, K.: Increased radioresistance is triggered by damage from high-LET negative pi-mesons. 18th L.H. Gray Conference. April 10-14, 1994. University of Bath, UK.
15. Koch, C.J. and Giandomenico, A.R.: Do low doses of radiation induce a response modulating induction or repair of DNA single-strand breaks? *Radiat. Res.* 138: S89-92, 1994.
16. Koch, C.J. and Giandomenico, A.R.: The alkaline elution technique for measuring DNA single strand breaks: Increased reliability and sensitivity. *Analytical Biochemistry* 220: 58-65, 1994.
17. Ikushima, T.: Chromosomal responses to ionizing radiation reminiscent of an adaptive response in cultured Chinese hamster cells. *Mutation Research* 180: 215-221, 1987.
18. Ikushima, T.: Radio-adaptive response: characterization of a cytogenetic repair induced by low-level ionizing radiation in cultured Chinese hamster cells. *Mutation Research* 227: 241-246, 1989.
19. Koch, C.J. and Skov, K.A.: Comparisons of cellular radiation response using absolute rather than relative parameters. *Radiat. Res.* 132: 40-49, 1992.
20. Skov, K.A., Marples, B. Adomat, H. and Zhou, H.: DNA damaging agents and increased radioresistance (IRR). 18th L.H. Gray Conference. April 10-14, 1994. University of Bath, UK.
21. Marples, B., Joiner, M.C., and Skov, K.A.: The effect of oxygen on low dose hypersensitivity and increased radioresistance in Chinese hamster V79-379A cells. *Radiat. Res.* 138: S17-20, 1994.
22. Marples, B. and Skov, K.: Pre-treatment with cisplatin increases the resistance of cells to x-irradiation below 0.5 Gy. Conference on "Gene Induction and Adaptive Responses", Montreal, June 3-4, 1994 (In preparation).
23. Painter, R.B., Drew, R.M., Hughes, W.L., Science 127: 1244-5, 1958
24. Olivieri, G., Bodycote J., and Wolff, S.: Adaptive response of human lymphocytes to low concentrations of radioactive thymidine. Science 223: 594-597, 1984.
25. Ribeiro, P.L., Martinez, M.C. and Mitra, R.S.: A possible correlation between the growth rate and the extent of DNA damage induced by radiodecay in mouse lymphoma cells. *Biochem. Biophys. Res. Comm.* 128: 204-210, 1985.

Author Index